RESTORATIVE REDEVELOPMENT

OF

DEVASTATED ECOCULTURAL LANDSCAPES

Integrative Studies in
Water Management and Land Development

Series Editor
Robert L. France

Published Titles

Boreal Shield Watersheds: Lake Trout Ecosystems in a Changing Environment
Edited by J. M. Gunn, R. J. Steedman, and R. A. Ryder

The Economics of Groundwater Remediation and Protection
Paul E. Hardisty and Ece Özdemiroğlu

Forests at the Wildland–Urban Interface: Conservation and Management
Edited by Susan W. Vince, Mary L. Duryea, Edward A. Macie, and L. Annie Hermansen

Handbook of Regenerative Landscape Design
Edited by Robert L. France

Handbook of Water Sensitive Planning and Design
Edited by Robert L. France

Land Use Scenarios: Environmental Consequences of Development
Allan W. Shearer, David A. Mouat, Scott D. Bassett, Michael W. Binford, Craig W. Johnson, Justin A. Saarinen, Alan W. Gertler, and Jülide Kahyaoğlu-Koračin

Porous Pavements
Bruce K. Ferguson

Restoration of Boreal and Temperate Forests
Edited by John A. Stanturf and Palle Madsen

Restorative Redevelopment of Devastated Ecocultural Landscapes
Robert L. France

Wetland and Water Resource Modeling and Assessment: A Watershed Perspective
Edited by Wei Ji

RESTORATIVE REDEVELOPMENT
OF
DEVASTATED ECOCULTURAL LANDSCAPES

Robert L. France *and Contributors*

CRC Press
Taylor & Francis Group
Boca Raton   London   New York

CRC Press is an imprint of the
Taylor & Francis Group, an **informa** business

Front cover photographs are of the Euphrates and Tigris Rivers in Syria, and that for the back cover is of the Tonle Sap Great Lake in Cambodia (*R. France*).

CRC Press
Taylor & Francis Group
6000 Broken Sound Parkway NW, Suite 300
Boca Raton, FL 33487-2742

© 2011 by Taylor and Francis Group, LLC
CRC Press is an imprint of Taylor & Francis Group, an Informa business

No claim to original U.S. Government works

Printed in the United States of America on acid-free paper
10 9 8 7 6 5 4 3 2 1

International Standard Book Number: 978-0-415-95225-5 (Hardback)

Library of Congress Cataloging-in-Publication Data

Restorative redevelopment of devastated ecocultural landscapes / editor, Robert L. France.
    p. cm.
    "A CRC title."
    Includes bibliographical references and index.
    ISBN 978-0-415-95225-5 (alk. paper)
    1. Restoration ecology. 2. Landscape protection. 3. Cultural property--Protection. I. France, R. L. (Robert Lawrence)

QH541.15.R45R555 2010
333.71'53--dc22
                                      2010006331

Visit the Taylor & Francis Web site at
http://www.taylorandfrancis.com

and the CRC Press Web site at
http://www.crcpress.com

To the once and future dwellers in the Iraqi marshlands.

With the river filled with flowing waters,
the marshes stocked with life …
Abutting heaven.
Its awe and glory
cast upon the country …
Grown up 'twixt heaven and earth.

**—The Cylinders of Gudea
(an ancient Mesopotamian cuneiform inscription)
In France, R.** *Wetlands of Mass Destruction: Ancient Presage for
Contemporary Ecocide in Southern Iraq.*
Green Frigate Books, 2007.

Contents*

Prologue: Countdown to 2012: Iraq and Other Landscapes Devastated by
Conflict and Natural Disasters xiii

Preface: Learning from Experience: Relevant Applications from around
the World for Sustaining the Once and Future Iraqi Marshlands....xvii

About the Editor..xxi

Contributors xxiii

Introduction to the Paradigm: Ecocultural Restorative Redevelopment as
a Guiding Principle in Rebuilding Devastated Landscapes in
Uncertain Times 1

Section I Background

Overview: Intellectual and Pragmatic Context for the Restorative
Redevelopment of the Iraqi Marshlands 11

Environmental Restoration Theory and Practice

Chapter 1 Restoration: Philosophical and Political Contexts 15

Elizabeth V. Spelman

Chapter 2 Restoration, Ecological Citizenry, and Ziggurat Building in Iraq..... 21

Chapter 3 Restoring Babylon: Leaving a Better Iraq for This Century of
Revitalization 29

Storm Cunningham

Environmental Planning Theory and Practice

Chapter 4 Before the Master Plan: A Framework and Some Difficult
Questions for Regional Land Use Planning 39

Carl Steinitz

---

* All non-ascribed chapters or other sections are authored by Robert L. France

Chapter 5 Control Models of Tourism Development and Conservation
Management with Respect to Indigenous Culture 47

Shiau-Yun Lu

Desert Wetland Restorations

Chapter 6 Wetlands Lost and Found in the Levant.. 65

Chapter 7 Rehabilitation of a Historic Wetland: Jordan's Azraq Oasis 85

Chapter 8 Integrated Restoration and Adaptive Management of the Las
Vegas Wash 117

Section II   Wetlands and Nature Reserves

Overview: Ecological and Cultural Context for the Restorative
Redevelopment of the Iraqi Marshlands 127

Ecosystems

Chapter 9 Hydrology and Wetland Restoration for Human Subsistence
and Regional Biodiversity: The Challenge to Restore the Living
Landscape of Iraq's Mesopotamian Marshes................................... 131

Steven I. Apfelbaum and James P. Ludwig

Chapter 10 Rebuilding Wetlands and Waterfowl Resources in North
and South America: Importance of Capacity Building and
Landscape Perspective 157

People

Chapter 11 Preservation, Rehabilitation, and Management of Heritage
Wetlands in Mexico..173

Chapter 12 Involving People in Science and Sustainability in the Pantanal
Wetlands of Brazil..211

Chapter 13 Planning and Development of a Desert Wetland Park in the
United States 219

Chapter 14 Living over Water: Introduction to Ecotourism Development
Concerns for the Tonle Sap, Cambodia, and Lake Titicaca, Peru ...227

Robert L. France and Evi Syariffudin

Ecology and Economics

Chapter 15 Ecotourism, Local Economy, and Ecological Strategies in
Planning Nature Reserves in Panama..297

Chapter 16 Sustainable Development and Nature Conservation in Jordan 305

Section III   Innovative Approaches and Technologies

Overview: Habitation and Livelihood Context for the Restorative
Redevelopment of the Iraqi Marshlands 343

Transportation Planning and Design

Chapter 17 Transportation Corridors as Vehicles for Cultural and
Ecological Regeneration..347

Ilze Jones, René Senos, Ints Luters, and Robert L. France

Alternative Wasterwater Treatment, Reuse, and Infrastructure

Chapter 18 Linking Water Treatment with Wetland Restoration:
Engineering Challenges and Associated Benefits............................ 375

Chapter 19 Advanced Ecotechnology for Decentralized Wastewater
Treatment and Reuse 391

David Austin

Chapter 20 Infrastructure Models for Ecological Wastewater Management...... 413

Scott Wallace

Agriculture and Water Management

Chapter 21 Salinity Management in Arid Regions: Lessons for Iraq from
the Western United States 429

Chapter 22 Managing Scarce Water Resources in Agriculture in West Asia
and North Africa 451

Theib Y. Oweis and Ahmed Y. Hachum

Epilogue   Cambodia's Institutional Frameworks for Ecocultural Restorative
Redevelopment Introduction to Optimistic Management Models
Useful for Iraq... 485

Index.. 495

Series Statement: Integrative Studies in Water Management and Land Development

Ecological issues and environmental problems have become exceedingly complex. Today, it is hubris to suppose that any single discipline can provide all the solutions for protecting and restoring ecological integrity. We have entered an age where professional humility is the only operational means for approaching environmental understanding and prediction. As a result, socially acceptable and sustainable solutions must be both imaginative and integrative in scope; in other words, garnered through combining insights gleaned from various specialized disciplines, expressed and examined together.

The purpose of the CRC Press series Integrative Studies in Water Management and Land Development is to produce a set of books that transcends the disciplines of science and engineering alone. Instead, these efforts will be truly integrative in their incorporation of additional elements from landscape architecture, land-use planning, economics, education, environmental management, history, and art. The emphasis of the series will be on the breadth of study approach coupled with depth of intellectual vigor required for the investigations undertaken.

Robert L. France
Series Editor
Integrative Studies in Water Management and Land Development
Associate Professor of Watershed Management
Department of Engineering, NSAC
Science Director of the Center for Technology and Environment
Harvard University
Principal, W.D.N.R.G. Limnetics

Prologue

COUNTDOWN TO 2012: IRAQ AND OTHER LANDSCAPES DEVASTATED BY CONFLICT AND NATURAL DISASTERS

The persistently threatened and increasingly degraded state of the planet's environment has attracted its cult of doomsayers. There seemed to be a flurry of pessimistic books that were produced in the years following the first Earth Day celebration (Wali 1992), as, for example, *Ecocide and Thoughts toward Survival, Ecocatastrophe, The Doomsday Syndrome*, and *The Eco-Spasm Report*. Then, as the second millennium was drawing to a close, amid all the New Age Chicken Littles squawking about our imminent end due to computer-related glitches, environmental books also followed suit: for example, *The End of Nature; The Death of Nature; Our Final Hour: A Scientist's Warning: How Terror, Error, and Environmental Disaster Threaten Humankind's Future in This Century—on Earth and Beyond*; and *Countdown to Apocalypse: A Scientific Exploration of the End of the World*. Obviously, though, somehow we and nature seemed to muddle our collective way through.

Reflecting on the grim legacy of natural disasters that have occurred over the last decade (e.g., Indonesia, Thailand, Sri Lanka, and the Asian tsunami; New Orleans and Hurricane Katrina; the Greater Antilles and Hurricane Jeanne; Bam [Iran], Bhuj [India], Pakistan-administered Kashmir, Java and Sumatra [Indonesia], China, Haiti, and their respective earthquakes; and Burma and the cyclone), which together have killed almost a million people and laid waste to thousands of square kilometers of urban and rural landscapes, one could almost believe that "the end *is* nigh." Indeed, all this takes on an ominous, surreal overtone when one considers that for the Mayans, ancient history's greatest and most accurate cosmologists and timekeepers, the end of time does actually have a precise date, namely, December 21, 2012. And not surprisingly, a new batch of Cassandras have hit the bookstores with their latest warnings (*2012: Mayan Year of Destiny; The Mayans and Doomsday 2012; Maya Cosmogenesis 2012: The True Meaning of the Maya Calendar End-Date; Apocalypse 2012: A Scientific Investigation into Civilization's End; The World Cataclysm in 2012; Gaia 2012: The Earth's Coming Great Change*; etc.). Optimistically, however, a handful of the titles suggest that some of us might still be around in 2013 to pick up the pieces of our devastated global environment (e.g., *How to Survive 2012; 2012 and the Ring of Light: When Mankind Finally Grows Up; Beyond 2012: A Shaman's Call to Personal Change and the Transformation of Global Consciousness; 2012—Year of Apocalypse: The Destruction and Resurrection of Earth*; and *2012: You Have a Choice! Archangelic Answers and Practice for the Quantum Leap*).

The Earth is indeed a precarious place, disasters being very much the status quo of our collective history (Kondratyu, Krapiuin, and Varostos 2006). As a species, we've scraped by several times when confronted by pandemics such as the Black Death in the fourteenth century, which killed as much as one-third of Europe's population, or the Spanish Influenza Epidemic of 1917–1918, which killed up to 100 million people, about one-twentieth of the world's population (including one-fifth of that of India). The closest we have probably come to extinction, however, was the cataclysmic eruption of the Toba volcano seventy thousand years ago, which was three hundred times stronger than the 1815 eruption of Tambora, which itself produced the "year without a summer" around the world. It is estimated that Toba might have reduced the entire world population of *Homo sapiens* to as few as thirty to forty thousand individuals. On a smaller scale, natural disasters can be related to the destruction of past civilizations (Keys 2000) such as, for example, the eruptions of Thera in the Mediterranean and Hekla in Iceland and the "Bronze Age collapse" in the Near East and the end of the Shang Dynasty in China.

Disasters needn't be such instantaneous cataclysms, however, to effect massive change. The widespread famine in Europe during the early fourteenth century killed a quarter of the population and forever altered the physical and conceptual landscapes.

And finally, as the recent sad fate of New Orleans has reminded us, disasters needn't be completely natural to be decimating. In this regard, the worst single anthropogenic disaster in recorded history was the willful breaching of the raised levees of the Yellow River in 1931, which led to as many as 4 million deaths. And famines such as that in nineteenth-century Ireland have long been known to have been caused as much by humans as by harsh climates or disease outbreaks. In this regard, conflict-related famines are estimated to have killed as many as 60 million people during the twentieth century alone. And, of course, the physical act of war itself exerts many direct impacts on the physical environment (Sirimarco 1993; Richardson 1995; Hulme 2004; Austin and Bruch 2000).

All these topics play out on the physical and historical landscapes of Iraq (France 2012). Following the great melting at the end of the last ice age eight thousand years ago, sea levels rose 120 to 130 m back to their former elevations. During the previous period of glaciation, many of the spreading peoples of the world had settled, as they still do today, near coastlines. Global inundation, therefore, produced our worldwide legacy of flood myths, nowhere more prominently than in the creation of the Persian/Arabian Gulf and destruction of the original marshes (and "Garden of Eden") between the Tigris and Euphrates Rivers (France 2012). What followed were thousands of years of Mesopotamian history in which the overwhelming roles played in the lives of commoners by despotic leaders and harsh environmental conditions are reflected in their cuneiform texts (France 2007).

The meteor impact that left behind the Umm al Binni Lake in the southern Iraqi marshes is thought to be related to the Bronze Age collapse of the Assyrian Empire in the twelfth century bce (Hamacher 2005). The "Plague of Justinian" pandemic that swept the world between 541 and 700 ce killed about half of Europe's and the Middle East's populations and weakened the Byzantine and Persian Empires, thereby paving the way for the ascendancy of Islam (Little 2006). The sacking of Baghdad by Hulagu Khan and the Mongols in 1258 resulted in the deaths of from two hundred thousand to 1 million citizens and brought an end to the Abbasid Caliphate (delivering a blow to Islam from which it has never recovered; Kennedy 2004). Further, the ancient irrigation canal system was destroyed and due to the massive concomitant casualties was never able to be restored, thus bringing about substantial changes to the Iraqi landscape (France 2012). The next sacking of the city—by Timur the Lame (Tamerlane) in 1401, when he built a pyramid of over ninety thousand decapitated skulls—furthered the desolation of people, spirit, and place.

Fast-forward to modern Iraq and Saddam Hussein, and we have yet again an environment ravaged by conflict (El-Baz 1994; Sadiq and McCain 1993). And following the first Gulf War, we have what might be regarded as history's most severe ecocide (France 2008), that is, the deliberate destruction of an environment, in this case the southern marshlands, for the purposes of ethnically cleansing it of a specifically targeted population, as described in detail in such chapters as "Assault on the Marshlands" by C. Mitchell and "A Personal Testimony" by A. Hayder in Nicholson and Clark (2002), and "Witness to a Lost Landscape: The Marshes in the Mid-Seventies" by N. Wheeler, "Human Rights Issues in the Iraqi Marshlands: A Case for Genocide" by E. Nicholson, and "Experiences and Hopes of the People of the Al-Ahwar Marshes" by R. Al-Khayoun in France (2007).

The destruction and restoration of the marshlands will be comprehensively described in a companion volume to this book, *Restoring the Iraqi Marshlands: Potentials, Perspectives, Practices* (France 2011), and are reviewed in the aforementioned books by Nicholson and Clark (2002) and France (2007). As such, they will not be reiterated here besides the following brief description from the front-cover flap in the latter book:

> Paradise Found and Lost—The Mesopotamian marshes, located between the Tigris and Euphrates rivers in southern Iraq, were historically one of the world's most important wetland environments. The area—once over twenty thousand square kilometers and thought of by some to be the original Garden of Eden—has provided habitat for millions of migrating birds and had been inhabited since the time of Sumerians by thousands of people living on artificial islands of mud and reeds and depending on sustainable fishing and farming. In the early 1990s, however, this important ecological and unique cultural jewel was destroyed by Saddam Hussein's Ba'thists through a series of constructed dams and water diversions designed to eradicate the remaining marsh dwellers who had escaped previous massacres.

This, then, is the lens through which the case studies in the present book will be examined.

REFERENCES

Al-Khayoun, R. 2007. Experiences and hopes of the people of the Al-Ahwar marshes. In *Wetlands of mass destruction: Ancient presage for contemporary ecocide in southern Iraq*, ed. R. L. France. Winnipeg, MB: Green Frigate Books.

Austin, J. E., and C. E. Bruch. 2000. *The environmental consequences of war: Legal, economic, and scientific perspectives*. Cambridge: Cambridge University Press.

El-Baz, F. 1994. *Gulf War and the environment*. Boca Raton, FL: Taylor & Francis.

France, R. L. 2008. Ecocide: A definition. *Archeology* (Spring), p. 10.

———. 2011. *Restoring the Iraqi marshlands: Potentials, perspectives, practices*. Sussex, UK: Sussex Academic Press.

———. 2012. *Back to the garden: Searching for Eden in the Mesopotamian marshes*. Cambridge, MA: Harvard University Press.

———, ed. 2007. *Wetlands of mass destruction: Ancient presage for contemporary ecocide in southern Iraq.* Winnipeg, MB: Green Frigate Books.

Hamacher, D. W. 2005. The Umm al Binni structure and Bronze Age catastrophes. *The Artifact: Publications of the El Paso Archeological Society* 43:115–38.

Hayder, A. 2002. A personal testimony. In *The Iraqi marshlands: A human and environmental study*, ed. E. Nicholson and P. Clark. London: Politico's.

Hulme, K. 2004. *War torn environment: Interpreting the legal threshold.* Leiden: Brill.

Kennedy, H. 2004. *When Baghdad ruled the Muslim world: The rise and fall of Islam's greatest dynasty.* New York: Da Capo.

Keys, D. 2000. *Catastrophe: An investigation into the origins of the modern world.* New York: Ballantine.

Kondratyu, K., V. Krapiuin, and C. Varostos. 2006. *Natural disasters as interactive components of global-ecodynamics.* Berlin: Springer.

Little, L., ed. 2006. *Plague and the end of antiquity: The pandemic of 541–750.* Cambridge: Cambridge University Press.

Mitchell, C. 2002. Assault on the marshlands. In *The Iraqi marshlands: A human and environmental study*, ed. E. Nicholson and P. Clark. London: Politico's.

Nicholson, E. 2007. Human rights issues in the Iraqi marshlands: A case for genocide. In *Wetlands of mass destruction: Ancient presage for contemporary ecocide in southern Iraq*, ed. R. L. France. Winnipeg, MB: Green Frigate Books.

Nicholson, E., and P. Clark, eds. 2002. *The Iraqi marshlands: A human and environmental study.* London: Politico's.

Richardson, M. 1995. *Effects of war on the environment: Croatia.* Boca Raton, FL: Taylor & Francis.

Sadiq, M., and T. C. McCain. 1993. *The Gulf War aftermath: An environmental tragedy.* Berlin: Springer.

Sirimarco, E. 1993. *War and the environment.* Milwaukee, WI: Gareth Stevens.

Wali, M. K. 1992. *Policy issues*, vol. 1 of *Ecosystem rehabilitation.* The Hague: SPB Academic.

Wheeler, N. 2007. Witness to a lost landscape: The marshes in the mid-seventies. In *Wetlands of mass destruction: Ancient presage for contemporary ecocide in southern Iraq*, ed. R. L. France. Winnipeg, MB: Green Frigate Books.

Preface

LEARNING FROM EXPERIENCE: RELEVANT APPLICATIONS FROM AROUND THE WORLD FOR SUSTAINING THE ONCE AND FUTURE IRAQI MARSHLANDS

The destruction of the marshlands in southern Iraq (Nicholson and Clark 2002) is one of the most dramatic environmental cataclysms to have occurred in recent times. The promise and possibilities of restoring the marshes and their surrounding agricultural landscapes (hence the more accurate reference to the "marsh*lands*" rather than the more limiting "marshes") have captured worldwide attention. Searching for some positive news to report from postwar Iraq, the press in particular has embraced the concept of restoring the imagined Garden of Eden (France 2007a). In 2004, I organized an international conference titled "Mesopotamian Marshes and Modern Development: Practical Approaches for Sustaining Restored Ecological and Cultural Landscapes" (Werthmann 2005; Reed 2005; Fink 2005; Bauer 2005). Previous conferences attended by the usual assortment of governmental and consulting experts had examined the feasibility of restoration efforts and the expected products that might ensue. Instead, I was interested in exploring different directions in that the conference was largely concerned with examining practical approaches for sustaining the *process* of those restoration efforts, both during and *after* the reparation work. In this respect, I was less interested in the immediate steps necessary to restore the marshlands than I was in the long-term approaches that would be required to *sustain* all those important efforts. Furthermore, whereas previous conferences had focused primarily on either the natural or cultural aspects of restoration, but not on both, I was well aware that by its very concept and application, restoration effectively blurs the boundaries between what is "natural" and what is "cultural" (France 2007b, 2008). As such, I built the conference around the unified theme of *ecocultural* landscapes, particularly in relation to moving from restoration per se to the much more useful term "development," as indicated in the conference title and as endorsed by the Iraqis in charge of managing the marshlands (for example, the Iraqi Minister of Water Resources, Abdul-Latif Jamal Rasheed, who also opened the conference). As quoted in Reed (2005),

> "Some of the surveys of [marshland] refugees living in camps in Iran have shown that if you ask them, 'Would you come back?' many say, 'No. Life is just as good in this horrible refugee camp as it was in the marshes,'" said France in an interview before the conference. "We mustn't romanticize their [the marsh Arabs'] old way of life. We mustn't want people to live in a museum to benefit ecotourism. Since Saddam drained the marshes, people have been doing dry-land agriculture, and as hard as that is, it was worse in the marshes. So they say, fine, bring back the marshes, but don't do it where I've been farming for a decade. Some of these people do want to get back and live on their artificial islands in the middle of the marsh, but they want Internet access. And of

course they want, and are entitled to, good healthcare. And education, and roads, and proper wastewater disposal. That's why development is the theme of the conference."

In this respect, the specific goals of the conference (as laid out on the webpage and in the advertising brochure) were as follows:

- Present practical approaches for sustaining the process of restoration efforts, both during and after the reparation work has been accomplished.
- Introduce designed and built projects from around the world that have achieved a high degree of success and that, either in their entirety or in part, can be adapted to the situation in the southern Iraqi marshlands.
- Examine actions that might be designed and planned to offer possible solutions to the sustainable development of the region.

So this was a very different type of conference altogether since it was based on something that is counterintuitive to the worlds of government environmental management and environmental consulting—where the former, if targeted with addressing a particular landscape or ecosystem, will only study that location, and where the latter often earn their keep by reinventing the wheel each and every time in an effort to appease their funders. In both cases, there is a glaring absence of taking a synoptic or cross-system approach to problem solving. In reality, environmental management could learn much from empirical ecology as practiced by the late Robert Peters of McGill University (Peters 1991), wherein often the best insights about problems at hand come through looking at systems and approaches located or implemented elsewhere (as in France 2008, which deals with Venice). To this end, I cast the net wide and sought out individuals to present (and later write about) case studies with direct applicability to the evolving restoration situation in Iraq from such wide-ranging locations as Poland, the United States, Taiwan, Israel, Jordan, Brazil, Mexico, Cambodia, Peru, Panama, Canada, and Algeria, among other countries. Mostly I wanted to bring in fresh perspectives and new players to contribute to the complex situation of the Iraq marshlands, thereby avoiding having to rely on the same old global trouble-shooting "call girls," to use Arthur Koestler's (1973) apt parlance from his novel of that name.

As I concluded in the book *Wetlands of Mass Destruction: Ancient Presage for Contemporary Ecocide in Southern Iraq*,

> The bottom line is that it is impractical, unrealistic, and condescending to expect the returning Ma'dan [marsh Arabs] to assume a museum lifestyle to satisfy our own romanticism [fueled by such books as those by Maxwell (1968) and Thesiger (1964)]. The challenge in the comprehensive restoration of the Iraqi marshes then becomes how best to interface a desire by some marsh Arabs to return to a semblance of their former lifestyle and at the same time keep the benefits of modernity to which they are entitled (such as communication, sanitation engineering, education, transportation, industry, agriculture, and tourism to name but a few). In the end, the restoration of the marshes and their inhabitants will be judged a success only by approaching that restoration within a framework of sustainable development. (France 2007a)

And this is the precise message of the present book, captured in that integration of restoration and sustainable development that I have since come to refer to as "restorative redevelopment." The present book then highlights interdisciplinary case studies of demonstrable success from around the world that could be used as models for sustaining marshland restoration efforts with a balanced focus on sustaining both restored nature *and* lifestyles.

ACKNOWLEDGMENTS

I would like to thank A. Tabernier and her small support army for helping to organize and run the conference, including B. Kenet. The generous sponsors for the conference, which would not have occurred otherwise, included Applied Ecological Services, Banrock Station Winery, Cambridge Brewing Company, Canadian International Development Agency, CH2MHILL, Design Workshop, Dharma Living Systems, Ducks Unlimited, Iraq Foundation Eden Again Project, Jones & Jones, Michael Baker Corporation, Montgomery Watson Harza, North American Wetland Engineering, United States Agency for International Development, in addition to Harvard University. Harvard's Milton Fund and Asia Fund are thanked for enabling the author's travel to Jordan, Syria, and Cambodia.

REFERENCES

Bauer, J., ed. 2005. The marsh Arabs of Iraq: The legacy of Saddam Hussein and an agenda for restoration and justice. *Carnegie Council Insider* (March–April): www.cceia.org/resources/transcripts/5102.html.

Fink, S. 2005. Saving Eden: Can the ecology and the economy of Iraq's once-glorious wetlands be restored? *Discover Magazine* 26(7): http://m.discovermagazine.com/2005/jul/saving-eden.

France, R. L., ed. 2007a. *Wetlands of mass destruction: Ancient presage for contemporary ecocide in southern Iraq*. Winnipeg, MB: Green Frigate Books.

———, ed. 2007b. *Healing natures, repairing relationships: New perspectives on restoring ecological spaces and consciousness*. Winnipeg, MB: Green Frigate Books.

———, ed. 2008. *Handbook of regenerative landscape design*. Boca Raton, FL: CRC Press.

Koestler, A. 1973. *The call girls*. New York: Random House.

Maxwell, G. 1968. *People of the reeds*. Boynton Beach, FL: Pyramid Books.

Nicholson, E., and P. Clark. 2002. *The Iraqi marshlands: A human and environmental study*. Berkeley, CA: Counterpoint Press.

Peters, R. H. 1991. *A critique for ecology*. Cambridge: Cambridge University Press.

Reed, C. 2005. Paradise lost? What can—or should—be done about the environmental crime of the century? *Harvard Magazine* 107(3): http://harvardmagazine.com/2005/01/paradise-lost.html.

Thesiger, W. 1964. *The marsh Arabs*. New York: Penguin.

Werthmann, C. 2005. Mesopotamian marshes and modern development: Conference overview. *Landscape Journal* 24:223–24.

About the Editor

Robert L. France is associate professor of watershed management in the Department of Engineering at Nova Scotia Agricultural College (NSAC). Dr. France has conducted research in regions from the High Arctic to the tropics, on subject areas from bacteria and algae to whales, as well as on chemistry and environmental theory. He has taught at the universities of McGill, Ca'Foscari Venice, and Harvard. France is an acquisition editor for CRC Press, where he runs the Integrative Studies in Water Management and Land Development series, and is also on the editorial board of the independent environmental press Green Frigate Books. He has published over two hundred articles and is the author or editor of over a dozen books of both a technical nature as well as general public interest. France conducts research on the environmental restoration of postagricultural and postindustrial landscapes, integrated watershed management and water-sensitive planning and design, the use of stable isotope analysis to trace material flow in aquatic foodwebs, the impacts of clearcutting on land–lake linkages, landscape modifications at the suburban-agricultural interface, agricultural urbanism, environmental biography, and immersion into historic agricultural and utilitarian landscapes.

Contributors

Steven I. Apfelbaum
Applied Ecological Services
Brodhead, Wisconsin
steve@appliedeco.com

David Austin
Natural Treatment Systems
Mendota Heights, Minnesota
david.austin@ch2m.com

Storm Cunningham
Resolution Fund, LLC
Washington, DC
storm@resolutionfund.com

Ahmed Y. Hachum
International Center for Agricultural
    Research in Dry Areas (ICARDA)
Allepo, Syria
A.HACHUM@cgiar.org

Ilze Jones
Jones & Jones
Seattle, Washington
ijones@jonesandjones.com

Shiau-Yun Lu
National Sun Yat-Sen University
Kaohsiung, Taiwan
shiauyun@faculty.nsysu.edu.tw

James P. Ludwig
607 Canard Street
Port Williams, Nova Scotia, Canada
 jpludwig@xcountry.tv

Ints Luters
Jones & Jones
Seattle, Washington
iluters@jonesandjones.com

Theib Y. Oweis
International Center for Agricultural
    Research in Dry Areas (ICARDA)
Aleppo, Syria
ICARDA@cgiar.org

René Senos
Belt Collins Northwest LLC
Seattle, Washington
RSenos@beltcollins.com

Elizabeth V. Spelman
Smith College
Northampton, Massachusetts
espelman@email.smith.edu

Carl Steinitz
Graduate School of Design, Harvard
    University
Cambridge, Massachusetts
steinitz@gsd.harvard.edu

Evi Syariffudin
Graduate School of Design, Harvard
    University
Cambridge, Massachusetts

Scott Wallace
Naturally Wallace
Minneapolis, Minnesota
scott.wallace@naturallywallace.com

Introduction to the Paradigm
Ecocultural Restorative Redevelopment as a Guiding Principle in Rebuilding Devastated Landscapes in Uncertain Times

One of the modern founding fathers of environmental restoration considered landscape degradation "a universal concomitant of human societies" (Bradshaw 2002). Indeed, estimates are that over 40 percent of the earth's terrestrial surface may have a diminished capacity to sustain human life due to recent impacts of degradation (Daity 1995). No wonder, then, that pioneering restorationist and environmental ethicist Aldo Leopold would make reference to our damaged "world of wounds."

Philosopher and contributor to the present volume, Elizabeth Spelman, has written about the human imperative to repair our broken world (Spelman 2007). Therefore, the concept of healing, applied to both nature and culture, is very much a part of comprehensive, ecumenical restoration (France 2007a,b). The publication *Healing Wounds: How the International Research Centers of the CGIAR Help to Rebuild Agriculture in Countries Affected by Conflicts and Natural Disasters* (Varma and Winslow 2003), produced by the same organization headquartered in Aleppo, Syria, where chapter 22's contributors are based, also links these themes. With a focus on agriculture, Varma and Winslow (2003) include sections on "Rebuilding Seed and Food Systems," "Safeguarding and Restoring Agrobiodiversity," "Rebuilding Human and Institutional Capacities," "Reducing Future Vulnerability to Conflicts and Disasters," "Making Relief Aid More Effective and Efficient," and "Returns on Investments." Another recent book, *Rebuilding Urban Places after Disaster: Lessons from Hurricane Katrina* (Birch and Wachter 2006), includes a major chapter titled "Restoration," which includes sections on "Decreasing Vulnerability," "Returning Economic Viability," "Resolving Social Issues," and "Recreating a Sense of Place." And it is here where we reach a disjuncture, for the subject areas in these two books seem to have much more in common with that part of sustainable development that concerns ecosystem and human well-being (Millennium Ecosystem Assessment

1

2005) than they do with ecological restoration, at least as the latter is portrayed in theoretical or practical books about the discipline/profession. Ecological restoration's shortcomings in contributing to sustainable development are at least fourfold.

First, ecological restoration still remains preoccupied with concerns about historical fidelity (France 2007a). Returning to original "natural" conditions for severely or widespread degraded landscapes or regions is difficult if not impossible to imagine. Every few years, the American Society for Ecological Restoration retools its definition of "ecological restoration" largely in response to concerns about the importance of a return to original-state conditions. For others, wariness of an ability to achieve this aspiration has led them to avoid the term "restoration" altogether in favor of other, less exacting but of course variably meaning, expressions such as "reclaiming" (Schaller and Sutton 1978; Berger 2008), "rehabilitating" (Wali 1992), "reconstructing" (Bradshaw 1983; Buckley 1989), "repairing" (Whisenant 1999), "rewilding" (Hall 2005), "regenerating" (France 2008a,b), "reinventing" (Gross 2005), "redesigning" (Higgs 2003; France 2007b), and "resynthesizing" (Jordan, Gilpin, and Aber 1990) natures, any of which might be more appropriate to use for sustainable development.

Second, practitioners of ecological restoration predominantly operate inside a belief system where humans are regarded as interlopers rather than imbedded members in ecosystems (France 2007a). Some sociological theoreticians of restoration have, however, posited divergent views (Jordan 2003; Gross 2005), even going as far in one case (France 2007b) as to purposely advance the emerging paradigm of restoration *design* as being distinct from that of restoration *ecology*. Shifting focus from restoring "nature" (a loaded term if ever there was one; Evernden 1992) to regenerating "landscapes" (a term that implies human presence; Krinke 2002) promotes a concept of restoration (using the term in its form as a convenient overarching, umbrella expression; Bradshaw 2002) in which humans matter (France 2008a; see also the illuminating discussion about this in Hall's 2005 book). This is obviously a much more accurate view of the world to hold for sustainable development. It is not surprising, therefore, that Iraqi Water Minister, Abdul-Latif Jamal Rasheed, prefers the word "development" over "restoration" when referring to the rebuilding of the marshland ecosystems and communities (again, along the lines of the directions outlined in Millennium Ecosystem Assessment 2005 rather than those in any descriptive text of ecological restoration).

Third, review most of the case studies described in books or journals about ecological restoration, and one would come away with a belief that the profession focuses, for the most part, on individual sites of under a hundred or so hectares in size. Due to inherent difficulties, both technical and financial, case studies concerned with "restoring" landscapes on the scale of, for example, the Everglades (Davis and Ogden 1994) or the Venetian lagoon (Fletcher and Da Mosto 2004; France 2010) are much rarer. A few attempts have been made, however, to situate ecological restoration within a framework of landscape or ecosystem processes (Wali 1992; Hobbs 2002; Whisenant 1999), this being the physical scale at which sustainable development customarily operates.

Finally, ecological restoration is primarily concerned with repairing the bits and pieces of nature through creating improved habitat templates to support restocked or recolonized wildlife. Economic issues, according to Edwards and Abivardi (1997), have been largely neglected. Recently, however, adopting the concept of "natural capital" that is becoming part of conservation biology (Daily and Ellison 2002) and watershed management (Abbott 2005), holistic-thinking restorationists are approaching, conducting, and interpreting case studies from a viewpoint of "economics as if nature matters, and ecology as if people matter" (Aronson, Milton, and Blignaut 2007). Given the integral role that natural capital plays in sustainable development (Daly 1996), its new and exciting inclusion in restoration studies offers promise as the vehicle needed to move that profession forward into the twenty-first century (Boyce, Narain, and Stanton 2007).

So, to summarize and expand upon the above, the newly evolving social, economic, and ecological concerns in recent environmental restoration studies seem to suggest that the profession could contribute to sustainable development given the latter's oft-stated mandate in dealing with social or cultural perspectives in reference to the quality of life, economic perspectives in reference to steady-state systems, and ecological perspectives in reference to supporting essential life processes. In fact, some early writers coined the expression "ecodevelopment" (Sachs 1980; Riddell 1981) to refer to just this sort of concept.

But is development itself the answer to today's concerns in a world where resources are being alarmingly diminished at the same time as landscape effluvia are rapidly increasing? Certainly the inclusion of restoration into sustainable development implies that the latter should not continue business as usual. Urbanska, Webb, and Edwards (1995) appear to have been among the first to recognize that "sustainable development can only be achieved through reversing environmental degradation by developing and using knowledge of how to restore ecosystems." Restoration economist and chapter 3 author Storm Cunningham is correct in his criticism that Urbanska, Webb, and Edwards's concept of restoration was limited to traditional ecological restoration, with its implicit limitations as described above, rather than to the more useful approaches of "restorative development." Cunningham (2002) argues that there are three modes of the development life cycle; that the time for new development and for maintenance and conservation development, regardless of their perceived or realized sustainability, is outdated; and that only restorative development can provide the framing vision to navigate our way through the twenty-first century in a way that considers both nature and people (Birch and Watcher 2006; France 2008b).

The need, therefore, is to move beyond Urbanska, Webb, and Edwards's (1995) perception of ecological restoration as merely being "a supporting device for strategies of sustainable development" to a mind-set that I refer to as "restorative redevelopment"—preserving the important "re" prefix in front of "development" as in Ferguson et al. (1999), the Center for Watershed Protection (2001), France (2002), and Pinkham and Collins (2002)—as being the only real solution to address and manage our inherited "world of wounds." So where does restorative redevelopment fit in relation to the disciplines of restoration ecology, restoration design, and sustainable development?

Figure I.1, based on extending the graphical interpretations of Bradshaw (2002), shows the restoration–development framework. The vertical axis indicates the degree

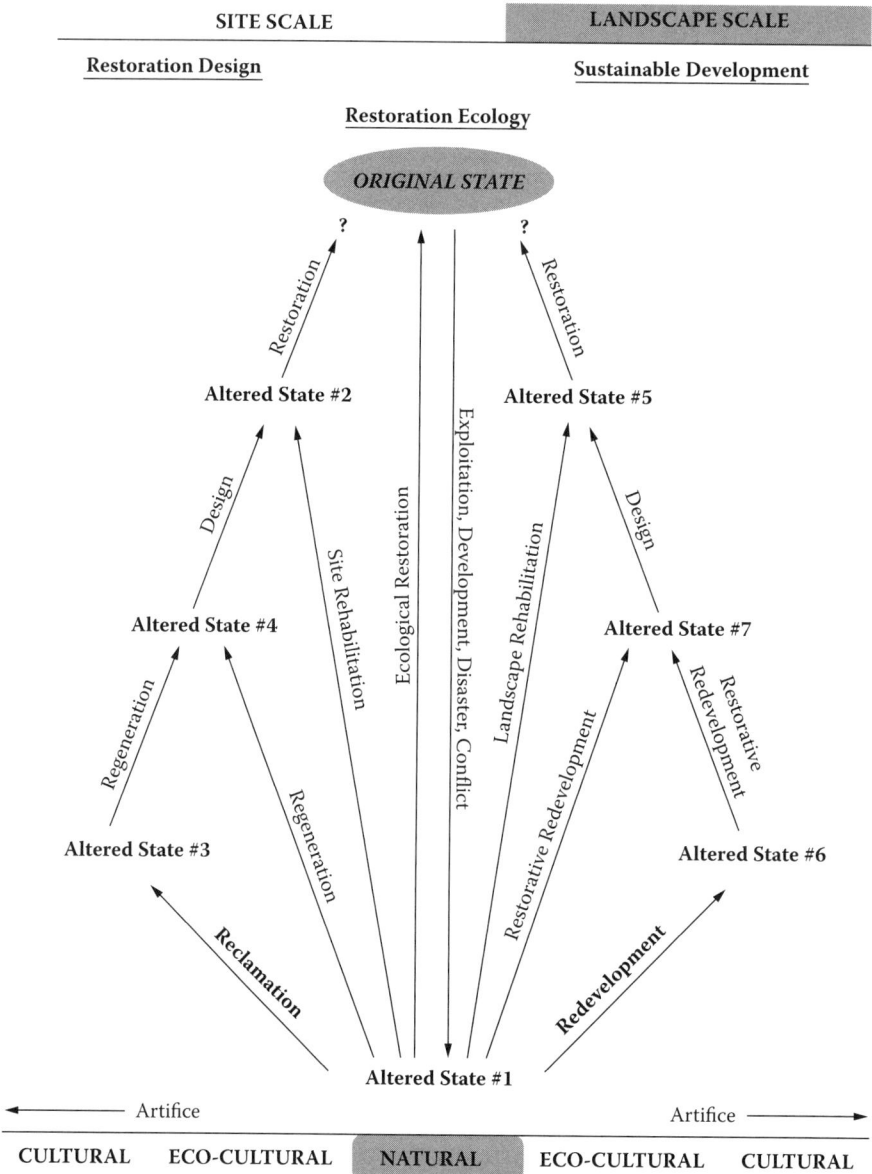

FIGURE I.1 Degradation: recovery subdisciplines and trajectories. "Restoration design" is the paradigm described in France (2007b); "regeneration" is short for "regenerative landscape design," which is the paradigm described in France (2008b); and "restorative redevelopment" is the paradigm described in the present book. The terms "ecological restoration," "reclamation," "design" (short for "ecological design"), and "sustainable development" have all been described previously in the literature.

of degradation or recovery: the more pristine a location, the closer to the top of the figure; and, conversely, the more disturbed a location, the closer it is to the bottom of the figure. The horizontal axis denotes the extent of artifice in locations, proceeding from natural in the center through ecocultural to cultural locations on either side. The distinction of one side or the other reflects the scale of the location, with smaller sites on the left and large landscapes on the right. Together these two "X-axis" variables are reflected in the three major disciplines of restoration ecology in the center of the diagram with restoration design toward the left side and sustainable development toward the right side. Using these waymarkings, it is possible to situate the various subdisciplines and degradation–recovery trajectories within this matrix.

Degradation of a location from its original-state conditions through exploitation, development, disaster, or conflict results in a shift to altered state 1. Through ecological *restoration*, it may be possible to return the location to its original state. More realistic is the recovery of the ecological conditions of a site (dealing with small-scale locations on the left of the diagram) to a new altered state (state 2) that approaches but does not actually achieve an explicit return to the original ecological conditions. This is the process of site *rehabilitation*. Transformation of a site to a completely new and artificial altered state (state 3) for human utilitarian purposes and without any attempt to mimic natural conditions is the process of *reclamation*. The partial recovery of the environmental conditions of a site to a new altered state (state 4) that is achieved for the mixed benefits of humans and the environment is the process of *regeneration*. Such sites are neither as artificial as reclaimed sites nor as natural as rehabilitated ones. As well, such regenerated sites are further along the way to recovery than reclaimed sites but are not as preoccupied with fidelity to original-state conditions as rehabilitated sites. Finally, it may be possible to move from altered reclaimed state 3 to altered regenerated state 4 through regeneration, and from there to altered rehabilitated state 2 through ecological *design*. Beyond that, though it might be possible, realistically it seems doubtful that such a site can ever be truly returned to original-state conditions. All these processes can be considered to be within the new paradigm of restoration design (France 2007a,b).

Moving to the right or landscape side of the diagram, a parallel group of trajectories and states can be described. Recovery of a landscape to environmental conditions approaching the original state (altered state 5) occurs through landscape *rehabilitation*. In contrast, transformation of a landscape to a completely new and artificial altered state (state 6) primarily for the benefit of humans and with little attempt to return to original-state conditions occurs through *redevelopment*. The partial recovery of a landscape to a new altered state (state 7) that is achieved for the mixed benefits of humans and the environment is the process of *restorative redevelopment*, the subject of the present book. Such landscapes are neither as artificial as redeveloped landscapes nor as natural as rehabilitated ones. As well, such restorative redeveloped landscapes are further along the way to recovery than simply redeveloped landscapes, but are not as preoccupied with fidelity to original-state conditions as rehabilitated landscapes. Finally, as was the case on the other side of the diagram, it may be possible to move from altered redeveloped state 6 to altered restorative redeveloped state 7 through restorative redevelopment, and from there to altered rehabilitated state 5 through ecological *design*. Beyond that, though it might be possible, it seems doubtful that such a landscape can ever be truly returned to original-state conditions. All these processes can be considered to be within the paradigm of sustainable development.

"Restorative redevelopment," then, is based on the reuse of landscapes to improve the value and livability of a location for humans at the same time as effectively reinstating natural processes and functions. Restorative redevelopment, therefore, links concepts of ecosystem and human well-being (Millennium Ecosystem Assessment 2005) with ecosystem and societal design, health, and sustainability (Steedman 2005; Abbott 2005; Williams 2002).

REFERENCES

Abbott, R. M. 2005. Into the great wide open: Rethinking design in an era of economic, social, and environmental change. In *Facilitating watershed management: Fostering awareness and stewardship*, ed. R. L. France. Lanham, MD: Rowman & Littlefield.

Aronson, J., S. J. Milton, and J. N. Blignaut. 2007. *Restoring natural capital: Science, business, and practice*. Washington, DC: Island Press.

Berger, A. 2008. *Designing to reclaim the landscape*. Boca Raton, FL: Taylor & Francis.

Birch, E. L., and S. M. Wachter, eds. 2006. *Rebuilding urban places after disaster: Lessons from Hurricane Katrina*. Philadelphia: University of Pennsylvania Press.

Boyce, J. K., S. Narain, and E. A. Stanton. 2007. *Reclaiming nature: Environmental justice and ecological restoration*. London: Anthem Press.

Bradshaw, A. D. 1983. The reconstruction of ecosystems. *Journal of Applied Ecology* 20:1–17.

———. 2002. Introduction and philosophy. In *Principles of restoration*, vol. 1 of *Handbook of ecological restoration*, ed. M. R. J. Perrow and A. J. Davy. Cambridge: Cambridge University Press.

Buckley, G. P. 1989. *Biological habitat reconstruction*. London: Belhaven Press.

Center for Watershed Protection. 2001. *Redevelopment roundtable: Smart site practices for redevelopment and infill projects*. Ellicott City, MD: Center for Watershed Protection.

Cunningham, S. 2002. *The restoration economy*. San Francisco: Berrett-Koehler.

Daily, G. C., and K. Ellison. 2002. *The new economy of nature: The quest to make conservation profitable*. Washington, DC: Island Press.

Daity, C. G. 1995. Restoring value to the world's degraded lands. *Science* 269:350–54.

Daly, H. 1996. *Beyond growth: The economics of sustainable development*. Boston: Beacon Press.

Davis, S., and J. C. Ogden 1994. *Everglades: The ecosystem and its restoration*. Boca Raton, FL: CRC Press.

Edwards, P. J., and C. Abivardi. 1997. Ecological engineering and sustainable development. In *Restoration ecology and sustainable development*, ed. K. M. Urbanska, N. R. Webb, and P. J. Edwards. Cambridge: Cambridge University Press.

Evernden, N. 1992. *The social creation of nature*. Baltimore: John Hopkins University Press.

Ferguson, B. K., with others. 1999. Re-evaluating stormwater: The Nine Mile Run model for restorative redevelopment [11 pp.]. Snowmass, CO: Rocky Mountain Institute.

Fletcher, C., and J. Da Mosto. 2004. *The science of saving Venice*. Torino, Italy: Umberto Allemandi.

France, R. L. 2002. Post-industrial restorative redevelopment: Rethinking the urban environment [brochure for series of executive education seminars]. Cambridge, MA: Harvard Design School.

————. 2007a. Landscapes and mindscapes of restoration design. In *Healing natures, repairing relationships: New perspectives on restoring ecological spaces and consciousness*, ed. R. L. France. Winnipeg, MB: Green Frigate Books.

————, ed. 2007b. *Healing natures, repairing relationships: New perspectives on restoring ecological spaces and consciousness*. Winnipeg, MB: Green Frigate Books.

————. 2008a. Environmental reparation with people in mind: Regenerative landscape design at the interface of nature and culture. In *Handbook of regenerative landscape design*, ed. R. L. France. CRC Press.

————, ed. 2008b. *Handbook of regenerative landscape design*. Boca Raton, FL: CRC Press.

————. 2010. *Veniceland Atlantis: The bleak future of the world's favorite city*. Libri, London.

Gross, M. 2005. *Inventing nature: Ecological restoration by public experiments*. Lexington, MA: Lexington Books.

Hall, M. 2005. *Earth repair: A transatlantic history of environmental restoration*. Charlottesville: University of Virginia Press.

Higgs, E. 2003. *Nature by design: People, natural process, and ecological restoration*. Cambridge, MA: MIT Press.

Hobbs, R. J. 2002. The ecological context: A landscape perspective. In *Principles of restoration*, vol. 1 of *Handbook of ecological restoration*, ed. M. R. J. Perrow and A. J. Davy. Cambridge: Cambridge University Press.

Jordan, W. R. 2003. *The sunflower forest: Ecological restoration and the new communion with nature*. Berkeley: University of California Press.

Jordan, W., M. E. Gilpin, and J. D. Aber. 1990. *Restoration ecology: A synthetic approach to ecological research*. Cambridge: Cambridge University Press.

Krinke, R. 2002. Overview: Design practice and manufactured sites. In *Manufactured sites: Rethinking the post-industrial landscape*, ed. N. Kirkwood. London: Spon.

Millennium Ecosystem Assessment. 2005. *Ecosystems and human well-being: Synthesis*. Washington, DC: Island Press.

Pinkham, R. D., and T. Collins. 2002. Post-industrial watersheds: Retrofits and restorative redevelopment (Pittsburgh, Pennsylvania). In *Handbook of water sensitive planning and design*, ed. R. L. France. Boca Raton, FL: CRC Press.

Riddell, R. 1981. *Ecodevelopment*. Farnham, UK: Gower.

Sachs, I. 1980. *Strategies de l'ecodeveloppment*. Paris: Ouvrieres.

Schaller, F. W., and P. Sutton, eds. 1978. *Reclamation of drastically disturbed lands*. Madison, WI: American Society of Agronomy.

Spelman, E. 2007. Embracing and resisting the restorative impulse. In *Healing natures, repairing relationships: New perspectives on restoring ecological spaces and consciousness*, ed. R. L. France. Winnipeg, MB: Green Frigate Books.

Steedman, R. J. 2005. Buzzwords and benchmarks: Ecosystem health as a management goal. In *Facilitating watershed management: Fostering awareness and stewardship*, ed. R. L. France. Lanham, MD: Rowman & Littlefield.

Urbanska, K. M., N. R. Webb, and P. J. Edwards, eds. 1997. *Restoration ecology and sustainable development*. Cambridge: Cambridge University Press.

Varma, S., and M. Winslow. 2003. *Healing wounds: How the international research centers of the CGIAR help to rebuild agriculture in countries affected by conflicts and natural disasters*. Washington, DC: Consultative Group on International Agricultural Research.

Wali, M. K., ed. 1992. *Policy issues*, vol. 1 of *Ecosystem rehabilitation*. The Hague: SPB Academic.

Whisenant, S. G. 1999. *Repairing damaged wildlands: A process-oriented, landscape-scale approach*. Cambridge: Cambridge University Press.

Williams, D. 2002. The design of regions: A watershed planning approach to sustainability. In *Handbook of water sensitive planning and design*, ed. R. L. France. Boca Raton, FL: Lewis.

Section I

Background

Overview: Intellectual and Pragmatic Context for the Restorative Redevelopment of the Iraqi Marshlands

Section 1 comprises eight chapters that deal with background topics that are important to address prior to undertaking the comprehensive restorative redevelopment of the Iraqi marshlands and other landscapes devastated by conflict or natural disasters. Together these chapters address fundamental concerns about the nature and culture of environmental restoration and environmental planning, as well as demonstrate how these themes have been approached in other successfully implemented projects. The lessons here are important: restoration is possibly the most rewarding yet most intellectually challenging of all forms of environmental management, land use planning is really just as much about managing mindscapes as it is about managing landscapes, and much can be learned from other desert wetland restoration projects that will be useful for the reparative work needed in Iraq.

In terms of environmental restoration theory and practice, the chapter by Elizabeth Spelman (chapter 2) grapples with the difficult question about just what kind of repaired landscape is to be sought for in Iraq and how this end goal might be shaped by our shared or imposed concepts that are in turn influenced by the philosophical and political forum in which the work is conducted. Chapter 2 considers the intriguing possibility that the most important result of any restorative redevelopment project may very well be the participatory process of communities of individuals involved in undertaking that restoration rather than any final physical products that might ensue from the work. Chapter 3 by Storm Cunningham discusses how restorative redevelopment can be related to economic and ecologic (in other words, "sustainable") regional revitalization and nationwide rebuilding.

For environmental planning theory and practice, chapter 4 by Carl Steinitz advances a logical sequence of questions that must be addressed prior to developing any regional master plan and posits that one integrative approach with an established record of answering such questions in a quantifiable manner is alternative futures scenario modeling. In chapter 5, Shiau-Yun Lu provides an informative case study contrasting top-down and bottom-up control models of tourism development and environmental conservation with particular reference to indigenous communities (and hence direct application to the situation in southern Iraq).

Considering other desert wetland restoration studies from around the world, chapter 6 reviews the Hula Swamp project in Israel (which shares many interesting parallels to the situation of the Iraqi marshlands) and describes the scientific and sociological problems that had to be circumvented there. Chapter 7 describes the destructive history, constructive reparation, and fledgling ecotourism activities involved with rehabilitating the Azraq Oasis in Jordan, the most successful wetland restoration project in the Arab Middle East and one that is becoming regarded as a model for undertaking other such activities in the region. Finally, the history of environmental degradation and the successful technical restoration efforts for the Las Vegas Wash are introduced in chapter 8 as another case study to possibly emulate for the Iraqi marshlands.

*Environmental Restoration
Theory and Practice*

1 Restoration
Philosophical and Political Contexts

Elizabeth V. Spelman

The restoration projects of human beings are not limited to natural habitats such as the marshlands of southern Iraq and other wetlands around the globe. Even the most cursory glance at newspaper headlines reminds us of frequent efforts to restore human artifacts such as valuable works of art (Michelangelo's *David*, to take a recent and controversial example) and historically significant buildings (such as the British Museum). There is an almost daily recitation of attempts to restore broken ties between individuals or peoples (think, for example, of South Africa's Truth and Reconciliation Commission, to take another recent and controversial example). Indeed, so numerous and ubiquitous are the repair and restoration activities of human beings that it seems quite fitting to think of *Homo sapiens* as *Homo reparans* (Spelman 2002, 2007): to the venerable and captivating portraits of *Homo sapiens* as the rational animal, the political animal, the social animal, the animal that really-and-not-just-apparently uses language, the only thinking thing that also has emotions, the only thinking thing that worries about whether it is the only thinking thing, we should add the portrait of the human being as the repairing animal (or, in any event, *a* repairing animal). And the human as repairing animal, *Homo reparans*, is called upon to develop and exercise not only the technical skills to carry out repair, but also the capacity to judge what is reparable and what is not, and also to decide, among those things judged reparable, what is worth fixing and what is not.

The English language, among others, is generously stocked with words for the many preoccupations and occupations of that bipedal tinkerer, *H. reparans*: repair, restore, rehabilitate, renovate, rebuild, recreate, reconstruct, regenerate, reconcile, redeem, heal, fix, and mend—to name a few. Such linguistic variety is not gratuitous. These are distinctions that make a difference. Do you want the car simply to be repaired, so that you can use it again to get where you need to go? Or do you want it restored to the bright shininess it had as it left the factory floor? Is a bold patch on your jacket adequate, or do you insist on invisible mending, on having it look as if there never were a rip to begin with? Should the work of art be restored, or simply conserved? Why do some ecologists want to preserve an environment rather than try to repair the damage done to it? Does forgiveness necessarily restore a ruptured relationship, or simply allow a resumption of it? What does an apology achieve that monetary reparations cannot—and vice versa? What was thought to be at stake for citizens

15

of the new South Africa in the contrast between restorative justice and retributive justice—between the healing promised by a Truth and Reconciliation Commission, and the remedy that might be procured by legally imposed punishment? In short, there typically is considerable aesthetic, economic (see chapter 3), moral (see chapter 2), or political significance attached to the difference between repairing and restoring, between restoring and preserving, and so on. One authority in charge of the treatment of Michelangelo's *David* hoped to defuse criticism of the methods that he endorsed by insisting that critics "made a mistake in talking about restoration. It's not a restoration. It's pure and simple maintenance" (D'Emilio 2003).

Important as such differences among the family of repair activities are, however, any reparative project is likely to involve many of the members of this family. For example, in his widely used book *Historic Preservation: Curatorial Management of the Built World*, James Marston Fitch (1990) first goes to considerable length to clarify some of the differences among the following: preservation (which "implies the maintenance of the artifact in the same physical condition as when it was received" by those in whose hands its care has been placed), restoration ("the process of returning the artifact to the physical condition in which it would have been at some previous stage"), and reconstruction ("the recreation of vanished buildings on their original site"). Fitch then goes on to point out that what is called the "restoration" of Colonial Williamsburg, Virginia, is in fact "a beguiling mixture of the preserved, the restored, and the reconstructed" (and that hearty brew includes "evasion": when Fitch published his book, Williamsburg had still managed to ignore the role of slavery in the historical moment to which the restoration is supposed to take us; Fitch 1990). Moreover, preserving, rehabilitating, or reconstructing a building may simultaneously be part of a project to restore a community's historical identity, its wounded pride, or its tattered hope (think, for example, of what is thought to be at stake in what happens at New York City's Ground Zero).

So while it is crucial to keep in mind the differences among repairing and restoring and preserving and reconstructing, it is also necessary to note that many of the projects undertaken by *H. reparans* in fact involve a mix of these different activities. But note, too, that there is no reason to think that people who have the knowledge, skill, and judgment to execute a repair job of one kind thereby have the knowledge, skill, and judgment needed to carry out a repair job of another kind. The point is not just that being good at repairing a car does not insure being good at restoring it, that an automobile mechanic who knows how to get your old Mustang running again probably doesn't have the skills to return it to its 1966 glory. Even a good mechanic who is also a good restorer isn't necessarily good at mediating conflict, and vice versa. Otherwise, the people who repair your appliances would be interchangeable with the damage control folks who work in customer service. Of course, some people may be good at both, but that's not because being skilled in one area of repair automatically translates into being skilled in another.

It's no wonder, then, that what appears to be even the simplest repair job can succeed at one level and misfire at another. For example, James has broken one of Sarah's favorite keepsakes, a flower vase given to her by her mother, now long deceased. A skilled restorer, James hopes that carefully mending the vase will also help repair Sarah's ever more fragile trust in James's sense of responsibility. The restoration is technically a great success: the fact that the vase had been broken will be noticeable only to the professionally trained eye. But as it turns out, James has only deepened the rift between him and Sarah. Not uncharacteristically, he failed to consult with her, in this case about the proper treatment of the broken vase; for a variety of reasons, she would have much preferred to preserve the vase in its brokenness. A technically successful restoration can, therefore, also be a socially unsuccessful restoration. A project meant to restore an object and shore up a relationship achieved the former at the price of doing further damage to the latter.

Given this brief reminder of the enormous family of activities of which restoration is but one member; of the fact that nonetheless most of what are called "restoration" projects involve a variety of such family members—restoration, preservation, reconstruction, and so on; and of the reality that good repairers in one domain are not necessarily thereby good repairers in another, it should be no surprise that proposals for the restoration of the marshlands of southern Iraq, for example, involve much more than restoration and much more than the marshlands. Indeed, "restoration" may not even be the best term to use in conjunction with such a project—not so much because damage has been so great as to render impossible full restoration (though that seems to be true) as because marshlands are by their very nature changeable in ways that works of art and buildings are not; we expect a well-restored painting to look just like it did before the damage, but we don't expect a restored river to cut through the riverbed just the way it did before being diverted or drained.

More to the present point, some proposals may explicitly or implicitly call for improvement rather than mere restoration: not just improvement of the current damaged condition, but also improvement upon the earlier, more desirable condition. That is, to the extent to which any proposal calls for making the marshlands not only better than they are now but also in some respects better than they've ever been, it may well be an improvement project more than a restoration project. Does that matter? Well, in some contexts the difference between restoration and improvement certainly does matter: art restorers, for example, are not supposed to take it upon themselves to improve upon a damaged work of art, to try to make it better than it ever was. A professional conservator isn't supposed to even think to herself, "Hmmm, that Van Gogh would look a lot better if I make the sky a bit lighter." In the context of the Iraqi marshlands and other wetlands with a history of human habitation, the difference between restoration and improvement certainly becomes significant to the extent that the question is not only about the land but also about a way of human life intimately tied to that land. Do those who were torn from their way of life when the marshes and marshlife were so severely damaged wish that the land and their way of life be put back together the way they were, say, two decades ago? How much is what they desire not a restoration of, but an improvement upon, those days, in terms of both the condition of the marshes and the conditions of their lives? They obviously didn't get to decide in the 1990s what was to become of the land or their lives; will they get to decide now, or in any event be partners in such decisions? Decisions about whether or not to restore (and, if to restore, how to restore), decisions about whether or not to improve (and if to improve, how to do so), and decisions about how to measure the success of the restoration or improvement are inherently political; they reflect the result of struggles over who has the power and authority to make crucial determinations of where precious resources will go and to evaluate how well they have been used. Though perhaps this is especially clear in the case of something so obviously politically charged as the restoration of the Iraqi marshlands, it is no less true in the case of the restoration of major works of art (e.g., *David* and *The Last Supper*) and in the mundane case of James and Sarah (James, you recall, simply preempted Sarah's role in the decision to restore).

So we can ask of any proposal for the restoration of the Iraqi marshlands or the other landscapes under discussion the same questions we need to ask of any repair or restoration project: (1) what is the nature of the project—is it restoration? Reconstruction? Preservation? Improvement? Or some combination of these? To whom and to what does it matter which approach is taken? And (2) what are among the proposed objects of repair, restoration, rehabilitation, improvement, or the like? Artifacts? Natural habitats? Relations among people? Cultural identity? Human dignity? One measure of the depth of the political charge to any reparative undertaking is the extent of the material and symbolic consequences of the project, the variety of domains in which repair of some sort is being attempted. Part of what appears to be at stake in the restoration, reconstruction, improvement, and so on of the marshlands, as we've already seen, is the repair not only of the land but also of a way of life. But even more: when asked by Congresswoman Ileana Ros-Lehtinen (R-FL) why the United States and its allies should provide economic and logistical support for efforts to restore the marshlands, Gordon West of the *U.S. Agency for International Development* (USAID) referred not only to the restoration of ecological conditions but also to the need to remedy the injustices done to the marshdwellers and more generally under Saddam Hussein (U.S. House of Representatives 2004). Moreover, as Professor Fernando Miralles-Wilhelm pointed out to Ms. Ros-Lehtinen and her congressional colleagues, among the serious rents in the social fabric of Iraqi society under Saddam was the deterioration of the educational system, "not only physically with the aging infrastructure, but also morally, with academic isolation taking its toll on the capacity of the country to provide solutions in the area of knowledge" (U.S. House of Representatives 2004). Ms. Ros-Lehtinen herself indicated that she understands the return of life to the marshes and its human and nonhuman inhabitants to be part of an effort to "erase the scars of a dictator" (U.S. House of Representatives 2004). (Whether any particular restorative project involves imposition of the will of one country upon another in the name of healing such scars is another but not irrelevant matter.) If you agree to one layer of repair, do you thereby agree to all?

To come around, in conclusion, to the broad theme of the "Background" section of this book, what might be dubbed "restoration writ large" (*sensu* France 2004):

1. I have suggested that we think of the restoration of the Iraqi marshlands and other landscapes as further instances of the ubiquity and variety of repair projects undertaken by the very familiar creature we call *Homo sapiens*, that wily language-using social and political animal who also might aptly be called *Homo reparans*. Repairing is a crucial skill for beings like ourselves, who are limited by the resources at our disposal, who are subject to the ever-present possibility of error and decay, who are capable of terrifying acts of destruction, who seek continuity with the past, and who face

the necessity of patching up relationships with their neighbors. But *Homo sapiens* is not now the repairing animal, now the social animal, now the political animal, and so on. The repairing animal is at one and the same time the social animal, the political animal, and the like, and that is why the restoration of the Iraqi marshlands and other landscapes, were they to be undertaken, can never be simply a matter of the right science, engineering, and technology.

2. But if thinking about repair and restoration writ large helps to situate and illuminate some of the reasons why wetlands restoration is such a complex reparative project, it is also true that the complexity to which we must attend in the case of the restoration of the marshlands drives home a particularly vivid lesson for students of repair and those who reflect on "restoration writ large." And that is that visible damage to the natural and artifactual world (if we assume this overly tidy distinction) around us is hardly ever simply that. We are embodied beings. It's not just that we are dependent for life on the land and water and air around us. We come to be and to know ourselves, we come to know and to create community with others, in part through our intimate relation to the physical world we inhabit and the material objects we create. We don't just live on or live near other living and nonliving things; our lives are in them, and they are in our lives. That is why it is not simply a coincidence that our spirits may be broken when the land on which we live or the objects we have created are shattered against our will.

This general point about the necessity of our intimacy with the physical and material world cannot, of course, be cited in defense of every instance of such intimate relation between us and that world. For one thing, to recognize such deep connection is not yet to say anything about the means by which any particular person or community comes to be in such intimate relation to a particular patch of the earth or their lives so intertwined with particular structures and objects. There are just and unjust ways of coming to occupy, live off, and live with land, water, buildings, and other objects. Moreover, this kind of intimacy, however achieved, hardly rules out selfish manipulation, exploitation, and defilement. Indeed, some of the most vicious forms of exploitation are possible only in conditions of intimacy. This surely is clear to us in the case of relations between human beings. Under what conditions such questionable forms of intimacy exist between humans and other living beings, other living systems, is one of the defining questions in ecological restoration, as discussed in chapter 2.

REFERENCES

D'Emilio, F. 2003. Amid criticism, "David" to get a cleaning. *Boston Globe*, September 16.
Fitch, J. M. 1990. *Historic preservation: Curatorial management of the built world.* Charlottesville: University Press of Virginia.
France, R. 2004. Mesopotamian marshes and modern development: Practical approaches for sustaining restored ecological and cultural landscapes. www.gsd.harvard.edu/mesomarshes.
Spelman, E. V. 2002. *Repair: The impulse to restore in a fragile world.* Boston: Beacon Press.

———. 2007. Embracing and resisting the restorative impulse. In *Healing natures, repairing relationships: New perspectives on restoring ecological spaces and consciousness*, ed. R. L. France. Winnipeg, MB: Green Frigate Books.

U.S. House of Representatives. 2004. Testimony before the Subcommittee on the Middle East and Central Asia of the Committee on International Relations. February 24. http://wwwc.house.gov/international_relations/108/92186.pdf.

2 Restoration, Ecological Citizenry, and Ziggurat Building in Iraq[*]

CONTENTS

Introduction 21
The Importance of Ecological Restoration 22
Moral Issues Intrinsic to Restoration 23
Restoration and Relationships... 24
Conclusions... 25
Acknowledgment 26
References.. 26

INTRODUCTION

Ecological restoration may be perhaps the most intellectually stimulating and ethically troubling of all forms of environmental action. Restoration design (*sensu* France 2007a), as opposed to restoration ecology (the assembly of the broken bits and pieces of damaged ecosystems), is a comprehensive and integrative way in which to look upon both the world and our role within it. Restoration design can be defined as the activity through which "reparation-minded individuals directly and creatively find ways in which to engage nature by establishing deeper physical and intellectual relationships with their world during the process of re-imagining, reconfiguring, and 'restoring' the ecological spaces in which they live and their consciousness in terms of how they live" (France 2007a). The restoration of the Iraqi marshlands has the potential to become one of the most important cases in the next few years of how this full potential of restoration can be realized, a way of not simply restoring the damage we have done to the physical environment, but also restoring the relationship that people have had to that special environment. The full potential of restoration, then, as Andrew Light (2004) sees it, is not simply the technical restoration of the environment but also the restoration of the cultural relationship with nature.

Almost immediately after the American tanks rolled northward past the remnants of the marshes on their way toward Baghdad, locals began to dismantle the dams and

---

[*] Adapted by Robert L. France from Light, A., 2004. Conceptual and moral issues in ecological restoration. Paper presented at the Mesopotamian Marshes and Modern Development: Practical Approaches for Sustaining Restored Ecological and Cultural Landscapes conference, Cambridge, MA, October.

channels that had for years strangled the region of water in their attempt to restore the wetlands. And soon after the initial toppling of the Ba'athist regime in Baghdad (and before the counterinsurgency), hundreds of native Iraqis were mobilized by the U.S. Army Corps of Engineers to help clear away debris from the canals around the city as a first attempt to restore the water infrastructure. Today, restoration efforts are being planned for the Iraqi marshlands by American, Italian, Canadian, and Japanese government agencies and nongovernmental organizations (NGOs; France 2011) in association with the Iraqi government (see the foreword). The ultimate success of the restoration efforts there will almost certainly be dependent upon the mobilization of the local community through the physical act of repairing the ecosystem (Light 2004). It is through participating in the healing of nature that the participants' relationships to that nature are, in turn, healed as well (France 2007a). This chapter explores some of the conceptual and moral issues underlying restoration.

THE IMPORTANCE OF ECOLOGICAL RESTORATION

With every practice referred to as "restoration," there is always the question whether it is really just that or is some other form of reparation (see chapter 1). Generally speaking, "ecological restoration" is defined as the human practice of restoring natural ecosystems, particularly those damaged by anthropogenic causes (but also occasionally those damaged by natural catastrophes as well). The science of restoration as it evolved in North America in the early 1970s dealt with human-damaged systems (Higgs 2003). Some of the first large-scale real-world experiments were conducted on tall-grass prairies in the Midwest and derived from the early work of pioneering conservation biologist Aldo Leopold at the University of Wisconsin.

Two of the indicators of the growing importance of restoration in our contemporary world involve ecology and economy. Ecology is normally given as the justification for restoration in terms of delivering ecosystem services. The defining issues addressed by ecological restorationists in this regard were habitat reconstruction, pollution mitigation, biodiversity sustenance, and exotic species limitation (Light 2004). Restoration has also become a big business (Cunningham 2002; and see chapter 3), and the federal government now spends more on this than almost any other type of environmental endeavor (e.g., the Clinton administration's multibillion-dollar appropriation budget for restoring the Florida Everglades).

Higgs (2003) defined three types of criteria for judging the worth of a good restoration project. Effective criteria are based on ecological reasons such as historical fidelity with respect to structural composition and replication, and functional success in terms of whether the replicated system can exist and be durable through time. Efficient criteria are the economic considerations for completion of a restoration project in terms of cost-effectiveness in relation to time and person-hours to complete the task. Expanded criteria represent the most interesting and challenging horizon issue in restoration (Light 2004).

The science of restoration is evolving at a very rapid pace with many controlled tests examining the veracity of restoration in a wide variety of physical landscapes. The expanded criteria are harder to assess as they deal with the historical, cultural, social, political, moral, and aesthetic dimensions of restoration. In a situation like

Iraq and the aspirations of the one-time marshland dwellers toward their restored former homeland (Al-Khayoun 2007), it is these expanded criteria that become paramount. Only in a place where one had the illusion that humans and nature existed in entirely separate realms of morality and practice (see Cronon 1996 to understand the reality of the situation) could one make a case that restoration only involved technical criteria (Light 2004; France 2007b). The expanded criteria speak more to the *process* of restoration rather than the product, and this is exactly what comprehensive restoration design is all about (France 2007a).

The Chicago Wilderness Project (Gobster and Hull 2000) is one of the most engaging and important restoration projects. Debates that surfaced concerned different opinions about what is "natural" and what is manipulated, and to what lengths restoration should go in order to obtain an impression of former times—as, for example, the controversial issue of removing oak forests to set the clock back to a time of grasslands. Perhaps the most important development of the restoration was the fact that much of the work was done by volunteers, not professionals (Light 2004). Indeed, thousands of individuals participated in the project, creating an incredible network of motivated advocates. The act of restoring the grasslands became an extension of nature in their urban lives, especially for those who did not have backyards. By getting their hands literally dirty, these people became more closely connected to the land around them. There is a certain kind of value that is captured in these sorts of voluntary intensive endeavors in contrast to professional restorations where a hired firm comes in from afar (Light 2000, 2007). So even in terms of Higgs's efficient criteria, if a restoration project takes longer to be completed through the use of volunteers, the benefits outweigh any potential disadvantages (Light 2004). In this regard, thinking about the process rather than the product as being the most important attribute resulting from any restoration project can lead one to reach different conclusions regarding the success of that project (France 2007a).

MORAL ISSUES INTRINSIC TO RESTORATION

For all environmental endeavors in general, there are issues of professional ethics prescribed by the organization to which one is a member, individual and social ethics concerning how one makes personal decisions in the public and private realms, and ethics between the species such as those having to do with the possibility of destroying certain parts of nature in order to restore other parts (as, for example, the loss of agricultural plants when restoring the Iraqi marshes for migratory waterfowl). Light (2004) asked what specific moral issues were involved with ecological restoration.

It is first important to ask the difficult question as to what actually is meant by "restoration." There is a big difference between restoration, which arguably has to be targeted to a specific moment in the past, and gardening, which some have likened to restoration. But the question is whether restoration has to go back *exactly* to the situation that was there before; or can it be just a typology rather than an actual token of the ecosystem that existed previously (Light 2004)? Also, is restoration the same as mitigation, as for example constructing replacement wetlands for those destroyed? In the end, is restoration really anything that makes you feel happy? Recognizing that broad definitions are problematic, the Society for Ecological Restoration tried to define ecological restoration more precisely, but this creates its own problem because it becomes very difficult to come up with a single definition that captures all the likely cases.

The potential of restoration to reconnect humans with nature is profound. The view of restoration theorist William Jordan (2003) is grounded in existentialism, claiming that the very act of being alive entails a necessary component of destruction which results in a shame. Jordan and Turner (2007) believe that we need rituals to help us overcome this shame and use this as a metaphysical tool in order to enable us to become better citizens.

Ecological restoration is important for creating a cadre of ecological citizens (Light 2000, 2007; France 2007b). Survey work has shown, for example, that many of the volunteers who would not have called themselves environmentalists before they became involved in various restoration projects now do so. Feelings of neighborhood confraternity and a sense of learning ecological history (that can develop very quickly, sometimes within only a few weeks, through engagement in restoration projects) can help to build this connectivity. And once this connectivity is in place, other unforeseen environmental benefits can ensue. Chicago, for example, now has the most ambitious green roof program in the country, avidly supported by the mayor, which might be directly related to the success of the wilderness restoration project and its created community of environmental advocates that cut across ideological differences (Light 2004). In contrast, New York's attempts to create a regional green network have been more difficult because, unlike in Chicago, there has not yet been a major restoration project for them to undertake together which would build that feeling of community and environmental citizenry.

RESTORATION AND RELATIONSHIPS

Light (2004) asked the following questions: if one of the values of restoration is its potential to get people physically connected with the land around them, to form relationships with the land, then how are we to understand these relationships and what are their qualities? And what does that mean in terms of our future relationships with the land if we do restoration as opposed to something else?

Relationships are not just defined by reciprocity and obligation. Normative relationships, for example, do not have to be justified in their creation of a form for interaction which results in actions. So what kind of guidance can this give us? We can have this sort of special relationship with a place based on accumulated personal history, as for example the marshland dwellers have in relation to the marshes (Al-Khayoun 2007). Light (2004) continued to ask a series of questions and posited answers based on this idea of relationship building:

- Do members of a particular marsh Arab family have presumptively decisive reasons to object to the destruction of the marshes? Absolutely yes.
- Do these same individuals have presumptively decisive reasons not to wantonly destroy the marshes? Yes, of course.
- Do these marsh dwellers exhibit virtues and vices in relation to their interaction with the marshes? Yes, as well.

- And, finally, can ecological restoration serve the same kind of connection between people and community? Absolutely.

Surveys from Chicago have, for example, found high sources of satisfaction in restoration that participants based on feelings of undertaking meaningful action; an increased fascination with nature; a belief in making life better for coming generations through connecting the past to the future; and an impression that they were doing the right thing for the larger human community, not just for nature alone (Light 2004).

So in addition to the goals of repairing the broken bits of nature, the participants in the restoration of Chicago grasslands were interested in establishing these types of normative relationships in terms of at least three outcomes: knowledge of local environmental history, environmental education about local and global environmental problems, and promotion of a sense of ecological citizenship.

CONCLUSIONS

People are not called to restoration work out of mere propinquity because it is simply another thing for us to do. People who become involved in restoration see it as another way of taking responsibility for their relationships with each other, with future humans, and with other biota in the environment (Light 2004). Restoration, therefore, becomes something that knits us together in a profound way (Jordan 2003; Light 2007). And the morality of restoration can be best described in terms of its virtue in building character. Higgs (2003) tried to look at what counts as a good ecological restoration. Light (2004), in turn, considered that ecological restoration can truly help us to define what counts to be a good human. Virtues have to be captured through actions, so the best restorations are going to simply be those that we actually do. As a result, there may be no more virtuous endeavor in the entire controversial arena of activities concerned with Iraq than that of assembling a healing confraternity of restorers to work toward healing an environment (France 2007c). So the value of ecological restoration is that it has a great democratic potential by maximizing public participation. In contrast to other environmental activities such as preservation, restorationists can therefore become value makers (Light 2004).

As France (2007c) describes it, ecological restoration then is a positive process of offering hope for a better future while at the same time acknowledging a sometimes shameful past. By addressing and correcting the sins of history (as discussed in chapter 1), restoration becomes an act of reciprocity, important for improving not only the quality of the outside environment of nature, but also that of the internal environment of human nature. Therefore, more than simply being a collection of repaired end products, restoration is a healing process of both ecological spaces and consciousness (France 2007a), both in obvious need of fixing in Iraq (France 2007c). And given that some practitioners have likened ecological restoration to a form of therapeutic gardening, what better place to engage in such than in the original garden—Eden?

The idea that community and shared identity with a place can be created through shared experience goes back to one of the sociological theories explaining the construction of the pyramids and the unification of Upper and Lower Egypt. Mendelssohn (1974) believed that the primary purpose of building the pyramids was a secular, not a religious, one. By bringing together formerly warring but recently unified peoples, pyramid building operated like a gigantic make-work project similar in scope to those implemented in the New Deal by Franklin D. Roosevelt during the troubled times of the 1930s. In this respect, it may not really be as accurate to refer to ancient Egyptians as building the pyramids as it is to say that it was the pyramids that actually built Pharaonic Egypt (Shaw 2003). Could marshland restoration in the form of ziggurat building (to use the vernacular monumental architecture as a model) play the same unifying and healing role in a present-day fractious Iraq? It is an intriguing idea. In this regard, the concept put forward by Light (2004) that restoration can lead to democratization is a direction worth pursuing much further.

For example, the Alexander River Restoration Project in Israel and Palestine (Brandeis 2007) demonstrates that restoration can indeed provide positive, unifying benefits to a war-torn, fragmented society. This restoration project served as a bridge between the different groups and was the recipient of an important international award. Significantly, the success of this project came about through nonsectarian cooperation among locals on both sides of the border. Today, river festivals unite Jews, Arabs, and Druze who enjoy the common benefit of a restored river channel, and plans exist for a cross-border peace park. People unite behind a desire to combat their common enemy—mosquitoes—and all military were banished from the opening ceremonies. Could such a strategy provide a unifying salvo against the sectarian violence now paralyzing Iraq?

ACKNOWLEDGMENT

This chapter was adapted from Light (2004).

REFERENCES

Al-Khayoun, R. 2007. Experiences and hopes of people of the Al-Ahwar marshes. In *Wetlands of mass destruction: Ancient presage for contemporary ecocide in southern Iraq*, ed. R. France. Winnipeg, MB: Green Frigate Books.

Brandeis, A. 2007. River of hope: The Alexander River Restoration Project: A unique Israeli-Palestinian effort to restore one polluted river flowing across both sides of the border. Lecture at the Harvard Kennedy School of Government.

Cronon, W. 1996. *Uncommon ground: Rethinking the human place in nature*. New York: Norton.

Cunningham, S. 2002. *The restoration economy: The greatest new growth frontier*. San Francisco: Berrett-Koehler.

France, R.L. ed. 2007a. *Healing natures, repairing relationships: New perspectives on restoring ecological spaces and consciousness*. Winnipeg, MB: Green Frigate Books.

———. ed. 2007b. *Handbook of regenerative landscape design*. Boca Raton, FL: CRC Press.

———. 2007c. *Wetlands of mass destruction: Ancient presage for contemporary ecocide in southern Iraq*. Winnipeg, MB: Green Frigate Books.

———. 2011. *Restoring the Iraqi marshlands: Potentials, perspectives, practices.* Sussex, UK: Sussex Academic Press.

Gobster, P. H., and R. B. Hull. 2000. *Restoring nature: Perspectives from the social sciences and humanities.* Washington, DC: Island Press.

Higgs, E. 2003. *Nature by design: People, natural processes, and ecological restoration.* Cambridge, MA: MIT Press.

Jordan, W. R. 2003. *The sunflower forest: Ecological restoration and the new communion with nature.* Berkeley: University of California Press.

Jordan, W. R., and A. Turner. 2007. Ecological restoration and the uncomfortable, beautiful middle ground. In *Healing natures, repairing relationships: New perspectives on restoring ecological spaces and consciousness,* ed. R. France. Winnipeg, MB: Green Frigate Books.

Light, A. 2000. Restoration, the value of participation, and the risks of professionalism. In *Restoring nature: Perspectives from the social sciences and humanities,* ed. P. H. Gobster and R. B. Hull. Washington, DC: Island Press.

———. 2004. Conceptual and moral issues in ecological restoration. Paper presented at the Mesopotamian Marshes and Modern Development: Practical Approaches for Sustaining Restored Ecological and Cultural Landscapes conference, Cambridge, MA, October.

———. 2007. Restorative relationships: From artifacts to "natural" systems. In *Healing natures, repairing relationships: New perspectives on restoring ecological spaces and consciousness,* ed. R. L. France. Winnipeg, MB: Green Frigate Books.

Mendelsohn, K. 1974. *The riddle of the pyramids.* Westport, CT: Praeger.

Shaw, M. 2003. Who built the pyramids? *Harvard Magazine* 4:42–49.

3 Restoring Babylon
Leaving a Better Iraq for This Century of Revitalization

Storm Cunningham

CONTENTS

Background 29
What Iraq Can Learn from Warsaw 30
Reblindness 31
From Pioneers to Long-Term Residents 32
Restoration: The Great Integrator 33
References 36

BACKGROUND

We're at a historic global turning point. The three crises (contamination, corrosion, and constraint) resulting from sprawl (from population growth) and from basing societies on natural resource extraction have always been local in nature: we've long had polluted, dilapidated, and overcrowded cities, regions, or even nations. Never before, however, have these three crises reached *global* levels, and now all three are coming to a head simultaneously in what can be referred to as "the restoration economy" (Cunningham 2002).

In the past, humanity dealt with these crises in one of three ways: the city or region was allowed to decline or even die, the people packed up and moved to an area that wasn't affected by the three crises, or the committed residents shifted to a restorative mode. Those three options exist when the crises hit at a local level, but two of them disappear when the crises become global: humanity can't yet move to another planet, and letting the entire global economy go into a death spiral isn't a viable option. Therefore, put simply, we either restore and revitalize, or decline and die.

In the 1960s, we started addressing the environmental problems of the world—contaminated land, air, and water; crashing fisheries; dwindling biodiversity; exhausted farmlands; destroyed farmlands (from sprawl); and so on—by trying to slow down the rate of destruction and pollution, and by trying to save a few examples of pristine ecosystems for future generations. All well and good. But that rearguard approach only delays inevitable economic and environmental disaster. It hardly lives up to the goal of leaving things better than we found them.

Here's the good news: we're now spending approximately $2 trillion annually on restoring the world, and it's the fastest growing of the three modes of development: new development (sprawl), maintenance and conservation, and restorative development (Cunningham 2002). Children have every reason to believe that the world they inherit will be healthier, wealthier, and more beautiful than the one their parents grew up in.

Obviously, renewing, restoring, and revitalizing the entire planet will have to be accomplished by a collaboration of industry, government, and nonprofit professionals. But where will this vast army of restoration–remediation–redevelopment professionals come from? Which universities offer degrees in restorative architecture, restorative ecology, or restorative engineering? Which ones turn out planners, public policy professionals, or economists with specific training in the theory and practice of revitalization? And where do students go to learn how to leave the world better than we found it … how to restore the world for a living?

More to the point of this book: as we labor under a worldwide plague of degraded watersheds, exhausted farmlands, polluted waterways, collapsing fisheries, dying ecosystems, contaminated urban land, decrepit historic assets, and failing infrastructure, what chance does Iraq have of being restored, ecologically, economically, or culturally?

WHAT IRAQ CAN LEARN FROM WARSAW

I submit that the sheer level of destruction that has taken place in Iraq, both pre- and post-U.S. invasion, makes it a candidate for (eventually) becoming the "Silicon Valley" of the Arab world's restoration economy. The only way an area can quickly build a critical mass of restorative skills at all levels—white collar, blue collar, technical, scientific, academic, and so on—is to have a huge quantity and diversity of restoration challenges in its own backyard. Preinvasion Iraq showed a strong inclination toward engineering and the sciences, so there should be no cultural resistance to moving to the leading edge of these disciplines, which is revitalizing communities and nations via the restoration of their natural and built assets.

I refer to this as the "Warsaw effect" (Cunningham 2002). World War II left Warsaw the most thoroughly destroyed city in the world: its infrastructure, industries, and public buildings were almost completely dismantled and eradicated. After the war, there was no Polish economy to rebuild, so it was created from scratch via the rebuilding of Warsaw. After the bulk of the rebuilding and restoration of public services was complete, the local populace found itself imbued with skill sets much in demand throughout Europe. So, they went off to help other war-damaged cities rebuild. The money they sent home while working abroad was what kick-started a new Polish economy. It was unfortunate that heavy-handed Soviet mismanagement of that economy didn't allow it to follow its own evolutionary path, as it might have blossomed into something quite unique, given that restorative beginning.

Given the level of damage to Iraq's urban and rural infrastructure, its great marshes, its coastal areas, and its historic sites; given its vast areas contaminated by crude oil, industrial chemicals, and war materials; given its expanses of formerly fertile farmlands that are now barren, salinized, and arid; and given that its governance structures are being remade from scratch—given all this, it seems that Iraq has the potential for a similar Warsaw effect. The major factors likely to impede the emergence of a restoration economy are (1) the paucity of social unity (which the Poles had plenty of), (2) the political and economic distortions inherent to oil-based economies, and (3) the heavy-handed U.S. mismanagement of the rebuilding process, which is overly focused on urban infrastructure (thus leaving rural areas desperate and unstable).

REBLINDNESS

Most of us are blind to the highly lucrative industries, technologies, products, and services that are restoring our world. We're so used to being in pioneering (sprawl) mode—building and maintaining new stuff, and conserving what's left of nature—that we ignore most of the end-of-lifecycle activities. I refer to this as "reblindness," since so much of what we can't see starts with "re": restoration, remediation, renovation, replacement, revitalization, reuse, renewal, regeneration, redevelopment, and the list goes on. Our reporting systems are blind to this fast-growing economy, too. We hear government reports on "new housing starts," but where are the reports on housing renovations and restorations?

Most of the new commercial space that comes onto the market each year in the United States is renovated, adaptively reused, or restored, not built new. Add in the new office buildings that are built on remediated brownfield (and other infill) sites (which also counts as restorative development), and the percentage of commercial space attributable to new sprawl develops shrinks much further. The same applies to infrastructure, as indicated by this quote from Howard B. LaFever, PE, DEE, executive vice president, Stearns & Wheler Co. (a 250-employee engineering firm): "Almost all of our sewer and wastewater projects are rehabilitation these days. Even with urban sprawl, construction of new systems is a rarity."

Few government agencies have any idea of how to go about changing policies, codes, and tax structures so as to enable and encourage revitalization (most still encourage quite the opposite). Many environmental groups are still focusing on greening destructive practices rather than encouraging restorative practices. They put most of their efforts into protecting marginal remnant ecosystems, rather than restoring the areas around these ecosystems so they can expand and connect to other ecosystems (ecosystems don't like to be small, nor do they like to be severed from other ecosystems).

One might ask whether specific training in restoration is really necessary. After all, can't any competent biologist restore an ecosystem, can't any competent architect restore a historic building? Sure, it happens, but then, civilians with no police training also stop burglaries, and civilians with no medical degree also revive heart attack victims. But that doesn't mean they should try to make a living at it. In fact, restoration and revitalization, if done properly, usually require unique technologies, materials, processes, policies, training, and technical disciplines. Consider the following:

- Draining a marsh is a very different process from restoring life to a drained marsh.
- Designing, constructing, and maintaining a new building are very different processes from renovating, restoring, and/or adaptively reusing an old building, especially if it has features of historic significance or artistic beauty.
- Designing, constructing, and maintaining a new bridge or a new concrete sewer system are very different processes from restoring an old stone or steel bridge, and very different from renovating, redesigning, or relining a century-old brick sewer system.
- Pumping water out of a fossil aquifer and polluting what remains in the aquifer are very different processes from restoring the quantity and quality of water in an aquifer.
- Depleting fossil topsoil via tilling and overuse—while killing soil communities via herbicides, pesticides, and artificial fertilizers—is a very different farming process from the kind that increases the quantity and quality of topsoil with each agricultural cycle.
- Clearcutting an old-growth forest, strip mining a mineral deposit, and contaminating a property with industrial waste are very different processes from restoring a natural forest (not to be confused with the monoculture tree farms that we often call "forests"), from restoring a mining site to a healthy ecosystem, and from remediating and redeveloping a toxic old industrial site into a vibrant neighborhood.

FROM PIONEERS TO LONG-TERM RESIDENTS

> The nation behaves well if it treats the natural resources as assets which
> it must turn over to the next generation *increased* … in value.
>
> **—Theodore Roosevelt, 1910 speech entitled "The New Nationalism"**
> **(emphasis added)**

Restoring our world is a very different economic model from basing our wealth on plundering fossil resources and developing new communities. Thus, it's obvious that switching to an economy that leaves the world better than we found it is going to require a thorough overhaul of our higher education, our professional associations, our industries, our nonprofits, our sciences, our research agendas, our government policies, our community development models, and—above all—our expectations. This isn't a shift that requires excessive force. After all, restoration is a natural part of the lifecycle of everything, living or built (see chapter 1). Everything gets to a point where maintenance no longer suffices, where it needs redesign, replacement, or regeneration.

This is bad news for those who want to keep operating in the old "pioneer" model, which equates economic growth with the conquering of new lands and the extraction of virgin resources. Over the past fifteen years or so, we've raised our expectations. Now, we're no longer satisfied to reduce the rate of despoilment; we want to actually reduce the immense global inventory of despoiled lands, damaged resources, and worn-out built assets. As a result, a trillion-dollar "restoration economy" has arisen because there's just as much money to be made restoring our world as there is to be made from sacking it (Cunningham 2002).

Perversely, restoring damage from sudden catastrophes is far easier to fund than preventing politically connected companies from doing the damage in the first place, or restoring damage that took place over an extended period. Look at the hundreds of billions of dollars flowing toward the U.S. Gulf Coast after Hurricane Katrina: an expense (not to mention the human and wildlife suffering) that could have been largely prevented with a fraction of that amount spent on ecological restoration and infrastructure renovation that had been recommended since the mid-1990s. But even that entrenched political behavior is changing. Our new knowledge of the precise costs of restoration is empowering a new generation of legal tools to both prevent and restore damage. We're seeing a rise of restorative fines and deposits, and they are no longer based on arbitrary amounts, making them far more defensible in court.

The business, nonprofit, government, and academic leaders of this restoration economy have already proven beyond any shadow of a doubt that restorative development is the only possible path to sustainable economic growth on a finite planet with a growing population (or even with a static population) (Cunningham 2002). In fact, it's the only practical path forward even if we had a *shrinking* population, due to the vast global inventory of damaged and depleted natural resources, the vast global inventory of decrepit, demoralized communities that are plagued, in turn, with vast inventories of derelict buildings, crumbling infrastructure, dying (or fleeing) industries, and poisoned properties.

RESTORATION: THE GREAT INTEGRATOR

It would be presumptuous and foolhardy to prescribe the specific courses of action that Iraqi leaders of business, academia, citizen groups, and government should take in order to restore and sustainably develop their nation. But the right guidance and vision at the right time could work wonders.

One of Iraq's primary restoration agendas must, of course, be watershed restoration. As long as the nation remains largely dependent on Turkey, Iran, and Syria for its water, it can never achieve a meaningful restoration of the marshes, much less sustainable economic independence once the oil runs out. But watershed restoration is far easier to fund and accomplish if it's integrated with agricultural, infrastructure, ecosystem, brownfield, and fishery restoration (Cunningham 2002). What's more, taking large areas of arid land back to the savannahs, marshes, and forests they were before deforestation, tilling, and overgrazing ruined them (in many cases, many centuries ago) is both possible and practical today through integrated revitalization. Such renewal strategies integrate along two axes: the four stakeholder groups (business, academic, government, and citizen/nongovernmental organization), and the eight sectors of restorative development (ecosystem, watershed, fishery, agricultural, brownfields, infrastructure, heritage, and catastrophe damage) (Cunningham 2002). Tools devised by the Revitalization Institute, such as the *Integrated Revitalization Guide* (www.revitalizationinstitute.org), will help them collaboratively devise a sustainable renewal strategy that—while it might not please everyone—will produce the most revitalization in the shortest period for the least investment.

Integration actually has its own restorative effect. Cities often spend vast sums on redevelopment and restoration projects, without getting the level of revitalization they desire. This is due to "silo thinking," which produces a series of isolated, disconnected projects: historic, brownfields, watershed, and so on. With its natural efficiencies and synergies, integrated revitalization programs tend to yield a restoration effect far greater than the sum of their component projects.

The reverse also is true: restoration has an integrative value. Restoration is uniquely nonpartisan, and thus integrative, because everyone loves bringing things back to life. At least the Poles had the advantage of rebuilding their country in the absence of war. Rebuilding in the midst of armed conflict is a chancy proposition at best, but if anything can bring people together, restoration can (see chapter 2). For example, restoration brought Muslims and Christians back together in Mostar, Bosnia, where together they rebuilt the five-hundred-year-old Stari Most, an iconic Muslim bridge destroyed by the Christians during their war with each other (which had ended just a scant year or two earlier). And in 2004, Israelis and Palestinians together completed a ten-year project to restore the Alexander River, which they share. This once highly polluted, nearly dead little river has now been dramatically returned to health and beauty (see chapter 2).

For millennia, our heroes and people of wealth have gained prominence by discovering, conquering, and exploiting "new lands." English explorer James Cook, Spain's conquering general Francisco Pizarro, and U.S. oil magnate John D. Rockefeller each in their time gained public acclaim for their contributions toward the exploitation of newly found resources. The fact that other people were already living in the "new lands" was of little consequence, especially if they were "wasting" their resources (in other words, using them sustainably).

Companies based on sprawl and resource extraction are now regarded as villains, and the heroes of the twenty-first century are those undoing the damage of our earlier heroes. In some cases the future of entire organizations hinges on their ability to switch from new development to restorative development, and much of their restoration work will be undoing what they themselves did earlier. The U.S. Army Corps of Engineers is wrestling with that transition, as for example in their restoration of the Everglades. The Jewish National Fund spent most of the twentieth century draining marshes and straightening rivers (see chapter 6). They too are now undoing some of their own projects. In fact, they put up $12 million for the Alexander River restoration mentioned above.

Half a millennium ago, Spanish conquistadores destroyed the city of Tenochtitlan (now called Mexico City), in part because it was "too civilized." When the first Spaniards encountered the city, one wrote in his diary that they wondered if they had reached heaven. Their stinking, overcrowded, disease-ridden European cities were shamed by the sparkling white Aztec city, with its limestone structures, its hanging gardens of flowers on most buildings (reminiscent of Babylon?), and aqueducts, not to mention a restorative sewage treatment system at nearby Xochimilco—sort of a miniature version of Mesopotamia's marshes (see chapter 11)—that was state of the art (even by today's standards). Chinampas, an agricultural technique similar to floating gardens, were able to handle large influxes of organic waste, and could break down even the complex molecules of industrial toxins. It was also a restorative sewage system, in that it had the ability to return water to the watershed that was cleaner and more oxygenated than the original source water that was extracted from a river or lake. Such systems can thus contribute to river and lake restoration.

Tenochtitlan could not, and will not, rise again. Its culture and religion were merged (via the torture and execution of their priests) with European culture and religion (although it's said that the indigenous religion altered local Catholicism almost as much as Catholicism altered it). Likewise, the glory of ancient Babylon is unlikely to rise again, but this isn't for want of trying on Saddam Hussein's part.

From 1984 to his overthrow by the United States (which had helped him consolidate his dictatorial power in the first place), Hussein spent some $800 million rebuilding the ancient city of Babylon. Hussein's approach to "restoring" the city paid scant respect to the ancient ruins, which he literally built on top of, rather than excavating and properly restoring them. The United Nations Educational, Scientific and Cultural Organization (UNESCO) protested this destructive form of "restoration," as this immense project was obviously meant to be more a new monument to the restorer than a restoration of ancient Babylon: "Saddam Hussein assumed control of Iraq in 1979. Soon afterward, he allocated funds toward the restoration of Babylon. Over the past two decades sixty million bricks have been laid in the reconstruction of Nebuchadnezzar's fabled city.... He has taken great care to restore the glory of the ancient city. ...By rebuilding Nebuchadnessar's city, Hussein exploited the natural opportunity to portray himself as Nebuchadnezzar's modern successor" (Dyer 2003). Original bricks at the ancient site are inscribed, "I am Nebuchadnezzar, the king of the world." Saddam's bricks say, "In the year of President Saddam Hussein all Babylon was constructed in three stages. From Nebuchadnezzar to Saddam Hussein, Babylon is rising again."

It might well be that Babylon has more value as a ruins than as a rebuilt attraction. Maybe there's something of value that could be salvaged from that expensive effort. But if it were to be restored, and if peace were restored, and if the country's other built heritage were restored, and if Iraq's infrastructure (rural and urban) were restored, and if its agricultural lands were restored, and if its watersheds and rivers were restored, and if the marshes were restored, what a wonderful country it could be ... again.

That sounds like an impossibly expensive dream, but for one factor. The oil that has (at present) turned Iraq into the new Vietnam could fund all of this restoration, laying the groundwork for a country that could thrive well after the oil runs out. Of course, Iraq would have to own and be the primary beneficiary of its oil, which might not be one of the goals of the current U.S. agenda, but that's another story. While the United States has been acting more or less unilaterally in Iraq, the future of international support for Iraq looks good, because restoration has many political upsides and almost no political downsides. Japan, for instance, is funding an $11 million project to restore the Mesopotamian marshes, train 250 Iraqis in wetlands management, and provide clean drinking water and sanitation to about 100,000 marsh Arabs who are living in the marsh area during the restoration (France 2011).

Given a well-designed integrated revitalization strategy, along with the extensive restoration that could be funded by both international assistance and Iraq's vast-but-temporary oil wealth, the potential certainly exists for the peoples of what we now call Iraq to harness the Warsaw effect and rebuild a wonderful new civilization.

REFERENCES

Cunningham, S. 2002. *The restoration economy*. San Francisco: Berrett-Koehler.

Dyer, C. H. 2003. *The rise of Babylon*. Chicago: Moody.

France, R. L. 2011. *Restoring the Iraqi marshlands: Potentials, perspectives, practices*. Sussex, UK: Sussex Academic Press.

*Environmental Planning
Theory and Practice*

4 Before the Master Plan
A Framework and Some Difficult Questions for Regional Land Use Planning

Carl Steinitz

Before the objectives outlined in the master plan for the Iraqi marshlands (USAID 2004) can be met, there is a need to examine the methodology at arriving at that stage. I would like to focus in this chapter on the issue from the perspective of modern development, rather than the marshes. I believe that the marshes are part of the situation but won't, in the long run, drive the sustainable social ecology of that area.

If you see the issue as a modern development issue (see the foreword), you may end up asking different questions. No one has asked the question "What happens if there really is no water for these people in their territory?" And no one has asked the question "What's the smallest wetland that is sustainable, and what is the largest wetland that's sustainable?" These questions are hard questions. I'm not underestimating the difficulty of trying to do something positive in a devastated social ecology. That's an enormous task, and you have to admire the people who are doing it. But you also have to ask additional questions: what is the time horizon? What really is the best use of Italian and American money and Turkish, Iranian, and Iraqi water? And how long in the future and for whom are you really talking about?

I have organized around thirty regional studies (e.g., Steinitz et al. 1996, 2003; see also France 2006 for other examples as well as the case study described in chapter 15 of this book), and I'm not good at the end game of making decisions. I don't do that. I'm experienced in the beginning game of trying to figure out how change comes to a region and how that might be planned for. "Plan" isn't exactly the right word. Rather, how can change be managed as a process? I'm well aware that the Iraqi marsh situation is one that has had a decade or more of very serious people doing very important work on it. And in a sense, that trajectory is like a rolling train. And some fair questions might be "Is it aiming in the right direction?" and "Are there several directions and branches that have to come together, but that might not?"

If you step back from these questions, you might also say, "Wait a minute, the real issue is what's going to happen to these people in their territory, part of which is a marsh-related culture, but part of which isn't." You have to ask, "What's going to happen to three-quarters of the land area that's likely to stay dry? And what's going to happen to three-quarters of the people who are not interested in returning to live in the marshes? And where is the balance of effort and money between those two groups?" I'd like to approach this problem as though these were the main questions. I want to present the framework that we use in organizing regional studies and posit several other hard questions in the process.

First of all, there are some generalizations that can be made. There are two kinds of people, at least, involved at the stage of designing studies like this. There are the people called "stakeholders," which for the Iraqi marshlands include both the minister of water of Turkey and also the persons in the boats in the marshes themselves; in other words, it's not just intellectuals rhapsodizing about "Eden" or the minister of water of Iraq. There is also the technical research team. And our roles as part of this latter group, in my opinion, are very different; indeed, problems arise when they are the same.

The stakeholders have a dual perspective. The first is that they have essential input on the possible future of the region. And the assumption that you always should make is that they do not agree with each other. They are not going to agree with each other because they often see it as a zero-sum game. Once you believe that, then all the bottom-up community work, the handholding, is likely to take longer than the time span of the problem. It's not going to work. Certainly it's not going to work alone. Stakeholders have scenarios of what they think a desirable future might be; it may be partial, it may be speculative, it may be geographically limited, it may be private, it may be ecologically oriented, and/or it may be profit oriented. But they do have these ideas. And one problem is getting at those ideas. What we need to do is to organize information in their language—not in our language, but in their language. We need to model the processes that they're talking about, compare their scenarios, and present them with information for their second essential role. That is the public review of the process and findings as a component of local debate and decision making involving local, and in this case also regional and probably international, individuals. I think it's a mistake when the stakeholders and researchers are the same, because impartiality is absent. When co-joined, as a researcher you have ideas, you're trying to sell them, and you're trying to convince people. But that means that you're part of the problem and you're not reflecting the others' interests.

The question of what the master plan should be or what the decision support process should be is a complicated one. The likelihood of having a long-term master plan in a fluid situation like the Iraq marshes is essentially zero. It's not going to happen. And if it does happen, it's not going to be effective. Here we are in the present. It's very clear that the past has a role. The marshlands of Iraq have always been highly variability (France 2012). The past has therefore not been stable; it has ebbed and flowed spatially, functionally, hydrologically, and in all cases dramatically. And we are here in the present, yet need to do something for the future by looking forward in time. And one of the interesting questions is, how far forward in time? Does this need to be sustainable? For what: a year? Ten years? A hundred years? A thousand years? How far? And what is the periodicity of sustainability or even its overall effectiveness?

The problem is that we have billions of alternatives when we try to visualize all the restoration strategies that have been raised about the marshlands as a combinatory set—billions of alternatives which you want to avoid like the plague. Then there are possible ones. You can't do that either. Then there are plausible ones. "Plausible ones" are things that more than one person believes are possible. And then there is a much smaller set that may be feasible, meaning that we could actually do that or this, or this, or this. And what you want to be able to do is to narrow the range toward the feasible, knowing that you're not going to hit the best. Then you should hope that you can implement something, knowing that once you get there, the process starts again and people can change their minds. So what you're looking for is a short time plan with high adaptability: maybe ten years, maybe twenty years, but not longer than that because conditions will change, especially in a country like Iraq and especially in a traditional culture like that of the marshlands. How do you get to that point?

One way is to design an alternative future. Let's say that we want a large wetland; we want people to live in grass huts, we want ecotourism, we want this, we want that. If you're designing a master plan, the problem becomes how you get there, and the big question mark concerns the implementation strategy. And usually you fail. The second approach is to start with the present and then to design a set of programs and policies and ask in a simulation of some kind, "Where is that going to take us?" The problem in this approach is that you don't know where it's going to take you. So each of these two approaches has a serious and risky flaw.

Then there is the incredibly difficult problem of the chain of decisions. Here we are, and we know we have a history. And we know we have facts on the ground. And we know we have some constants. We know our geographical location. We know we're a part of the water system of Iraq and Turkey. We know we're ultimately subject to Turkish decisions. We know some things, such as assuming that we want to keep the people in the marshes despite their possibly inhospitable nature. And then we're going to embark on a set of assumptions. We're going to say, well, if this happens first, then this is assumption number one, then two, three, four, and so on. The master plan for the Iraqi marshlands (USAID 2004) probably had one thousand assumptions, each of which had its own budget. The problem is that if you embark on the most influential assumptions, the ones that drive the system, and if you have chosen the wrong primary assumption, you've lost. In other words, you've gone forward in one direction, whereas according to your original master plan you should have gone forward in another direction. Therefore, every combination of those assumptions leads you to a different master plan in space, time, consequences, and so on. So the idea that you can hit that ideal master plan and get it done is absurd. And this is especially true for a large, complicated, changing area such as the Iraqi marshlands.

So what do you do about that? The answer is to try to deal with the strategic policies and let the tactics go. It doesn't matter what the tactics are if you lose the strategy. And the questions, then, are "What are the leading issues?" and "Shouldn't you spend time assessing the likelihood of feasibility from the combinations of the leading issues before you go into the tactics?" The answer is "yes." And that's why it turns out, in my experience in the last two to three decades, that scenario-based studies of alternatives—before you even think of making a master plan—comprise a useful, and in some cases even essential, exercise.

Therefore, you need a methodology which is robust and which captures the basic issues. It doesn't matter whether the wetlands of Iraq flood this way or that way. The real issue is "Should they use scarce water to reflood the marshes in the first place?" The planning approach is to be flexible. It has to expand geographically. It has to be able to add components. It has to be easy to update in data and models. It has to be easy to change assumptions about land uses, scenarios, and the resulting alternative functions. A scenario, by the way, is a set of policies and plans. The alternative future is the result of those scenarios.

The planning and design framework that I and my colleagues have used for twenty years (Steinitz 1990) is based on answering six questions. And it has to answer them at least three times. These are the six questions: (1) How should the landscape be described? (2) How does it work? (3) Is it working well? (4) How might it change? (5) What predictable differences might the changes cause? And then, finally, (6) how should it be changed?

The first question is "How should the landscape be described?" This is essentially answered by representation models which include maps, pictures, photographs, stories, songs, and legends and memories. These models therefore include data such as dissolved oxygen measurements or maps of transportation systems, and they are neutral.

The second question is "How does the landscape work?" This is answered by process models which describe the systems that are at work in the area. These process models are ecological ones, hydrological ones, social ones, economic ones, political ones, and other kinds; they are not data but rather contain purposeful information.

The third question is "Is the landscape working well or not?" This is answered through the use of evaluation models based on cultural knowledge. What is considered a good life in Iraq in the marshes may not be considered a good life in Cambridge, Massachusetts. And the question is "Who decides whether it's working well?" Even in Iraq, there are two cultures: the Baghdad intellectual, academic, and political culture, and the uneducated farmer culture in the marshes. Whose decisions are going to determine what is good and bad?

These three questions must be asked of the existing conditions. The fourth question is "How might the landscape be changed?" This might be the simplest one to answer. This is because it is based on data that describe the following: what if we did this? Or this? And so on. This leads to the generation of many competing ideas as valid answers.

The fifth question is "What predictable differences might the changes cause?" This is answered through the use of impact models, which are essentially the same as the process models under different data conditions describing potential changes.

And, finally, how should the landscape be changed? These questions are answered through the use of decision models that are, once again, based on cultural knowledge.

The first time you have to ask these questions, you're doing it for the purpose of recognizing the context: (1) We've got to go to Iraq. We know where it is. (2) We know about the hydrological and social functioning. (3) It's working badly. Apparently, it used to work quite well. (4) It's going to change, but it could have a lot of water or no water. (5) What difference does it make? I'm not sure yet. (6) And decision making:

how should it be changed? I surely don't know that. I'm going to try and figure out who is going to need to make what decisions and when.

It is important to realize that the second time through, you are asking the questions in the reverse direction in order to design the method of the study itself. (6) Who needs to know what to make a decision, and what decisions are they going to make? (5) That is why you have to have impact models, and that is why you need process models. (4) And what changes are we going to propose and simulate? (3) And why make a change if things are working well? But we, in this case, know they're not necessarily working well. (2) And to do that, you need to understand the processes. (1) And to understand the processes, you need data. And only then, after you go though the framework in this direction, do you know what data you need to try to obtain.

And then you have to do the study: you get the data, you organize the models, you evaluate, you propose lots of changes, you compare them, and you present them for decision. And three things can happen. People can say "yes," "no," or "maybe." If they say "no," it might be because you need better data or more data. It might be because the evaluations are wrong. It might be because there are other changes that people want to propose. It might be because you might be able to mitigate or alter the impact. And it might be that you need to educate decision makers, or they need to educate you (which is more likely). It may be that the answer is "yes" and you've been lucky: you hit the nail on the head, as is said. And then you've got the problem of implementation. And you do the things you're supposed to do. And time moves forward, and the next generation will revisit the issue, and your changes and decisions will then be their data. The third thing that can happen is the answer "maybe." Maybe you are asking this question at the wrong scale, in which case you start the process over again. You ask the same questions, but the data are different and the models are different. You also have to realize that you're here, in a simultaneous matrix of scale and time. You're working at a regional scale, while somebody's working on their own little farm, and Turkey and Iraq are trying to figure out where the water is going to come from. This surely isn't the first plan for the marsh area. And somebody is surely going to do it again in the future. This is the framework.

I want to finish with a very quick overview of about thirty additional difficult overarching questions that have to be muddled through in regional land use planning. Who should participate and how? Regional experts, local residents, outsiders? What is the purpose? Scientific advancement or public action? What is the trade-off between faster study results and action versus possibly better research but later decisions? Could we solve the problem of a master plan by getting ten people around the table in one day? And would we be 70 percent right? And is it worth waiting three years and spending $20 million to be 90 percent right? That is a very serious question. Will the study product be a single effort, meaning a master plan? Or a continuing decision support process? What is the appropriate cost? How much time, money, and basic research are really needed?

Here are the questions asked on the framework's first pass:

1. *Representation*: where is the study area? What is its history? What are its past and present physical, economic, and social geographies?
2. *Process*: what are the area's major natural and social processes? How are they linked to each other? (By the way, several charts I've seen of the Iraqi departmental environmental structure show many organizations isolated in boxes with no arrows among them.)
3. *Evaluation*: are there important problems in the region? Are there important opportunities in the region? Which? Where?
4. *Change*: what major changes are foreseen for the region? One of them is the possibility that there is no water at all. Are the changes related to growth or decline? Are the pressures for change from inside or outside? (Personally, I think that for Iraq they're going to come from the outside.)
5. *Impact*: are foreseen changes seen as beneficial or harmful? Are they seen as serious? As irreversible?
6. *Decisions*: who are the decision makers? Who are the major stakeholders? Are they public or private? Are their positions known? Are they in conflict? You need to ask those questions the first day you go someplace.

So we've asked those very preliminary questions.

And now we're going to go up through the framework to define the methods:

6. *Decisions*: what do decision makers need to know? What are their bases of evaluation? Are these scientific evaluations? Cultural norms? Legal standards? Do the people who are going to make these decisions really know the difference between the water chemistry at pH 8.2 or 8.3? And do they care? Are there issues of public communication? Or visualization? Of education? A general rule is that the more complex the models, the less people understand and trust them, and they won't decide on their basis.
5. *Impact*: which impacts? Which costs and benefits are seen as good versus bad? How much, where, when, and to whom are these seen as good versus bad?
4. *Change*: is there one master plan, or several scenarios reflecting different assumptions, policies, and uncertainty? Who defines the scenarios for change, and how? Which scenarios are selected? Toward which time horizon? At what scales? Which issues are beyond the capabilities of the models? Are the alternative future outcomes simulated, or are they normative allocations? Are they the process of modeling the process of change, or are they designs?
3. *Evaluation*: what are the measures of evaluation? In development, economics, ecology, and politics?
2. *Process*: which models should be included, and which not? How complex should they be? How reliable must they be? In a special issue of *Ecological Systems* in 1971, there is a wonderful article about the special problem of modeling ecological systems that identified four models. First, they are systems with complex components and feedback mechanisms. Second, they react to historic events as well as to the present. Third, they are interlocking;

they have neighbor and other special effects. And finally, and most important, they are nonlinear. A small change can have a major impact.

1. *Representation*: which data are needed for which geography, at what spatial scale, at which classification, for which time, from which sources, at what cost, and in which mode of representation?

After all of this, perhaps now you know what data to collect. And then perhaps you can do a study that eventually leads to a feasible master plan, and then maybe you can get something positive implemented. I think that with respect to the marshlands of southern Iraq, it will be a messy and complicated process that will require many changes en route.

REFERENCES

France, R. L. 2006. *Introduction to watershed development: Understanding and managing the impacts of sprawl.* Lanham, MD: Rowman & Littlefield.

―――. 2012. *Back to the garden: Searching for Eden in the Mesopotamian marshes.* Cambridge, MA: Harvard University Press.

Steinitz, C. 1990. A framework for theory applicable to the education of landscape architects (and other design professionals). *Landscape Journal* 9:136–43.

Steinitz, C., and others. 1996. *Biodiversity and landscape planning: Alternative futures for the region of Camp Pendleton, California.* Cambridge, MA: Harvard Design School.

―――. 2003. *Alternative futures for changing landscapes: The Upper San Pedro River Basin in Arizona and Sonora.* Washington, DC: Island Press.

USAID. 2004. *Master plan for the Iraqi marshlands.* Washington, DC: USAID.

5 Control Models of Tourism Development and Conservation Management with Respect to Indigenous Culture

Shiau-Yun Lu

CONTENTS

Central Control Conservation Model and the Indigenous Culture 48
Co-Management Model and Indigenous Culture..50
Indigenous Sovereignty Model 52
Tourism Type and Control Power 53
Development Stages on Indigenous Land...54
The Relationship among Control Power, Development, and Tourism Type............56
References...57

The ecocultural landscape of the Iraqi marshlands has the potential to be a major tourist destination (as suggested in Chapters 12, 13, 15, and 16). In many parts of the world, the most biologically diverse lands and treasure grounds from the perspective of conservationists are homelands of indigenous peoples (such as the Marsh Arabs of southern Iraq), who have lived on the lands for generations and form their own culture, language, and identity (Abrahams 1994). The traditional worldwide conservation concept is to set aside these lands as nature reserve areas based on the advice of people outside the area and on strong control from a mainstream sovereign. In a good scenario, the indigenous people living in the reserved areas are told about the plan. In a worst-case scenario, these indigenous people would lose the right to practice their original lifestyle and would even have to move out of the reserve areas (Dasmann 1991). Thus, the conflict between conservation and indigenous community emerges, and similar situations can be found all around the world with different contexts and geographic environments.

The key issue of this conflict lies in the different interest groups, who have vested interests in the same piece of land and natural resources and different definitions of conservation. The interest group who gets the control power treats the land with their conservation objectives and consequentially affects the indigenous culture. Therefore, in most of the literatures, power relationships comprise one of the major aspects in the discussion about how indigenous culture is influenced by the world culture. This park and people relationship is usually divided by the degree of control power involvement (Stevens 1997; World Bank 1996; Igoe 2004). On one end lies entirely central control or a mainstream sovereign, and on the other end lies total indigenous control. And in between the two extremes are various instances of the government and indigenous people co-managing the same piece of land.

Apart from the issue of control power, tourism impact is another significant aspect in the discussion about the literature of indigenous culture transformation. Since these indigenous lands are targeted as tourism destinations because of their natural beauty and rich cultures, tourism would gradually become one of the most important incomes for the indigenous communities. Their life and culture are affected consequentially. These two aspects of influence, control model and tourism, should not be discussed separately. In fact, they relate to each other and shape the image of indigenous culture. In this chapter, I will begin with the discussion of how the indigenous culture is affected by control power, and subsequently I will add two other dimensions, tourism and development, to create a new relationship model.

CENTRAL CONTROL CONSERVATION MODEL AND THE INDIGENOUS CULTURE

The varieties of park–people relationships can be described by the degree of power involvement (Figure 5.1). The traditional national park model, which is entirely controlled by the central government without any involvement from the indigenous people, is the strictest conservation model. The most popular definition of "national park" that is cited as the basic principle internationally is that a national park is one that is "not materially altered by human exploitation and occupation … the highest competent authority of the country has taken steps to prevent or to eliminate as soon as possible exploitation or occupation" (International Union for the Conservation of Nature and Natural Resources 1980). Thus, conservation is often the main purpose of a national park conservation model, followed by research and tourism. From this definition, indigenous settlement and activities are not included in the nature reserve scene. Any utilization by indigenous people is strongly restricted in the protected area under the high conservation principle, and many indigenous people are eliminated from the park (West and Brechin 1991). The elimination process results in indigenous communities totally disappearing from the park or clustering along the park boundary.

There are three types of consequences for indigenous people who have been eliminated from the park. The first scenario is when an entire indigenous culture vanishes. The elimination process here is because of not only the physical relocation but also the cultural invasion. Many indigenous communities are assimilated by the mainstream culture and lose their cultural identity gradually. In this type of situa-

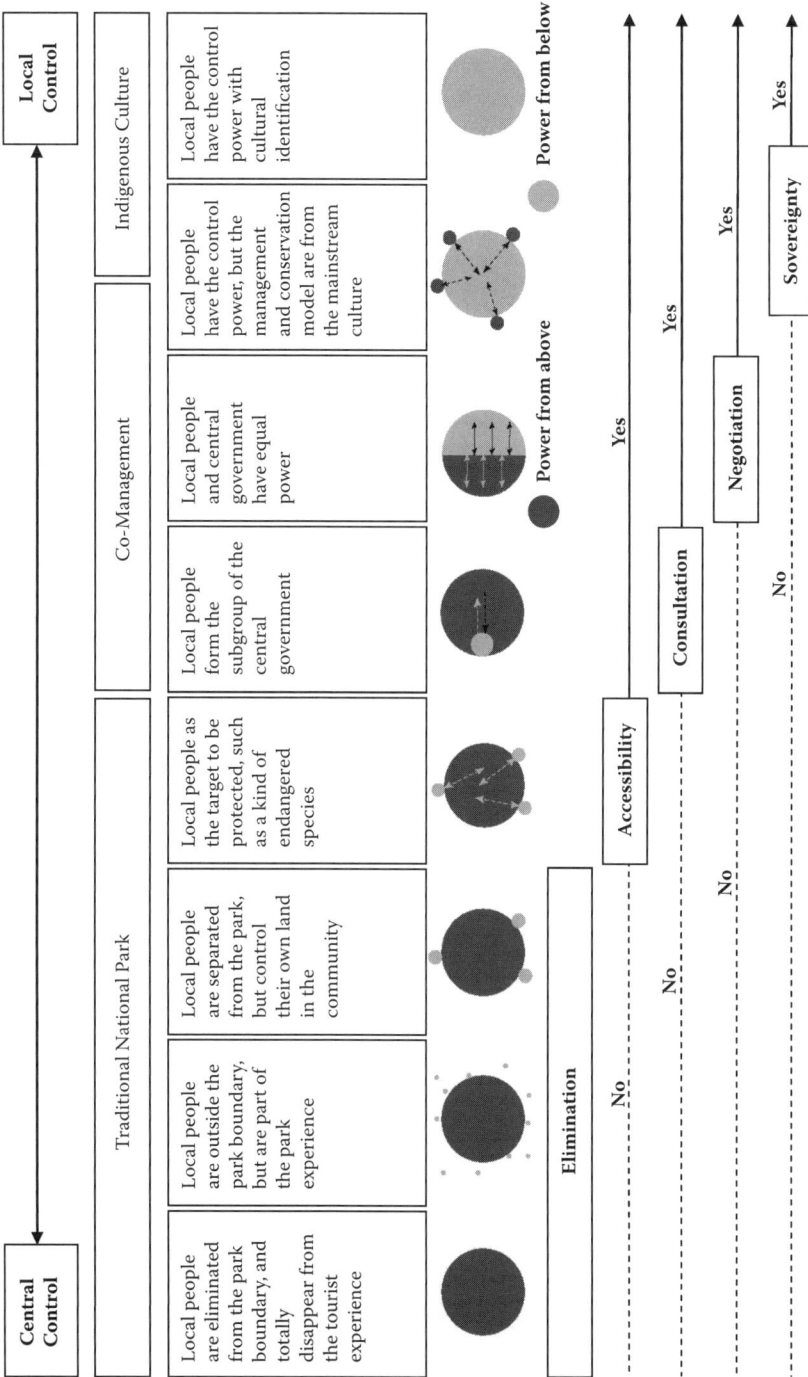

FIGURE 5.1 Park and people models divided by the control power relationship.

tion, indigenous culture totally disappears from the tourism experience, and is only seen on the history board in the museum or visitor center.

The second scenario is the attempt to eliminate the indigenous people from the park area, but the indigenous people will continue to live at the park boundary and be included in the tourism experience. It is a common scene to spot indigenous people wearing traditional cloth for pictures or selling crafts before entering and upon exiting the park. In some cases, one can find a strip of stores selling indigenous arts and crafts around the national park boundary. For the tourists, this will allow them to experience some aspects of the culture and bring back the local flavor with satisfaction. From the aspect of park authority, which has the entire control of the park, they assert that the national park can raise the local economy and job opportunities. However, the fact remains that the indigenous people have already lost their cultural identity since they are not living on their home land retaining their daily cultural practices. Although they may keep the skill of art and crafts alive, many indigenous people claim that the original motivation is missing, the craft work has become commercialized, and only a few indigenous people truly benefit from tourism (Hitchcock 1997). What is more, tourism increases the poverty gap and the inflation that results in the tourism area, increasing the hardship faced by the local people.

The third type of situation is the delimitation of the park boundary that physically excludes the indigenous settlement. The indigenous communities keep their own villages, but they do not have the right to utilize the resources inside the park areas. In this case, even though the indigenous people are separated from the park control and have part of their home land under their own control, their ways of life are changed. Their livelihoods are disrupted or destroyed by the establishment of national parks. People need to find other ways to live on. Many of the indigenous communities on the national park boundary are viewed as problems of poverty and a cause of the environmental degradation. It is hard for the indigenous culture to survive in this situation.

With the increase of recognition and appreciation of the different cultures, a new type of park–people relationship arises. Indigenous people are now viewed as the protected target. This type of relationship is still close to the end of the traditional national park model. Indigenous people have access to the park and resources inside the park, but they are not involved in the park management and governance. Under the central government control, the indigenous community is monitored as one of the endangered species. Although the indigenous people retain their own life, they lose their cultural autonomy.

CO-MANAGEMENT MODEL AND INDIGENOUS CULTURE

The discussion of the conflict between park and people usually revolves around two aspects. One is the overlay exploitation of the natural resources, and the other one is the insufficient involvement of local people in the process of decision making. The latter may cause fiercer resistance because of the neglect and disdain of indigenous rights. The conflicts obstruct indigenous people from cooperating with the conservation goal that the park authority sets out. When the traditional "top-down" approach suffers from resistance and criticism because it fails to deliver the promised environmental conservation and local development goal, a new approach emerges which is based on working with the local community to reduce the conflict. Many discussions have begun to recognize the importance of local knowledge and the necessity of cooperation between the local community and the conservationist (Western and Wright 1994). This is the start of a new type of conservation management model.

This co-management concept does not have a rigid definition, and it describes a wide range of possibilities based on the degree of local participation and control power involved. It marks the transition from the traditional central control model to the entire indigenous control model. The objectives of the traditional park plan try to identify what parks can do for the environment and local people. And the co-management strategy shifts the attitude to how the local people can be part of the control power and what they can do. In the past decades, cases of environmental and cultural conservation working with indigenous communities have happened around the world (Stevens 1997). The difficulty lies in co-managing the same resource with different meanings and definitions of natural resources. When these two groups, the central government and local community, co-manage the land, the group that gets more power will have more authority on the decision making. In the co-management process, indigenous people can be consulted and negotiated, and be treated as an autonomous entity from a different degree of power involvement. This is similar to the "co-management ladder" that Sneed has mentioned, which ranges from virtual nonparticipation at the bottom to almost full local control at the top (Sneed 1997).

Consultation by the conservationist and government marks the beginning of the involvement of indigenous people. In this stage, the indigenous people act as the local information source for decision making, but they do not have the real authority to change the decision. At the best-case scenario, the indigenous people form a subgroup in the central government organization. The purpose of this subgroup is to bring the local voice into the decision-making process and to show goodwill to the indigenous community. To the indigenous people, this type of co-management offers them some protection of their indigenous culture and environment, and promise of involvement in policy making. In reality, however, the involvement of the indigenous people is limited since not many of them are engaged in the process and they do not have an equal political position with the central government.

When indigenous people and government have equal power in management, negotiation will take place which will result in a strong, unified voice in environmental management. The case of co-management in Australia is one of the most influential models in this negotiation process, and it draws significant attention from the world's indigenous communities as an experience to learn from. One significant step in this case is the legislation of the Aboriginal Land Rights Act passed in 1976. It established the land trust system for indigenous people in Australia, and the indigenous people can claim their land back based on this land trust (Hill and Press 1994). Another important component during the negotiation process in Australia's case is the detriment clause, which permits indigenous people to terminate the co-management relationship if they feel their rights have been ignored (Lawrence 2000). This is a critical mechanism to force the negotiation process to keep running. The negotiation process involves the handing back of the national park land to the indigenous community, and the indigenous people in turn lease it back to the national park authority. This allows the indigenous people to continue to have their lives and livelihoods in the park, to have control of park management, and to share the benefits from the park. But under the co-management promise, the indigenous community still needs to compromise with some of the conservation objectives from the national park, and bear the accompanied tourism. Even though their homeland is still viewed as the tourism attraction (Cordell 1993), the role of indigenous people in the park–people relationship is changed in this stage. They can strive for their rights and home land instead of being observed and consulted.

When the co-management system consults or negotiates with the indigenous people, it is important who represents the local people. Many local groups and leaders emerge after an indigenous movement begins, and after many governmental and nongovernmental organizations help these local groups via funding and ideas. These organizations also bring mainstream conservation and management concepts to these local leaders. On top of the co-management model, there exists a type of relationship in which indigenous people have the control power, but the management or conservation model is from the mainstream culture. It is ambiguous whether this is co-management or indigenous sovereignty. Even though there is no violence put up by the indigenous community against the mainstream conservation model, the influence is tremendous and the conservation principle is not from below entirely.

INDIGENOUS SOVEREIGNTY MODEL

The indigenous sovereign is the final stage of community-based conservation, and it is the ultimate hope of indigenous groups to achieve this. Nietschmann called this type of park "conservation through self-determination" (Nietschmann 1992). Although this name still reflects viewing the park from a conservation purpose, it is a powerful tool for indigenous people to get the control power back. When indigenous people bring the issue of conservation to the common ground with mainstream culture, they get more recognition. Once recognized and treated as an autonomous entity, the premise that indigenous people can entirely manage natural resources on their territory and have the legal authority to exclude outsiders by treaties with their associated nations is established. The "human rights" and "indigenous conservation knowledge" concepts are effective weapons to seek the support from outside. It is debatable, however, whether getting help from outside to manage their land with the mainstream experience is truly indigenous sovereignty. The cultural influence should not be overlooked even though the physical sovereignty is returned. There exists a gray area between the top of the co-management model and the indigenous sovereignty model discussed above. For indigenous people, having sovereignty over their parks does not simply mean that they hold the power over their land. Meanings of cultural identity and freedom from the mainstream culture are important. There are only a few indigenous communities that can still keep their ways of life with a passive relationship with the outside world and that are not interested in the mainstream power (Igoe 2004). There are critiques about the survival of this form of island culture, because they cannot practice their own life in the traditional ways and cannot benefit from the surrounding culture either. However, this is the model which indigenous people are truly living in their culture, and it should not be undermined because of their abilities in dealing with environmental change.

TOURISM TYPE AND CONTROL POWER

The future tourism industry of the Iraqi marshlands will face decisions about how to share control power. Control power is not the only force that alters the indigenous culture. Governance changes indigenous cultures from within, while tourism and related development shape the culture's image externally. Because of the blooming of tourism in indigenous lands, the high attraction of income from the tourism industry changes the environment and culture accordingly to fit the tourism demand. Thus, different types of tourism affect the indigenous lands in various aspects. There are many ways to define tourism types. Smith divided tourism into five types, which are ethnic, cultural, historical, environmental, and recreational tourism (Smith 1989). He used destination attractions as the spectrum, which varies from involvement in the aboriginal life on one end to relaxing enjoyment on the other end. Applying this classification to a single destination, indigenous land, one end of the spectrum is fascination with the indigenous culture, and the other end is the attraction of the indigenous environment (Figure 5.2).

Ethnic tourism is marketed to people who are interested in the real indigenous life. They would like to see the traditional houses and ceremonies, and be involved in the cultural exchange, not just see the physical elements in the exhibition. *Cultural tourism* attracts people who come for the local color or traditional flavor. This type of tourism also expects to see a different lifestyle, but the involvement is not as deep as with ethnic tourism. The local meal, performance, and model villages are typical destination activities. Usually the local flavor means an old or even vanishing lifestyle for these visitors. They want to experience it, but do not

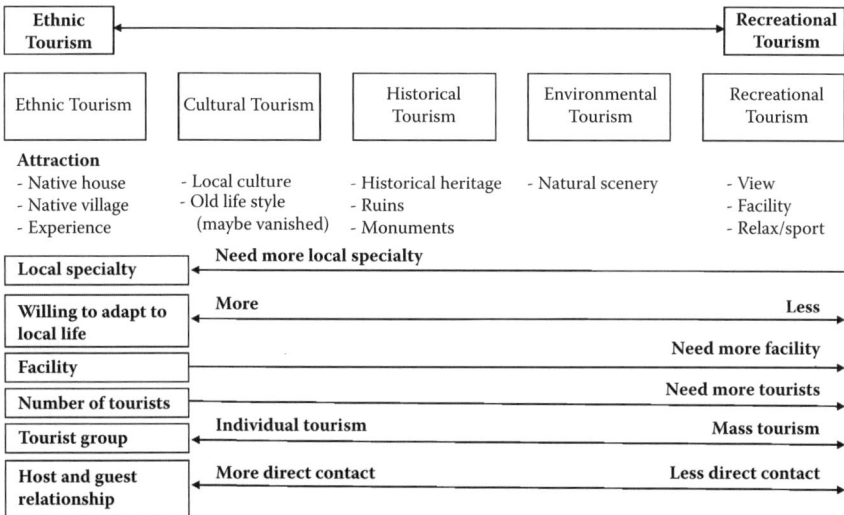

FIGURE 5.2   Tourism types divided by destination attractions.

want to get deeper into the culture. *Historical tourism* is for people to see traces of the past. They would like to see indigenous monuments and demonstrations on how the indigenous ancestors practiced their daily lives long ago. *Environmental tourism* brings people who want to enjoy the natural beauty and wildlife of the indigenous land. *Recreational tourism* attracts people who want to enjoy the colors of natural beauty and a luxurious accommodation package. Enjoyment of ease and comfort are the key features for the destination, which contains more general images without site specificity. For instance, the sun, beach, and sea compose the typical destination images for recreational tourism in many of the Pacific indigenous islands. In contrast, with ethnic tourism a destination is promoted with more site specificity.

When the number of tourists increases toward the recreational tourism end of the spectrum, it is accompanied by a decreasing willingness or ability of tourists to adapt to the local life. In addition, in order to meet the demand, more tourism facilities are needed, which usually differ from the indigenous people's way of life. On the other end of the spectrum, as mentioned, is ethnic tourism. This is marked by a decrease in the number of tourists and more emphasis is on tourists' willingness and desire to arrange their own trip. This is moving toward individual tourism, where contact with the indigenous community becomes more frequent and important.

Conservation policies controlled by different power groups affect the destination attractions. The national park conservation model encloses the highest biodiversity and a significant natural environment inside the park boundary. This then becomes the main target for environmental and recreational tourism. For example, safari tourism in East Africa's national parks is one of the hottest recreational tourism destinations currently. If we see the history of national parks, there are strong connections between parks and sport hunting and recreation by European elites in North Africa and North America. Even the name "national park" is already the attraction point in advertisement, because it means that nature is reserved. On the other hand, ethnic and cultural tourism promise tourists the experience of the indigenous life. This form of tourism is highest when the land is under indigenous sovereignty since culture authenticity and specialty are kept. With the decrease of the indigenous control involvement, the authenticity of the cultural experience lessens.

DEVELOPMENT STAGES ON INDIGENOUS LAND

Butler describes the development process in a tourism destination as a "growth-peak-decline" model (Butler 1980). This model characterizes the destination area in seven distinctive stages. They are the exploration, involvement, development, consolidation, stagnation, decline, and rejuvenation stages (Butler 1980). His model is probably the most cited model that discusses the development of tourism destinations, even though there are critiques about the model being overly simplified (Weaver 2000; Prosser 1995). When this model applies to an indigenous culture's land, it can be modified in the last three stages. The development sequence becomes the exploration, involvement, prime development, competition and decline, and regulated rejuvenation or unregulated rejuvenation stages (Figure 5.3).

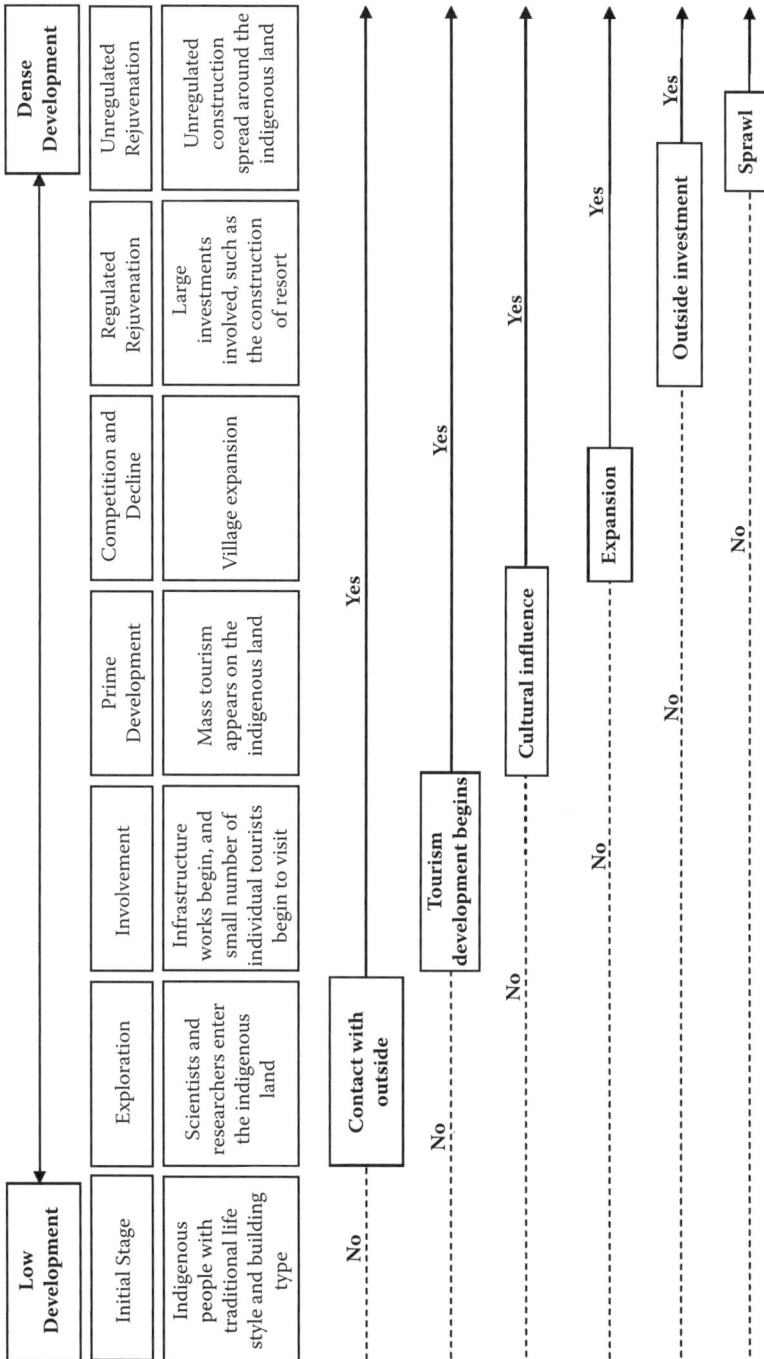

FIGURE 5.3 The development state in indigenous land.

The *exploration stage* is the beginning period in which indigenous people contact with the outside. In this stage, only a few visitors are attracted by the unspoiled nature and culture. Scientists and researchers comprise the major group of people staying in the indigenous land, and there are almost no tourism infrastructures and facilities in this stage. After the site is explored by researchers, it brings people's attention to the indigenous environment and its unique culture, and thus the involvement stage begins. In the *involvement stage*, the indigenous lands start to be promoted as tourism destinations and the basic infrastructures are constructed. Because indigenous people live in a different culture and different living conditions from those of the mainstream culture, indigenous communities are always viewed as impoverished and heathen lands from outside. Thus, development in this stage also comes from the involvement of mainstream governance and international organizations. Large numbers of visitors start to arrive at the indigenous land in the *prime development stage*. At the peak period, the number of tourists may even equal or become more than the indigenous population. In this stage, most of the tourism business is controlled by outside investments, and visitors mainly rely on making arrangements through travel agencies. This is also the stage at which indigenous people face tremendous cultural influence by tourists. Beyond this period of time, the rate of increase of tourists will stop or even decline. This is the stage that combines the development, consolidation, and stagnation stages in Butler's model.

After the tourism business operates in the indigenous land for a certain time period, there are instances in which indigenous people start to pick up the tourism game and run their own tourism business. They will capitalize on their advantageous position of using their traditional culture as attractions to compete with outside investment. But usually, it is also the tourism decline period when the destination is no longer fashionable. This is the stage at which local accommodations increase to compete for the tourism income, but this is also the stage where external investments start to decrease. After tourism declines and many indigenous people have started their businesses, there occurs a phenomenon in which the indigenous lands are repromoted with new attractions of their culture and environment a few years later. This begins as local tourism business initiatives, but very soon large external investments will roll in again with new tourist packages in the indigenous land if there is profit in the market. Depending on the political, social, and economic situation in the indigenous land, the rejuvenation stage may transfer from a regulated to unregulated situation with a spread of development. The rejuvenation stage may not happen in some of the indigenous lands, and in some cases, destinations may have more than one rejuvenation stage.

THE RELATIONSHIP AMONG CONTROL POWER, DEVELOPMENT, AND TOURISM TYPE

Control power, development, and tourism type are three basic elements that shape the indigenous culture and environment. These three factors are related to each other and result in numerous circumstances that will have to be faced by the indigenous Marsh Arab community in Iraq. Thus, a new model may be generated by

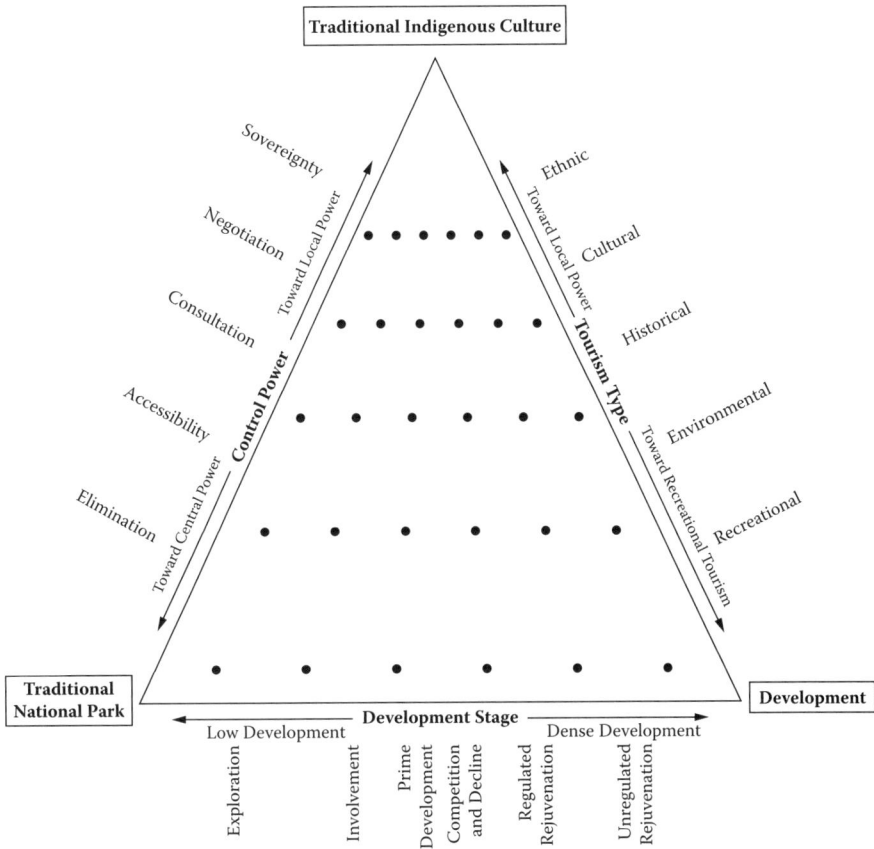

FIGURE 5.4   The relationship model among control power, development, and tourism.

these three factors (Figure 5.4). For the indigenous people, they would want to get close to the top of the model in Figure 5.4. On the other hand, the government expects to have control power (shown on the lower-left corner of the model) because of their highest conservation concern and (see the bottom right of the model) because of the highest profits from tourism. Is indigenous sovereignty for the Marsh Arabs the only option for them to keep their cultural and economic survival? Where is the compromise situation located in the model? Questions can be examined in the model depending on different geographic and cultural contexts. Using this model as a tool rethinks the relationship among government, tourists, Marsh Arabs, and their culture.

REFERENCES

Abrahams, Roger D. 1994. Powerful promises of regeneration or living well with history. In *Conserving culture: A new discourse on heritage*, ed. Mary Hufford. Urbana: University of Illinois Press.

Agrawai, Arun. 1995. Dismantling the divide between indigenous and scientific knowledge. *Development and Change* 26:413–39.

Asai, Erin. 1936. *A study of the Yami language: An Indonesian language spoken on Botel Tobago Island*. Leiden: J. Ginsberg.

Atte, O. 1992. *Indigenous local knowledge as a key to local level development: Possibilities, constraints, and planning issues*. Ames: Iowa State University.

Ayres, Ron. 2002. Cultural tourism in small-island states: Contradictions and ambiguities. In *Island tourism and sustainable development*, ed. Yorghos Apostolopoulos and Dennis J. Gayle. London: Praeger.

Beller, W. S., ed. 1986. *Proceedings of the Interoceanic Workshop on Sustainable Development and Environmental Management of Small Islands*. Washington, DC: U.S. Department of State.

Bender, Barbara, ed. 1993. *Landscape: Politics and perspectives*. Providence, RI: Berg.

Brady, Erika. 1994. The river's like our back yard: Tourism and cultural identity in the Ozark National Scenic Riverways. In *Conserving culture: A new discourse on heritage*, ed. Mary Hufford. Urbana: University of Illinois Press.

Briguglio, Lino, Brian Archer, Jafar Jafari, and Geoffrey Wall, eds. 1996. *Sustainable tourism in islands and small states: Issues and policies*. New York: Pinter.

Brunhes, J. 1920. *Human geography*. London: Methuen.

Butler, R. W. 1980. The concept of tourist area cycle of evolution. *Canadian Geographer* 24:5–12.

Butler, Richard, and Tom Hinch, eds. 1996. *Tourism and indigenous peoples*. Boston: International Thomson Business Press.

Canada, Department of Indian Affairs and Northern Development. 1991. *Building a new relationship with First Nations in British Columbia, Canada's response to the report of the BC Claims Task Force*. Ottawa: Supply and Service.

Carlquist, Sherwin. 1974. *Island biology*. New York: Columbia University Press.

Carneiro, Robert. 1960. Slash-and-burn agriculture: A closer look at its implications for settlement pattern. In *Men and cultures*, ed. Anthony F. C. Wallace. Philadelphia: University of Pennsylvania Press.

Chambers, Erve, ed. 1997. *Tourism and culture: An applied perspective*. Albany: State University of New York Press.

Chang, Wen-Jie. 1990. The livelihood and cultivation adaptation of Paiwan people and Yami people: Compare the Shiwen and Iraraley villages. Master's thesis, National Taiwan Normal University, Department of Geography.

Chen, Cheng-Hong, Tsung-Kwei Liu, Tsan-Yao Yang, and Yue-Gau Chen. 1994. *The geologic map and explanatory text of Taiwan: Sheet 71, LanYu*. Taipei: Central Geological Survey.

Chen, Yong-Long. 1999. Rethinking Taiwan indigenous culture and conservation. In *The utilization of indigenous preservation area and environmental protection*, ed. Shu-Ya Lin. Taipei: Chinese Institute of Land Economics.

Cheng, He-Wen, and Sheng-Yu Lu. 2000. *Botel Tabaco, Yami and plants*. Taipei: Lamper Enterprises.

Chiang, Daniel Bien. 1984. The development, move, and inheritance of Yami residence. *Bulletin of the Institute of Ethnology, Academia SINICA* 58:83–117.

Chou, Chong-Jing. 1996. *The folk song and culture of Yami people*. Taipei: Changming.

Coccossis, Harry. 2002. Island tourism development and carrying capacity. In *Island tourism and sustainable development*, ed. Yorghos Apostolopoulos and Dennis J. Gayle. London: Praeger.

Coccossis, Harry, and A. Parpairis. 1992. Tourism and the environment: Some observations on the concept of carrying capacity. In *Tourism and the environment: Regional, economic and policy issues*, ed. H. Briassoulis and J. van der Straaten. Dordrecht: Kluwer Academic.

Copper, C., J. Fletcher, D. Gilbert, and S. Wanhill. 1993. *Tourism: Principles and practices.* London: Pitman.

Cordell, John. 1993. Who owns the land? Indigenous involvement in Australian Protected Areas. In *Indigenous peoples and protected areas: The law of Mother Earth*, ed. Elizabeth Kemf. London: Earthscan.

Dasmann, Raymond. 1991. The importance of cultural and biological diversity. In *Biodiversity: Culture, conservation, and ecodevelopment*, ed. Margery L. Oldfield and Janis B. Alcorn. Boulder, CO: Westview.

Ellen, Roy, Peter Parkes, and Alan Bicker, eds. 2000. *Indigenous environmental knowledge and its transformations: Critical anthropological perspectives.* Amsterdam: Harwood Academic.

Esman, Marjorie R. 1984. Tourism as ethnic preservation: The Cajuns of Louisiana. *Annals of Tourism Research* 11:451–67.

Feld, Steven, and Keith H. Basso, eds. 1996. *Senses of place.* Santa Fe, NM: School of American Research Press.

Forman, R. T. T., and M. Godron. 1987. *Landscape ecology.* New York: John Wiley.

Ghai, D. P., and J. M. Vivian. 1992. *Grassroots environmental action: People's participation in sustainable development.* New York: Routledge.

Gilbert, F. S. 1980. The equilibrium theory of island biogeography: Fact or fiction? *Journal of Biogeography* 7:209–35.

Gilpin, M. E., and J. M. Diamond. 1980. Subdivision of nature and the maintenance of species diversity. *Nature* 285:567–68.

Government of Japan. 1985. *Treaty of Shimonoseki.* Tokyo: Government of Japan.

Greenwood, Davydd J. 1989. Culture by the pound: An anthropological perspective on tourism as cultural commoditization. In *Hosts and guests: The anthropology of tourism*, ed. Valene L. Smith. 2nd ed. Philadelphia: University of Pennsylvania Press.

Hache, J. D. 1987. The island question: Problems and prospects. *Ekistics* 54:88–92.

Hamnett, M. 1990. Pacific Islands resources development and environmental management. In *Sustainable development and environmental management of small islands*, ed. W. Beller, P. d'Ayala, and P. Hein. Paris: Parthenon/UNESCO.

Harrison, David, ed. 1992. *Tourism and the less developed countries.* London: Belhaven.

Hill, M., and A. Press. 1994. Kakadu National Park: An Australian experience in comanagement. In *Natural connections: Perspective in community-based conservation*, ed. D. Western and R. Wright. Washington, DC: Island Press.

Hitchcock, Robert K. 1997. Cultural, economic, and environmental impacts of tourism among Kalahari Bushmen. In *Tourism and culture: An applied perspective*, ed. Erve Chambers. Albany: State University of New York Press.

Hornborg, Alf, and Mikael Kurkiala, eds. 1998. *Voices of the land: Identity and ecology in the margins.* Lund, Sweden: Lund University Press.

Huang, Shu. 1995. *The living culture and its transition of Yami people.* Taipei: Daoshiang.

Huang, Yueh-Wen. 1997. An analysis of decision-making process of establishing Lanyu National Park. *Department of Geography, National Taiwan University, Journal of Geographical Science* 23:13–31.

Hunter, M. Jr., G. Jacobson, and T. Webb. 1988. Paleoecology and the coarse-filter approach to maintaining biological diversity. *Conservation Biology* 2:375–85.

Igoe, Jim. 2004. *Conservation and globalization: A study of national parks and indigenous communities from East Africa to South Dakota.* Belmont, CA: Thomson Wadsworth.

Inglis, Julian T., ed. 1993. *Traditional ecological knowledge: Concepts and cases.* Ottawa: International Program on Traditional Ecological Knowledge.

International Union for the Conservation of Nature and Natural Resources (IUCN). 1980. *World conservation strategy.* Gland, Switzerland: IUCN.

Kano, Tadao, and Kokichi Segawa. 1956. *An illustrated ethnography of Formosan aborigines*, vol. 1 of *The Yami.* Tokyo: Maruzen.

Keyser, Heidi. 2002. *Tourism development*. Oxford: Oxford University Press.

King, Russell. 1993. The geographical fascination of islands. In *The development process in small island states*, ed. Douglas G. Lockhart, David Drakakis-Smith, and John Schembri. New York: Routledge.

Kwan, Hua-Shan. 1988. Yami's living environment and religion. In *Proceedings of the conference on indigenous religious ceremony*. Taipei: SINICA.

Lack, D. 1976. *Island biology: Illustrated by the land birds of Jamaica*. Berkeley: University of California Press.

Lawrence, D. 2000. *Kakadu: The making of a national park*. Melbourne: Melbourne University Press.

Lee, Yi-Yuan, Lei Shi, Hei-Yuan Chu, Shin-Huang Shiao, and Guang-Hong Yu. 1983. *The evaluation report of the indigenous policy*. Taipei: SINICA.

Leones, Julie, and George B. Frisvold. 2000. Park planning beyond park boundaries: A Grand Canyon case study. In *National parks and rural development: Practice and policy in the United States*, ed. Gary E. Machlis and Donald R. Field. Washington, DC: Island Press.

Lin, Gang-Wei. 1986. The environmental adaptation and change of Yami people. Master's thesis, National Taiwan University, Department of Geography.

Lin, Shi-Juan. 1970. The living environment of Yami people. Master's thesis, *National Cheng Kung University*, Department of Architecture.

Liu, Pin-Hsiung. 1959. One example of Yami's funeral. *Institute of Ethnology Publications* 8:143–83.

Lusigi, W. J. 1981. New approaches to wildlife conservation in Kenya. *Ambio* 10(2–3): 87–92.

MacArthur, R. H., and E. O. Wilson. 1967. *The theory of island biogeography*. Princeton, NJ: Princeton University Press.

Machlis, Gary E., and Donald R. Field, eds. 2000. *National parks and rural development: Practice and policy in the United States*. Washington, DC: Island Press.

MaLai, GuMai. 2004. The establishment and vision of indigenous autonomy in Taiwan. Master's thesis, *National Sun Yat-Sen University*, Department of Public Affairs.

Moock, J., and R. Rhoades, eds. 1992. *Diversity, farmer knowledge, and sustainability*. Ithaca, NY: Cornell University Press.

Morris, A. 1996. Tourism and local awareness: Costa Brava, Spain. In *Sustainable development? European experience*, ed. G. K. Priestley, J. A. Edwards, and H. Coccossis. Wallingford, UK: CAB International.

National Taiwan University, Center of Urban Planning. 1984. *The research of Lanyu Island's natural and cultural resources*. Taipei: National Taiwan University.

National Taiwan University, Graduate Institute of Building and Planning. 1989. *Lanyu's development and national park plan*. Taipei: National Taiwan University.

National Taiwan University, School of Forestry. 1982. *The investigation and analysis of the natural resources and landscape in Lanyu Island and Green Island*. Taipei: National Taiwan University.

Nietschmann, Bernard Q. 1992. Conservation by self-determination. *Research and Exploration* 7(3):372–74.

Pearce, D. 1987. *Tourism today: A geographical analysis*. New York: Longman.

———. 1989. *Tourist development*, 2nd ed. Harlow, UK: Longman Scientific and Technical.

Pitt, David. 1980. Sociology, islands and boundaries. *World Development* 8:1051–9.

Prosser, G. 1995. Tourism destination lifecycles: Progress, problems and prospects. In *Proceedings of the national tourism and hospitality conference*, ed. J. Shaw. Melbourne: CAUTHE.

Ricci, F. P. 1976. *Carrying capacity: Implications for research*. Ottawa: University of Ottawa Press.

Ryuzo, Torii. 1911. *The investigate of Botel Tobago*. Tokyo: Taihoku.

Scoones, I., and J. Thompson. 1994. Knowledge, power, and agriculture: Towards a theoretical understanding. In *Beyond farmer first*, ed. I. Scoones, J. Thompson, and R. Chambers. London: Intermediate Technology.

Semple, Ellen Churchill. 1911. *Influences of geographic environment: On the basis of Ratzels system of anthropo-geography*. New York: Henry Holt.

Sen, G. 1992. *Indigenous vision: Peoples of India, attitudes to environment*. New Delhi: Sage.

Shelby, B., and Heberlein, T. 1986. *Carrying capacity in recreational settings*. Corvallis: Oregon State University Press.

Smith, Valene L., ed. 1989. *Hosts and guests: The anthropology of tourism*, 2nd ed. Philadelphia: University of Pennsylvania Press.

Sneed, Paul. 1997. National parks and northern homelands: Toward co-management of national parks in Alaska and Yukon. In *Conservation through cultural survival: Indigenous peoples and protected area*, ed. Stan Stevens. Washington, DC: Island Press.

Society of Wildlife and Nature. 1985. *The investigation of significant natural and cultural resources in Taiwan*. Taipei: Society of Wildlife and Nature.

Stanton, Max E. 1989. The Polynesian Cultural Center: A multi-ethnic model of seven Pacific cultures. In *Hosts and guests: The anthropology of tourism*, ed. Valene L. Smith, 2nd ed. Philadelphia: University of Pennsylvania Press.

Steinitz, Carl. 1995. A framework for landscape planning practice and education. *Process Architecture*, no. 127:42–53.

Stevens, Stan, ed. 1997. *Conservation through cultural survival: Indigenous peoples and protected area*. Washington, DC: Island Press.

Stringer, P. 1981. Hosts and guests: The bed and breakfast phenomenon. *Annals of Tourism Research* 8(3):357–76.

Syman, Rapongan. 1992. *The legend of BaDai Bay*. Taichung, Taiwan: Chen-Shin Press.

Tourism Bureau. 2005. *Visitors statistics report to the principal scenic spots in Taiwan*. Taipei: Tourism Bureau.

Van den Berghe, Pierre L., and Charles F. Keyes. 1984. Introduction: Tourism and re-created ethnicity. *Annals of Tourism Research* 11:343–52.

Wall, Geoffrey, and Veronica Long. 1996. Balinese homestays: An indigenous response to tourism opportunities. In *Tourism and indigenous peoples*, ed. Richard Butler and Thomas Hinch. Boston: International Thomson Business Press.

Warren, D. M. 1989. Linking scientific and indigenous agricultural systems. In *The transformation of international agricultural research and development*, ed. J. Lin Compton. Boulder, CO: Lynne Rienner.

Weaver, D. B. 2000. A broad contact model of destination development scenarios. *Tourism Management* 21:217–24.

Wei, Hwei-Lin, and Pin-Hsiung Liu. 1962. *Social structure of the Yami Botel Tobago*. Taipei: SINICA.

West, Patrick C., and Steven R. Brechin. 1991. *Resident peoples and national parks*. Tuscon: University of Arizona Press.

Western, D., and M. Wright, eds. 1994. *Natural connections: Perspectives in community-based conservation*. Washington, DC: Island Press.

Williamson, M. 1989. The MacArthur and Wilson theory today: True but trivial. *Journal of Biogeography* 16:3–4.

Wilstach, P. 1926. *Islands of the Mediterranean*. London: Geoffrey Bless.

World Bank. 1996. *The World Bank participation sourcebook*. Washington, DC: World Bank.

Yan, Ai-Jing, and Hong-Qian Yang. 1999. The land development and management in Lanyu's indigenous preservation area. In *The utilization of indigenous preservation area and environmental protection*, ed. Shu-Ya Lin. Taipei: Chinese Institute of Land Economics.

Yu, Guang-Hong. 1991. Ritual, society and culture among the Yami. PhD diss., University of Michigan.

———. 2004. *The Yami people*. Taipei: San-Ming.

Zheng, Shian-You. 2001. The ecological system in Lanyu Island: The influence from outside force. In *Indigenous Land Conference*. Taipei: Council of Indigenous Peoples.

Desert Wetland Restorations

6 Wetlands Lost and Found in the Levant*

CONTENTS

Introduction 65
History 66
Wetlands Lost 67
Wetlands Found 71
Sustaining Restoration 80
Lessons for the Iraqi Marshlands 82
References 83

INTRODUCTION

If one looks at ancient maps of the Levant or even at more recent ones of Palestine from the British Mandate Period, a large body of water called Lake Hula or the Hula Swamp is clearly visible in the region north of the Sea of Galilee. The Hula Valley is 25 km long and 7 km wide, bordered by the Golan Heights on its eastern side and the Levant highlands on the west, and at one time it contained a shallow lake and wetland complex with a surface area of 60 km^2 (Dimentman, Bromley, and Por 1992; Hambright and Zohary 1998) Similar to the Iraqi marshlands, Lake Hula is a snow-fed system, in this case beginning in the Lebanese mountains to the north. The extensive marshlands owe their origin to the presence of a large basalt plug situated about 30 km north of the Sea of Galilee which served to back up the southward flow of the Jordan River (Hambright and Zohary 1998).

The Hula lake–wetland complex in Israel, like the Azraq oasis wetlands in Jordan (see chapter 7), has been occupied for over fifty thousand years and is therefore a rich cultural landscape in addition to being an important hydrological and ornithological landscape. And of greatest significance to the marshland situation in southern Iraq, Lake Hula has become the most successful and highest visibility desert wetland restoration project anywhere in the Middle East—one that offers many lessons for Iraq.

---

* Authored by Robert L. France

65

HISTORY

The Hula area is one of the oldest documented wetland–lake systems in history, first being mentioned in about 1400 BCE in the Tel Amarna letters of Pharaoh Akhenaton (Dimentman, Bromley, and Por 1992). The region has been occupied since 73,000 BCE and contains Neolithic settlements as old as 10,000 BCE. The very important Bronze Age city of Hazor, frequently mentioned in the Old Testament as being a city of Solomon and contemporaneous with the marshland cities of southern Iraq, over-looked the marshes (Figure 6.1). Today the view from the Hazor Tell reveals agricultural lands where once marshes existed, and distant views to the south of the Sea

FIGURE 6.1   Ruins of the Bronze Age biblical city of Hazor, at the edge of the Hula wetlands.

of Galilee and the famous Horns of Hittin where Sal-al-din defeated the crusaders in one of history's most significant battles. Hazor is also historically tied to ancient Mesopotamia in that it was attacked and razed by the Assyrians.

The Hula Valley has a long history of irrigation manipulation extending back to Roman and Byzantine times (Dimentman, Bromley, and Por 1992). Notably in 1260, the Sultan Baibars constructed a bridge that narrowed the Jordan River and expanded the marshes. The area was explored and mapped at the end of the nineteenth century, revealing extensive wetlands referred to as the "Hooleh Morass" and a dense network of feeder streams for the Jordan River and Lake Hula. Maps from this period also show a group of small settlements scattered throughout the extensive wetland landscape.

The Iraqi marshlands are important as a staging area for migratory birds on their way for the Soviet Union, and Lake Hula provides the same function with respect to Europe (Dimentman, Bromley, and Por 1992). And as for all desert wetlands, the Hula region exhibited a rich biodiversity and abundance of wildlife, being regarded at one time as the finest hunting ground in all Syrio-Palestine, by supporting populations of large mammals such as panthers, leopards, bears, wolves, jackals, hyenas, gazelles, and, as in the Iraqi marshes, wild boars and otters. Like the Iraqi marshlands and Jordan's Azraq Oasis (chapter 7), water buffalo were introduced into the Hula marshes in the seventh century CE following their domestication in Pakistan and transport (through the Iraqi marshlands) to the Levant.

A final very interesting parallel to Iraq was the presence in the Hula region since the Middle Ages of a population of marsh Arabs referred to as *Ghawaina* who had originated as escaped slaves and deserters from the Egyptian Malmuk army and whose population had reached 12,000 by the 1950s. Imaginative mid-nineteenth-century lithographs show the explorer MacGregor Laird in his famous Rob Roy canoe and a paternalistic image of Hula marsh Arabs looking for all the world like equatorial Africans. In fact, lifestyle photographs of these marsh dwellers and their inhabitations, watercraft, and water buffalos from the Hula swamps during the first half of the twentieth century (Dimentman, Bromley, and Por 1992; Hambright and Zohary 1998) actually look identical to contemporaneous ones from the Iraqi marshes (France 2007c; Maxwell 1966). Both cultures were sustained by reeds, which in the case of Hula were a mixture of not only Phragmities, as in southern Iraq, but also papyrus. Under the British Mandate in the 1940s, a very high infant mortality rate of these marsh dwellers due to malaria outbreaks led to a widespread DDT-spraying program to eradicate the mosquito-borne disease.

WETLANDS LOST

Lake Hula was drained in the late 1950s by placing the upstream Jordan River into diversion canals and by increasing the size of the downstream river channel. The motivation behind the project was to increase arable land, and it soon became "the standard bearer of the entire Zionist movement of agricultural self-reliance and resettlement of the land" for the new country of Israel (Dimentman, Bromley, and Por 1992). Although it is easy to criticize Israel for the destruction of this important wetland habitat, draining the Hula swamps is merely one part of an often adopted strategy in nation building where the first step is to attack and control water resources as an act of sovereignty (e.g., Norway, on obtaining independence from Sweden, put Europe's most popular waterfall into a pipe for hydropower generation; and Quebec, which has its own independence dreams, has concrete plans to turn James Bay into a freshwater lake by diking it off from Hudson Bay and then using two nuclear reactors to send the water to southern California for profit). In short, grandiose (and often environmentally insensitive) water-engineering projects and nation building are often linked together in hubris. The drainage of the Hula swamps became a major tourist attraction for the new Israelis, and donations from Jews around the world were given to help drain the swamps (M. Ukleas, personal communication).

Today, all that remains of the once extensive Lake Hula are the vestigial wetlands in the nature reserve and the small wetlands restoration projects. The predominant land use is agriculture (Figure 6.2) through which the Jordan River makes its way, resembling not so much the majestic river of biblical fame as rather a modern insignificant drainage ditch which is largely filled with reclaimed wastewater (Figure 6.3).

As a result of drainage, agriculture did flourish, but for only a time. One hundred and nineteen animal species were lost to the region, of which thirty-seven were totally lost to the country (Hambright and Zohary 1998). Also very troubling is the fact that the one-time dwellers of the swamps, the indigenous marsh Arabs (who had had no role in the 1948 war with Syria), were "removed" or left on their own when their environment was destroyed.

The Hula Nature Reserve, an area of 4 km^2, was saved in 1964 as a token museum piece of what once was the extensive marshes. The lobbying by dedicated environmentalists to save this vestige led to the birth of the Israeli conservation movement.

FIGURE 6.2   Conversion of the historic Hula wetlands to agricultural fields.

FIGURE 6.3   The once-mighty Jordan River, now a channelized agricultural drainage ditch.

The Hula Nature Reserve includes a rare glimpse of the desert wetland landscape (Figure 6.4) and has an extensive network of trails, as shown in a map in the car park (Figure 6.5). The visitor center is not very architecturally inspiring (Figure 6.6) but functions as offices and provides some minor interpretation (only in Hebrew). Parking is arranged under an attractive glade of trees important in such hot Middle Eastern climates (Figure 6.7). Shaded areas are provided for visitors (Figure 6.8), as well as a buffalo-viewing hide (Figure 6.9) to observe the reintroduced buffalo (marsh Arabs have not been invited to return). There is an impressive network of raised boardwalks that wind through the dense reed beds (Figure 6.10) as well as viewing platforms,

FIGURE 6.4   Wetland remnants in the Hula Nature Reserve.

bird blinds, and observation towers (Figure 6.11). Overall, the strategy is to immerse visitors within the wetland where they can hear the cacophony of bird calls and observe such wildlife in addition to amphibians and fishes (Figure 6.12).

Drainage has led to a number of problems in addition to the obvious one of loss of a unique and important wildlife habitat (Hambright and Zohary 1998; Markel et al. 1998). Following water removal, the exposed peat became oxidized, which led to spontaneous underground fires. This in turn caused the land to subside (up to 3 m in places), which necessitated further drainage and with a consequent development of dust storms. Disappearance of natural predators led to skyrocketing rodent

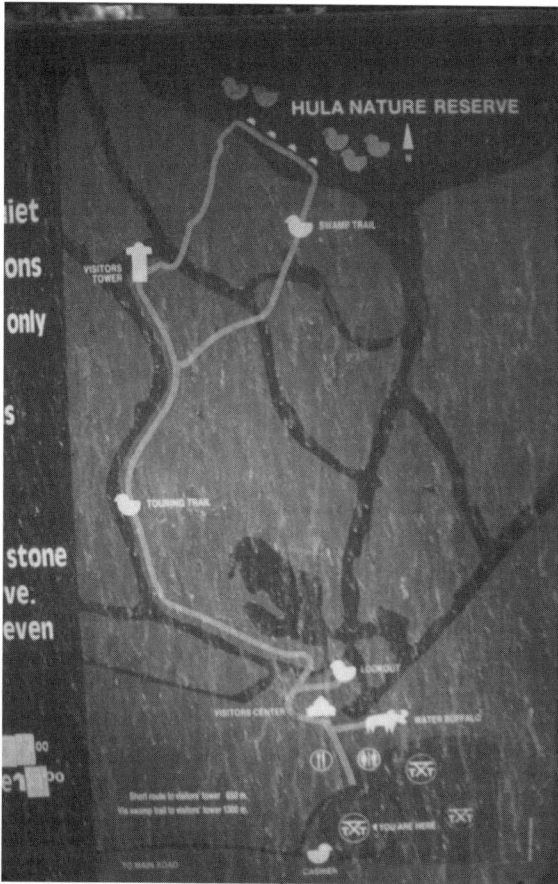

FIGURE 6.5   Signage showing the trail system in the Hula Nature Reserve.

populations. Over time the agricultural benefits fell and eventually led to large areas becoming neglected. The loss of the phytoremediation in the once extensive wetlands allowed nutrients to leak from the system and enter the Sea of Galilee, Israel's major water supply, where they were the trigger for accelerating eutrophication there. More than any other concern, it was this downstream water quality problem that became the major motivation for the Hula Restoration Project.

WETLANDS FOUND

There were six major goals of the Hula Valley Restoration Project:

- Prevent nutrient escape and transport to, and the resulting eutrophication of, the Sea of Galilee.

FIGURE 6.6   The uninspiring visitor center.

FIGURE 6.7   The lovely and functional green parking lot at the visitor center.

FIGURE 6.8 Shaded observation area.

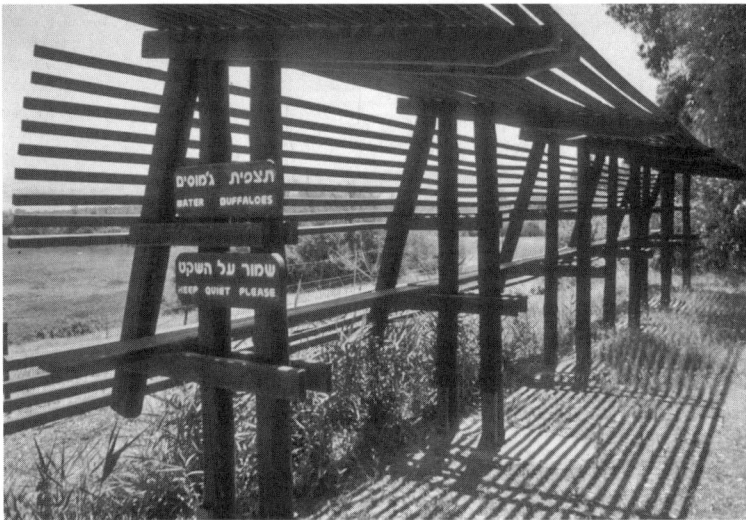

FIGURE 6.9 Screen from which to view water buffalo.

FIGURE 6.10   Impressive array of walkways to immerse visitors in the preserved wetlands.
(Continued)

- Control groundwater levels and develop year-round vegetative cover (keeping water levels high is necessary to prevent peat oxidation, decomposition, and subsequent erosion of peat).
- Reclaim abandoned agricultural soil in order to minimize erosion.
- Recreate aquatic habitats to restore biodiversity.
- Create an ecotourism industry as "a balance between nature and environmentally sensitive agricultural development" (Dimentman, Bromley, and Por 1992) in order to provide an alternative income for former farmers.
- Revive the spirit of an extinct ecosystem of worldwide ecological and historical significance (as was also a motivation behind the restoration of Xochimilco in Mexico, as described in chapter 11).

FIGURE 6.10 (*Continued*) Impressive array of walkways to immerse visitors in the preserved wetlands.

Excavation of the new lake and wetland complex, the 1 km² Lake Agmon (Hebrew for bulrush and also inferring a small lake), began in 1993, followed by reflooding due to release of water which had been constrained within the many small drainage channels (as has also been the case in southern Iraq). The ongoing supply of water is from the Jordan River and a major peat drainage canal through inflow channels of rock dissipaters (Figure 6.13). An underground barrier was constructed at the southern end to prevent downstream nutrient leakage (Hambright and Zohary 1998). Water in the restored lake–wetland complex is reused locally for agricultural irrigation. Attention was paid to the creation of wildlife habitat in the form of islands and

FIGURE 6.11   Observation tower and bird-viewing blind.

variable topography and planting (Shy et al. 1998; Ashkenazi and Dimentman 1998; Figure 6.14).

There was an amazing and very rapid return of visual and some ecological attributes as recently as two weeks following reflooding, leading to a very vibrant ecosystem. Over seventy plant species had colonized during the first two years (Kaplan, Oron, and Gophen 1998). The "build it and they will come" strategy paid off with the colonization of pelicans, herons, egrets, cranes, cormorants, ducks, and flamingos to such a degree that the area came to be regarded as "a new paradise for bird watching" (Ashkenazi and Dimentman 1998). In the case of cranes, numbers increased from levels of less than five thousand observed prior to restoration to levels exceeding

FIGURE 6.12 Numerous turtles and fishes are frequently seen from the boardwalks.

twenty thousand today. The restoration site is now a major tourist destination drawing hundreds of weekend visitors from around the country. Future plans exist for a wildlife safari park and an ecotourism center. Some species, however, had to be deliberately introduced as they had been extirpated from the local area and the seed source had been removed such that direct manipulation was required. And finally, and most importantly in the eyes of many, the new wetlands of restored Lake Agmon now operate as a nutrient sink, serving to maintain downstream water quality in the precious Sea of Galilee.

Israel has an active and highly regarded group of aquatic scientists, and given the near complete absence of desert wetland restoration projects anywhere in the

FIGURE 6.13   Rock inflow dissipators to receive water into the restored Lake Agmon wet-land complex.

world, it is interesting to note that they treated the Hula restoration project in the form of a large experiment. In order to investigate all the nuances of this pioneer project, researchers employed grazing exclusion cages, variable revegetation strat-egies, different riparian tree plantings, and the like, and wrote dozens of scientific papers, including a special issue of the international journal *Wetlands Ecology and Management* called "Destruction and Creation of Wetland Ecosystem in Northern Israel." In this same light, researchers established a vigorous long-term monitoring program to determine the successes and failures of the various restora-tion and conservation measures employed.

FIGURE 6.14 Images of the vibrant plant growth in the restored wetland. (*Continued*)

Much can be learned from the detailed scientific approach taken to the Hula Valley Restoration Project, and much of the methodology utilized in this Israeli case study is adaptable and applicable to the situation occurring less than a thousand kilometers away in southern Iraq. The times of prejudicially ignoring scientific insight (as, for example, with regard to Galileo and Giordano Bruno) are well behind us, and scientific knowledge presented in peer-reviewed international publications remains the best source of objective information accessible to all, and something which present and future Iraqi wetland scientists would be remiss not to turn to with respect to this important case study, regardless of their own personal political opinions.

FIGURE 6.14   (*Continued*) Images of the vibrant plant growth in the restored wetland.

SUSTAINING RESTORATION

Wetland restoration is not over when the newly excavated hole fills with water and plants; it needs certain monitoring and possible tending through time (France 2003). In the case of the Lake Hula wetland complex, some problems occurred as a result of what might be referred to as there being "too much of a good thing." Following the creation of Lake Agmon, thousands of birds returned to the Hula region for the first time in decades (Shy et al. 1998). Grain foraging made it difficult for nearby farmers trying to maintain their living. The large increase in the number of cranes (up to twenty thousand per day during the migration season) caused problems in terms of grazing and trampling on fields. The managerial solution was to provide supplemental food, which has led to a corresponding tourist boom with people driving up from as far away as the Negev Desert on weekends to see the cranes and perhaps buy promotional wine and other products. In addition, nearby commercial fish ponds were suffering from pelican overfeeding. Using the same strategy as for grain-eating waterfowl, wetland managers stocked the newly created lake–wetland with thousands of juvenile and adult fish (Degani et al. 1998) to attract the birds away from the fish ponds.

In terms of what might be referred to as being "too much of a (*perceived*) 'bad' thing," one unexpected problem that developed in the Hula restoration experience was the failure to convince environmental groups that elements of commercial tourism could in fact be integrated aesthetically and environmentally into the newly restored landscape. So here is an ironic situation whereby today's environmentalists, many of whose grandparents lobbied to preserve a small sliver of the Hula swamps a half century ago as the Nature Reserve and founded the national environmental group, were the most critical of attempts to work with the farmers in establishing ecotourism, believing such a fusion to be an impossibility. There certainly are bad examples in Israel of outdoor recreation facilities such as hedonistic water parks, and it is unfortunate that there appears to be little or no knowledge of what has transpired next door in Jordan with respect to its Royal Society for the Conservation of Nature's (RSCN) projects (see chapters 7 and 16) which demonstrate that it is certainly possible for nature and people to coexist (the whole subject of creating restored landscapes for people as much as for nature is the subject of *Handbook of Regenerative Landscape Design* [France 2007a]). As a result, when I first visited the restored Agmon lake–wetland complex in 2001, there were no interpretive signs nor a single person seen. Farmers had become angry because they had sold their land for the restoration project on the understanding that they would become part of the future ecotourism development, which, however, was being held in limbo by the green movement. When I next visited the site three years later, the situation had started to change: the supplemental feeding of cranes had raised national consciousness about the role of ecotourism, and planners were revisiting the concept in an improved light.

Finally, there was one major ecological problem that developed in the reflooding of areas that had been desiccated for long periods of time. Initially there was a massive growth of Typha recolonizing during the first three years. But, surprisingly, this was followed by a near complete die-off the following year. So in terms of restoration, things looked like they were progressing well until suddenly all that was present was vacant water. The reason turned out to be due to the complicated biogeochemistry characteristic of such desert wetland systems (Markel et al. 1998). The oxidizing peat affected the water chemistry of the shallow lake and restored wetlands. Gypsum, which had been formerly held by peat, became dissolved and produced sulfate levels as high as that of seawater. So in the end, the "restoration" project turned what had once been a freshwater wetland into an inland salt marsh. This was related to organic matter from the large blooms of filamentous algae which accelerated redox reactions such that the iron protection of plant roots failed due to the high hydrogen sulfide levels. The solution to this serious problem was to follow some Australian research showing the need to maintain low water levels in summer

FIGURE 6.15   Flooding of formerly desiccated peat in the restored wetland.

in order to allow the upper sediments to oxidize and get ready for another redox cycle during the period of winter flooding (Figure 6.15). This necessitates having good hydrological control in order to prevent the buildup of the carbon source that will affect the biogeochemical cycles. As a result of these water manipulations, Typha returned in 1998 and has maintained healthy levels thereafter.

LESSONS FOR THE IRAQI MARSHLANDS

There are a number of lessons from the Hula restoration project that are applicable to the situation in the Iraqi marshlands:

- Restoration is not over with reflooding; there is a need for ongoing supervision and monitoring (France 2003; and see chapter 8).
- It is essential to know and understand the biogeochemical cycles and hydrology for wetland restoration to be sustained through time (Markel et al. 1998; and see chapters 8 and 9).
- Restoration should be treated as a process rather than a product (France 2007b). In this way, restoration should be regarded as an evolving experiment with a full support team, including a soil scientist, agricultural engineer, hydrologist, biologist, tourism park planner, and more (Bays et al. 2002).
- To be successful, restoration projects must work with, not around or against, the local agricultural community (see chapters 10, 11, and 16).
- It may be necessary to accelerate species introductions and to promote alternative food sources to minimize local grazing impacts (see chapter 10).

- For all regenerative landscape designs (*sensu* France 2007a), it is important to be wary of environmental "allies"; in other words, sometimes those who one thinks will be most supportive of the restoration efforts may turn out to be the most difficult group to contend with. It is, therefore, often necessary to be innovative and adaptable in the various ecotourism options put forward in order to circumvent such confrontation (see chapters 13 and 15).

REFERENCES

Ashkenazi, S., and C. Dimentman. 1998. Foraging, nesting, and roosting habitats of the avian fauna of the Agmon wetland. *Wetlands Ecology* and *Management* 6:169–87.

Bays, J., J. Cormier, N. Pouder, B. Bear, and R. France. 2002. Moving from single-purpose treatment wetlands toward multifunction designed wetland parks. *In Handbook of water sensitive planning and design*, ed. R. France. Boca Raton, FL: CRC Press.

Degani, G., Y. Yehuda, J. Jackson, and M. Gophen. 1998. Temporal variation in fish community structure in a newly created wetland lake. *Wetlands Ecology* and *Management* 6:151–57.

Dimentman, C., H. J. Bromley, and F. D. Por. 1992. *Lake Hula: Reconstruction of the fauna and hydrobiology of a lost lake*. Jerusalem: Israel Academy of Sciences.

France, R. L. 2003. *Wetland design: Principles and practices for landscape architects and land-use planners*. New York: Norton.

———. ed. 2007a. *Handbook of regenerative landscape design*. Boca Raton, FL: CRC Press.

———. ed. 2007b. *Healing natures, repairing relationships: New perspectives in restoring ecological spaces and consciousness*. Winnipeg, MB: Green Frigate Books.

———. 2007c. *Wetlands of mass destruction: Ancient presage for contemporary ecocide in southern Iraq*. Winnipeg, MB: Green Frigate Books.

Hambright, K. D., and T. Zohary. 1998. Lakes Hula and Agmon: Destruction and creation of wetland ecosystems in northern Israel. *Wetlands Ecology* and *Management* 6:83–89.

Kaplan, D., T. Oron, and M. Gophen. 1998. Development of macrophytic vegetation in the Agmon wetland of Israel by spontaneous colonization and reintroduction. *Wetlands Ecology* and *Management* 6:143–50.

Markel, D., E. Sass, B. Lazar, and A. Bein. 1998. Biogeochemical evolution of a sulfur-iron rich aquatic system in a reflooded wetland environment (Lake Agmon, northern Israel). *Wetlands Ecology* and *Management* 6:121–32.

Maxwell, G. 1966. *People of the reeds*. Boynton Beach, FL: Pyramid Books.

Shy, E., S. Beckerman, T. Oron, and E. Frankenberg. 1998. Repopulation and colonization by birds in the Agmon wetland, Israel. *Wetlands Ecology* and *Management* 6:159–67.

7 Rehabilitation of a Historic Wetland

Jordan's Azraq Oasis*

CONTENTS

Introduction 85
The Place 87
Water in Jordan 87
Rehabilitation Efforts 90
Acknowledgment 115
References 115

INTRODUCTION

In 1996 I visited the Azraq Oasis in western Jordan for the first time. I had been attracted to the area to see the old Umayyad "castle" (really a desert fort) that had housed T. E. Lawrence during his 1917 campaign on Damascus. He had written quite favorably about the "magically haunted" area despite the harsh climatic conditions that were experienced by his army during their winter stay there: "The blue fort on its rock above the rustling palms, with fresh meadows and shining springs of water" (Lawrence 1999). From background reading, I knew that Lawrence was part of a long tradition of people (Romans, Byzantines, Crusaders, Mamlukes, Ottomans, Druze, Chechens, etc.) who had been attracted to this crossroad oasis that contained expansive freshwater pools and surrounding forests—the only permanent such oasis within 30,000 km² of desert—and that had apparently been inhabited since Paleolithic times, some two hundred thousand years ago (Figure 7.1). It was a shock, then, to see the extent of wetland devastation that greeted me on that first visit to this former cradle of life in the Syrian Desert through the near complete absence of standing water (Figure 7.2). A few years later, a guide book quite accurately described the situation: "Although ragged palms survive, the pools are stagnant, the buffalo are dead and the migrating birds now head for Galilee instead. Dust storms are more common today than ever before. The underground reservoirs, exploited almost to exhaustion, are slowly turning brackish" (Teller 1998). By the time I returned to the area more than a decade later, I was greeted by

---

* Adapted by Robert L. France from Irani, K. 2004. Death and rebirth of Jordan's Azraq Oasis. Paper presented at the Mesopotamian Marshes and Modern Development: Practical Approaches for Sustaining Ecological and Cultural Landscapes conference, Cambridge, MA, October.

FIGURE 7.1   Paleolithic artifacts found at the Azraq Oasis.

FIGURE 7.2   Desiccated wetlands and remnant trees.

a partially restored ecosystem and a wetland interpretive center run by the Royal Society for the Conservation of Nature (RSCN). This chapter briefly reviews the intriguing history of the Azraq Oasis, which I believe to be the most successful wetland restoration project in the Arab Middle East. The project has gained international attention (Pearce 2004; Mitchell 2005) and may offer many practical insights into how such future endeavors might be undertaken in neighboring Iraq.

THE PLACE

As described by Jordanian Minister of Environment, Khalid Irani (2004), the Azraq wetlands are an oasis ecosystem that is unique to the region and are located at the junction between basalt bedrock to the north and limestone–flint geology to the south. The oasis is an important area for migrating birds and a safe refuge for a variety of rare species, including an endemic killifish that only exists in these marshes; as a result, it is an official Ramsar site. The Azraq Basin is situated in the lowest depression of a great underground water system, 94 percent of which is in the eastern part of Jordan on the ancient crossroads leading to Iraq, with 5 and 1 percent being in Syria and Saudi Arabia, respectively. The oasis is fed by aquifers that drain rainwater filtered through the basalt into a large shallow basin in which are located five springs and ten seasonal rivers, or wadis. In Lawrence's day and depending on the season, up to 50 km^2 could be flooded to a depth of about a meter. At one time, more than a million migrant birds would visit the oasis each year, with representatives of over 85 percent of all the avian species that have ever been recorded in Jordan. Wild horses and particularly water buffalo similar to those found to the west in the Hula swamps (see chapter 6) and to the east in the Iraqi marshes (Young 1977) used to be common.

The oasis has a rich history of occupation (Figure 7.3) that has shaped neighboring local activities such as fishing, hunting (which was a very big problem in the 1970s and 1980s), water pumping for local use, limited farming, and the salt industry (which depended on evaporating rainwater). Throughout the world as well as in Iraq, the marginal landscape of the wetlands has always provided refuge for marginal ethnic groups. Bedouin have lived in the Azraq area for centuries, and for the last one hundred years Chechen and Druze minorities have also been present.

WATER IN JORDAN

Jordan is one of the ten poorest countries in terms of water resources, with 90 percent of the country being classified as desert. More than 780 million cubic meters (mcm) of water are used each year, with the country running an annual deficit of 19 percent (Irani 2004). In the recent past, Jordan used 70 percent of its freshwater resources for unsustainable agriculture, an amount that has recently decreased to 64 percent and that is targeted for further reductions in the future.

Following the war with Israel and the influx of tens of thousands of Palestinians, cities like Amman grew at a rapid rate and became very thirsty. The Azraq Basin became the first place to be selected for water extraction in order to satisfy the country's increasing demand for the resource. Large-scale pumping started in the

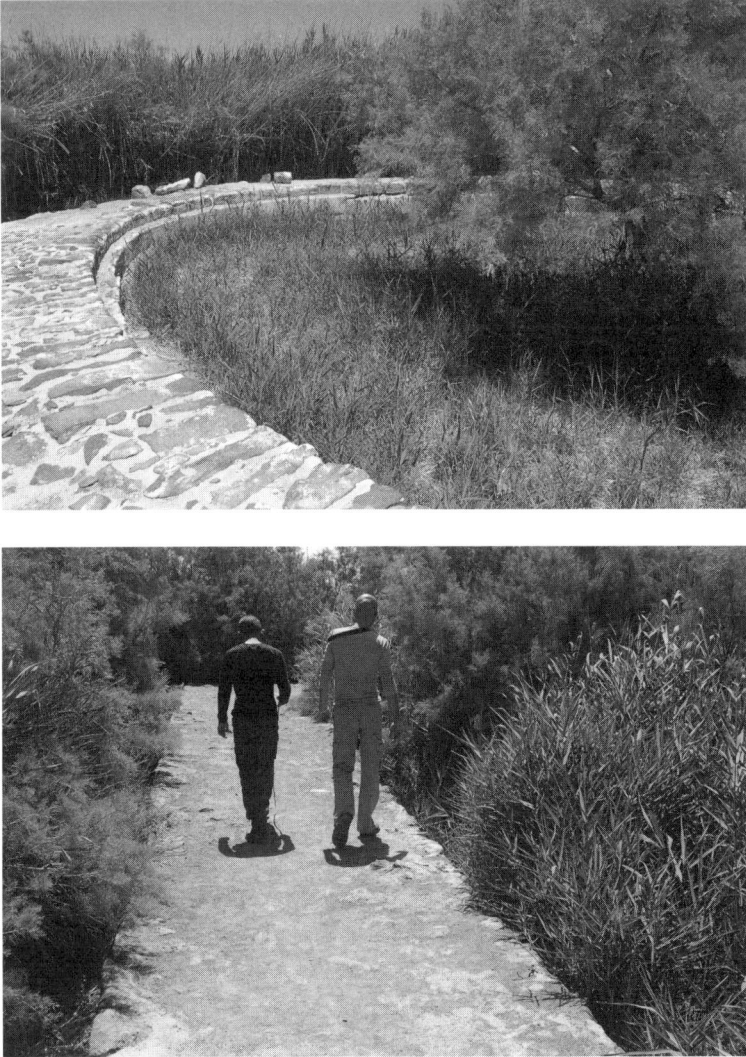

FIGURE 7.3 Building foundations from the Arab colonization period.

early 1980s to supply the major cities of Zarqa and Amman (Figure 7.4). During the period from 1980 to 2001, extraction rose from 15 to more than 55 mcm (Irani 2004). The most important factor to recognize is that today's annual pumping rate is actually now double the annual rate of recharge for the basin. At the same time, Syria dammed the major inflowing river, thereby holding back water which would have otherwise seeped into the aquifer. In short, the aquifer is now being mined like any nonrenewable mineral or petroleum resource.

In addition, many unlicensed wells sprang up within the oasis due to people from the city establishing agriculture plots (Figure 7.5). There are now almost a thousand of

FIGURE 7.4   Active pumps, transport pipelines, and storage reservoirs used for widespread aquifer mining.   *(Continued)*

these illegal wells, with most of the crops being grown, such as olive trees, having high rates of water consumption. Unregulated hunting for migrant birds and the introduction of exotic species (mainly fishes) have persisted since the days of the British Mandate.

When I first visited the oasis in the mid-1990s, there was little water for either birds or local people. Vast areas had begun to dry up, leaving behind nothing but cracked pans of hardened, sometimes salt-encrusted mud and drought- and salt-resistant plants (Figure 7.6). Astonishingly, in one year the wetlands actually dried up completely. It seemed that for all intents and purposes, the demise of one of the

FIGURE 7.4 *(Continued)* Active pumps, transport pipelines, and storage reservoirs used for widespread aquifer mining.

world's most ecologically unique and historically significant wetlands had occurred within the span of only a single human lifetime.

REHABILITATION EFFORTS

The rebirth of the Azraq Oasis began in 1994–1996 with a grant awarded from the Global Environmental Facilities (GEF), a fund developed following the Rio

FIGURE 7.5   Small-scale, localized groundwater withdrawal for agriculture crops.

Conference in 1992 as a capacity-building program to support activities relating to protected areas management. The goals of the project (Irani 2004) were as follows:

*   Rehabilitate and manage the area as the Azraq Wetland Reserve.
*   Establish an environmental impact assessment unit within the Department of Environment directed toward an improved implementation of the Ramsar Convention guidelines in Jordan.

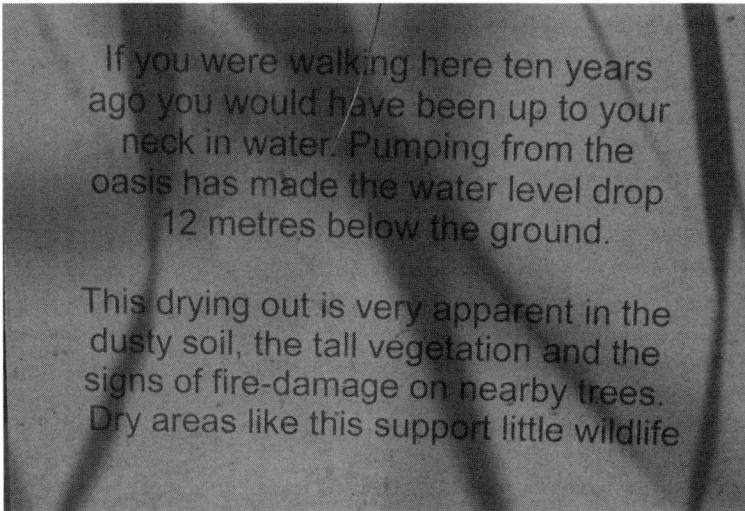

If you were walking here ten years ago you would have been up to your neck in water. Pumping from the oasis has made the water level drop 12 metres below the ground.

This drying out is very apparent in the dusty soil, the tall vegetation and the signs of fire-damage on nearby trees. Dry areas like this support little wildlife

FIGURE 7.6   Interpretive signage, destroyed wetland, mud pans, salt encrustation, and surviving xeric haleophyte plants.   (*Continued*)

FIGURE 7.6 (*Continued*) Interpretive signage, destroyed wetland, mud pans, salt encrustation, and surviving xeric haleophyte plants.

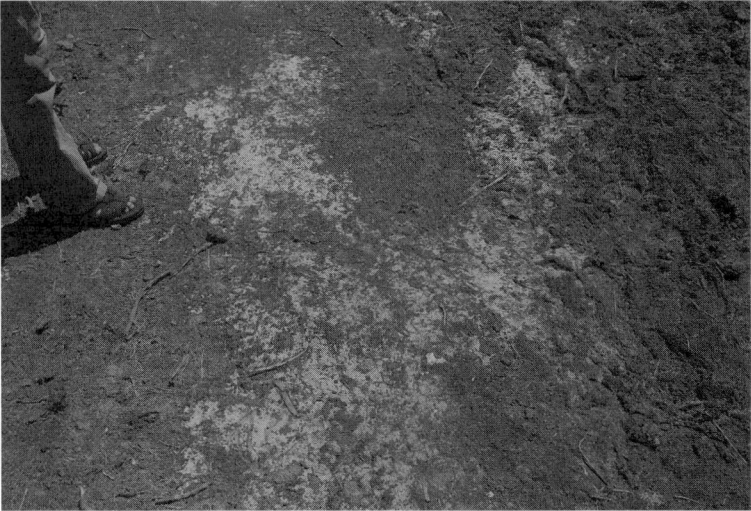

FIGURE 7.6 (*Continued*) Interpretive signage, destroyed wetland, mud pans, salt encrustation, and surviving xeric haleophyte plants.

FIGURE 7.6 (*Continued*) Interpretive signage, destroyed wetland, mud pans, salt encrustation, and surviving xeric haleophyte plants.

- Establish guidelines for agricultural development in the Azraq Basin in order to minimize the use of water through altering cropping patterns.
- Investigate the groundwater resources in the basin, particularly in relation to measuring the extent of salt intrusion into the aquifer consequent with water withdrawal (see chapters 18 and particularly 21 for more about arid-land water and salt management).
- To move beyond the short-term solution of simply pumping water back to the basin through providing support for long-term research on conservation and initiating a management plan for the basin in the style of those developed for other RSCN-protected areas in Jordan (see chapter 16).

The main operational objectives in the Azraq Management Plan of 1998–2002 were (1) to establish and maintain the diversity and abundance of key species characteristic of the area within the reserve, and (2) to restore and maintain a range of habitats characteristic of the area within the reserve through many different activities such as cleaning and enlarging the open-water bodies as well as through creating new areas of water. Rehabilitation has proceeded through stages: some water has already been pumped back (Figure 7.7) with the result being that 10 percent of the original area of the wetlands have returned (Figure 7.8), and alternative water sources have begun to be investigated to augment the return of water to the system.

Another objective was to help ensure land uses within the Azraq Basin that would be ecologically sustainable and hence help to replenish the upper aquifer. This included working with farmers to minimize their extraction of water and working with the government to try to reduce the degree of pumping to the major cities. As well, plans were formulated to conserve and study the archaeological history of the reserve.

FIGURE 7.7   Inputs of water to sustain newly established wetlands.

One major objective was to facilitate and encourage education and interpretation within the reserve (Irani 2004). For example, the RSCN worked with local children to begin to make them feel responsible for their area such that schools now use the visitor center (Figure 7.9) to discuss and plan options for developing sustainable practices (Figure 7.10). One marker of the success of the education program was that before the RSCN started, 70 percent of the locals thought that environmental protection and management were solely government responsibilities. Now, through implementing different interactive techniques such as role-playing and debates with the visiting national minister of water, and using music and theater to tell water

FIGURE 7.8   Rehabilitated wetlands in the Azraq Oasis.   (*Continued*)

FIGURE 7.8 (*Continued*) Rehabilitated wetlands in the Azraq Oasis.

FIGURE 7.9   The Azraq Oasis visitor center, designed in the architectural vernacular of the nearby "desert castles" (note the outdoor lecture amphitheater).

stories, students in Amman and other cities have also become engaged and feel just as passionate about the issues as those from the immediate area of Azraq itself.

Students from across the country, therefore, learned that the water that they drink or casually use originates from this beautiful wetland in the desert. The RSCN worked with the Ministry of Education to develop a new national curriculum about water for the schools which focused on its role in supporting ecosystems and societies, the unique qualities about water as a substance, the treatment and demand management of water resources, and the dynamics of water supply (Irani 2004).

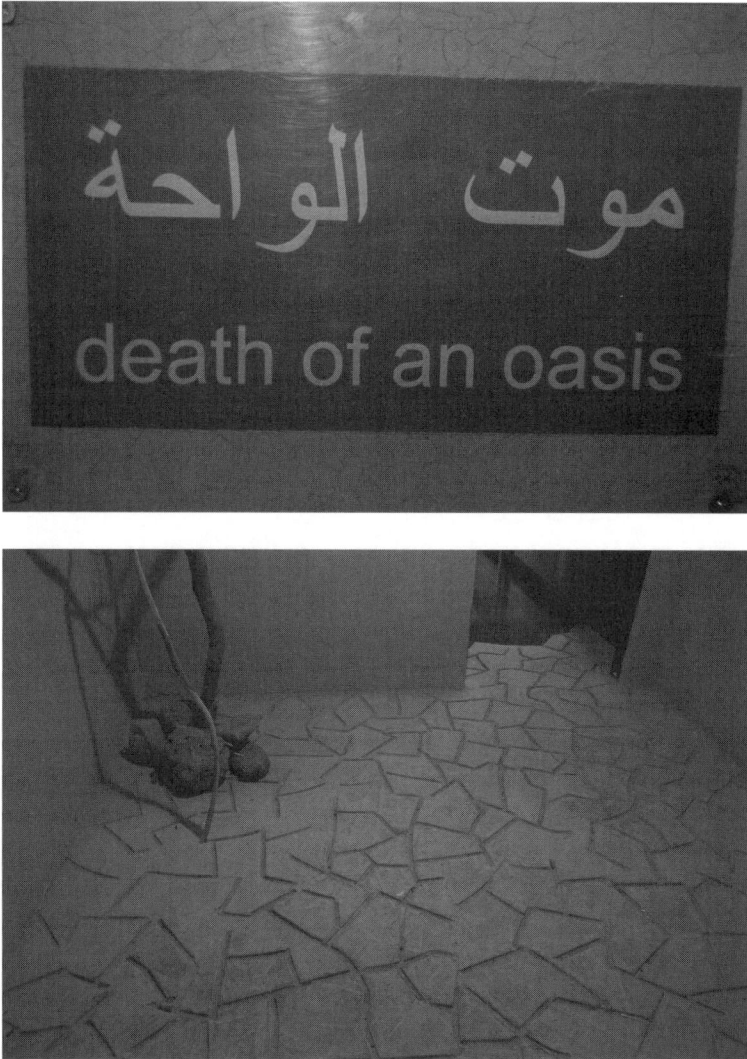

FIGURE 7.10 Pedagogic displays in the visitor center informing about the demise of the Azraq Oasis (note the nice detail of the tiles designed to resemble parched desert ground surfaces). Displays include those of the original piping used in the wetland destruction, information about Jordan's water use and how the wetlands were drained, and a board game concerned with land and water management decision making.   (*Continued*)

FIGURE 7.10 (*Continued*) Pedagogic displays in the visitor center informing about the demise of the Azraq Oasis (note the nice detail of the tiles designed to resemble parched desert ground surfaces). Displays include those of the original piping used in the wetland destruction, information about Jordan's water use and how the wetlands were drained, and a board game concerned with land and water management decision making.

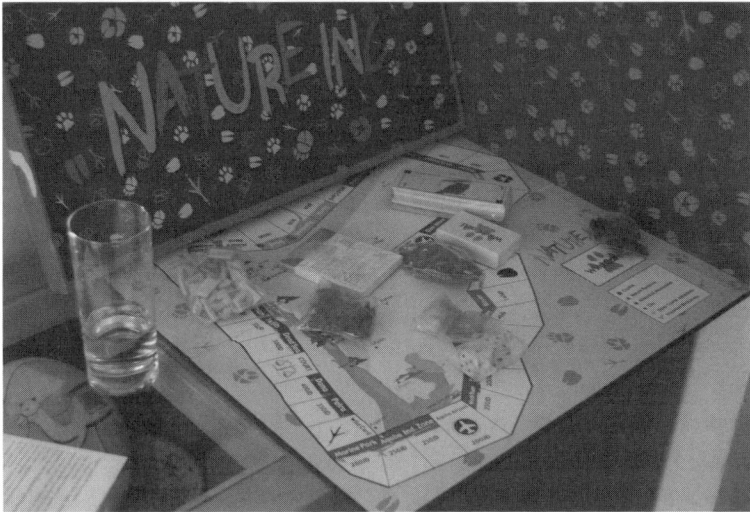

FIGURE 7.10 (*Continued*) Pedagogic displays in the visitor center informing about the demise of the Azraq Oasis (note the nice detail of the tiles designed to resemble parched desert ground surfaces). Displays include those of the original piping used in the wetland destruction, information about Jordan's water use and how the wetlands were drained, and a board game concerned with land and water management decision making.

Such "rehydration education" (*sensu* France 2003) is essential for fostering watershed awareness and stewardship (France 2005). In 2004 for the first time, the story of Azraq was mentioned in the country's textbooks. And attempts have even been made to reach urban children by developing electronic games about water management issues in Jordan, particularly the fate of the Azraq Oasis.

Other objectives outlined in the management plan were to enhance and maintain the publicity of the reserve nationally and internationally, to work with the Ministry

FIGURE 7.11   The Royal Society for the Conservation of Nature's Azraq Oasis Handicraft Center and various locally produced products.   (*Continued*)

FIGURE 7.11 (*Continued*) The Royal Society for the Conservation of Nature's Azraq Oasis Handicraft Center and various locally produced products.

of Water in exploring the use of alternative water sources, and to develop and maintain local community involvement in the reserve (Irani 2004). These locals are interested in developing sources of income that might arise from the wetland such as ecotourism and the creation of products other than those originating from unsustainable agriculture (Figure 7.11). The RSCN looks at these income-generating activities as tools for conservation that simultaneously reduce local poverty (as elaborated on in chapter 16).

One more objective was to establish and maintain a team of staff who would be able to implement the management plan effectively. This is the capacity-building idea, dealing with both institutional and technical issues, as the RSCN are not themselves wetland experts (see chapter 12 for a similar example in South America).

Ecotourism became another objective based on the establishment and maintenance of effective visitor management services and facilities such as boardwalks and bird-viewing hides within the reserve that were constructed by RSCN staff with local assistance (Figure 7.12), as well as the development of the long-term financial viability of the reserve. In this regard, related objectives included ensuring that management activities within the reserve would be effectively spatially zoned, and that the implementation of the management plan would be periodically reviewed.

A variety of achievement indicators demonstrate the success of the Azraq restoration efforts (Irani 2004). Ten percent of the oasis has been restored and endemic species, once threatened with extirpation, have been preserved. Others, such as the iconic water buffalos, have been reintroduced (Figure 7.13). Many of the birds have come back for feeding, resting, and nesting in the Azraq Wetland Reserve, and while they are not in the same numbers as before, they are at least the same species (raptors, kingfishers, herons, terns and gulls, and so on) that were once present. The Azraq Basin is

FIGURE 7.12 Visitor amenities include a shaded car park, Bedouin-style picnic shelters, trails and boardwalks, observation platforms and interpretive signage, and a birdwatching shelter in the style of the regional architecture.   **(*Continued*)**

FIGURE 7.12 (*Continued*) Visitor amenities include a shaded car park, Bedouin-style pic-nic shelters, trails and boardwalks, observation platforms and interpretive signage, and a birdwatching shelter in the style of the regional architecture.

FIGURE 7.12 (*Continued*) Visitor amenities include a shaded car park, Bedouin-style picnic shelters, trails and boardwalks, observation platforms and interpretive signage, and a birdwatching shelter in the style of the regional architecture.

FIGURE 7.12 (*Continued*) Visitor amenities include a shaded car park, Bedouin-style picnic shelters, trails and boardwalks, observation platforms and interpretive signage, and a birdwatching shelter in the style of the regional architecture.

FIGURE 7.12   (*Continued*) Visitor amenities include a shaded car park, Bedouin-style picnic shelters, trails and boardwalks, observation platforms and interpretive signage, and a birdwatching shelter in the style of the regional architecture.

becoming the gateway for a booming desert tourism industry (Figure 7.14) that capitalizes on its proximity to sites of historical interest such as the famous desert "castles" (Figure 7.15). The idea of Minister Irani and his colleagues is to bring people to the area not just for purposes of generating local income but also, just as importantly, for purposes of generating national awareness and support. An attractive visitor interpretation center was key in this regard in that it explains the quarter-million-year history of the Azraq Oasis (Figure 7.16). Already, design ideas from this Jordanian location are being

FIGURE 7.13   Young water buffalos in rearing pens.

FIGURE 7.14 The Azraq Tourist Lodge for overnight visits.

FIGURE 7.15   The nearby desert palace of Qusar Amra, constructed in the Umayyad period (seventh–eighth centuries CE) for purposes of retreat, agriculture, and/or local administration, and featuring some of the most spectacular (and unusual) examples of early Islamic art.

(Continued)

FIGURE 7.15 (*Continued*) The nearby desert palace of Qusar Amra, constructed in the Umayyad period (seventh–eighth centuries CE) for purposes of retreat, agriculture, and/or local administration, and featuring some of the most spectacular (and unusual) examples of early Islamic art.

FIGURE 7.15 (*Continued*) The nearby desert palace of Qusar Amra, constructed in the Umayyad period (seventh–eighth centuries CE) for purposes of retreat, agriculture, and/or local administration, and featuring some of the most spectacular (and unusual) examples of early Islamic art.

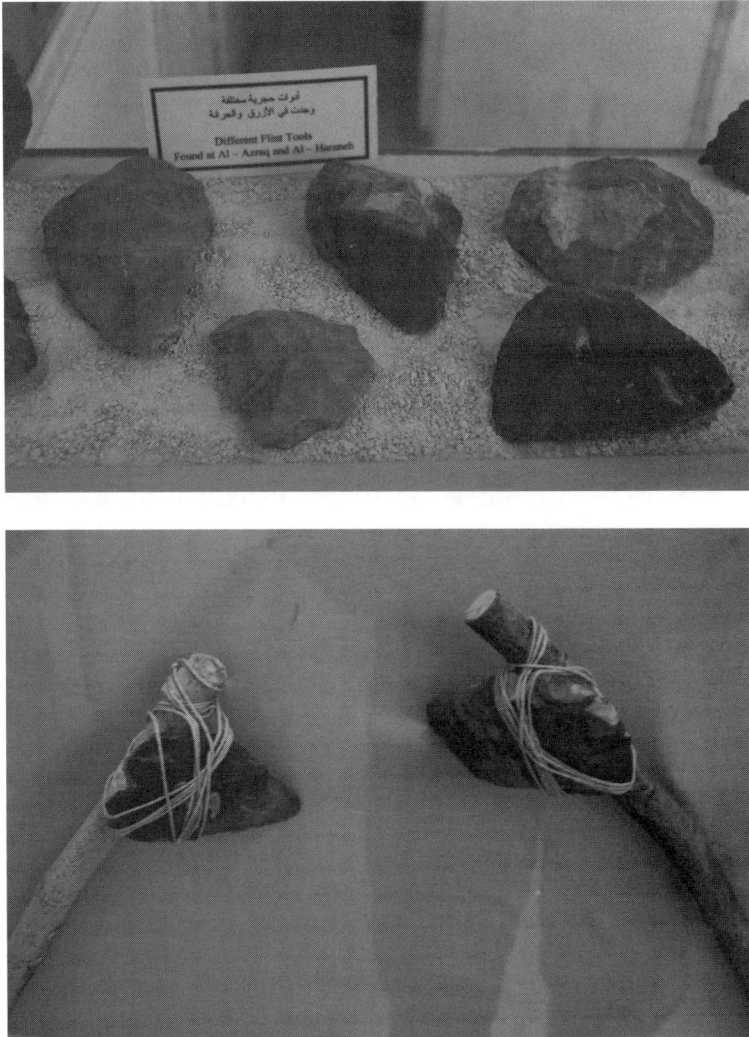

FIGURE 7.16 Exhibits of the extensive history of the Azraq Oasis include flint tools and axe heads, information about wetland wildlife, and pre-drainage photos of the wetlands, some showing swimming water buffalos in a scene reminiscent of those in both the Hula Swamp (see chapter 6) and the Iraqi marshlands.   **(Continued)**

FIGURE 7.16 (*Continued*) Exhibits of the extensive history of the Azraq Oasis include flint tools and axe heads, information about wetland wildlife, and pre-drainage photos of the wetlands, some showing swimming water buffalos in a scene reminiscent of those in both the Hula Swamp (see chapter 6) and the Iraqi marshlands.

adapted to another wetland interpretation center planned for Abu Dhabi (Abu Dhabi Urban Planning Council 2007), and they are certainly adaptable to what might develop in the future in Iraq. For the Azraq Oasis, it is important to emphasize that much attention was directed toward children in a belief that it is they who will become the best vehicles for advocacy.

So though the "rebirth" of these desert wetlands is still in its infancy, it is hoped that with maturity the Azraq Oasis will help the entire country of Jordan address its very serious water management concerns. At the same time, the environmental restoration efforts in Azraq, particularly with respect to the development of an increased feeling of connectivity to the ecosystem—an issue of paramount importance to the sustainable success of any restoration project (France 2007; see also chapter 2)—can offer a valuable lesson for how similar activities might unfold in Iraq.

ACKNOWLEDGMENT

This chapter is adapted from Irani (2004).

REFERENCES

Abu Dhabi Urban Planning Council. 2007. *Abu Dhabi 2030 urban structure framework plan.* Abu Dhabi: Abu Dhabi Plan Council.

France, R. L. 2003. *Deep immersion: The experience of water.* Winnipeg, MB: Green Frigate Books.

———, ed. 2005. *Facilitating watershed management: Fostering awareness and stewardship.* Lanham, MD: Rowman & Littlefield.

———, ed. 2007. *Healing natures, repairing relationships: New perspectives on restoring ecological spaces and consciousness.* Winnipeg, MB: Green Frigate Books.

Irani, K. 2004. Death and rebirth of Jordan's Azraq Oasis. Paper presented at the Mesopotamian Marshes and Modern Development: Practical Approaches for Sustaining Ecological and Cultural Landscapes conference, Cambridge, MA, October.

Lawrence, T. E. 1999. *Seven pillars of wisdom.* Ware, UK: Wordsworth Editions.

Mitchell, A. 2005. *Dancing at the Dead Sea: Journey to the heart of the environmental crisis.* London: Transworld.

Pearce, F. 2004. *Keepers of the spring: Reclaiming our water in an age of globalization.* Washington, DC: Island Press.

Teller, M. 1998. *Jordan: The Rough Guide.* London: Rough Guides.

Young, G. 1977. *Return to the marshes: Life with the marsh Arabs of Iraq.* London: Collins Books.

8 Integrated Restoration and Adaptive Management of the Las Vegas Wash*

CONTENTS

Introduction 117
The Place 118
Protection Efforts 120
Environmental Problems 121
Restoration Efforts 122
Adaptive Management 123
Conclusion 124
Acknowledgment 124
References 124

INTRODUCTION

After more than three decades of advocacy, what had long been a severely degraded landscape, the Las Vegas Wash, has been changed into something truly extraordinary: the Clark County Wetlands Park (see also chapter 3). The process and the ensuring products of this particular project represent the most successful and impressive restoration of a desert wetland that has been accomplished, a signature undertaking that offers lessons for not only the situation of the Iraqi marshlands, but also publicly supported environmental restoration projects of any type, anywhere in the world.

Before the Las Vegas Wash is examined in detail, it is important to review some of the most important things about any kind of successfully implemented ecosystem restoration project. Paramount is that all factions, personnel, and stakeholders move from their isolated spheres and their idiosyncratic agendas to a state of more interdependence with one another. Interdependence by its very nature can be very wrought with opportunity but also, it must be admitted, wrought with some forces of failure. But Vicki Scharnhorst (2004) believes this is the best strategy to adopt to what will

* Adapted by Robert L. France from Scharnhorst, V. 2004. Restoration of the Las Vegas Wash and other desert wetlands. Paper presented at the Mesopotamian Marshes and Modern Development: Practical Approaches for Sustaining Restored Ecological and Cultural Landscapes conference, Cambridge, MA, October.

117

eventually become (in every successful ecosystem restoration project) what can be called "integrated community stakeholder involvement."

The other items necessary for a successful restoration project are all structural (Scharnhorst 2004). The water balance absolutely and unequivocally has to be right. And this means not just the quality and the quantity of water, but also the timing and the distribution of its delivery (see chapter 18). One useful approach is to always plan for and ensure some early wins and successes. Overall success often depends, for example, on creating incremental products such as demonstration projects that can be celebrated by everyone (France 2005). And, finally, it is key to have a long-term implementation financing operations and maintenance strategy in place in order to develop a sustainable, long-term plan.

This chapter will briefly review the place that is the Las Vegas Wash. Next, the actions of the people and how they nearly destroyed the Wash will be covered, followed by how the people later became mobilized to restore the Wash. Issues concerning water will be examined throughout, and the protection of the Wash will be introduced, followed by the adaptive management plan and an assessment about how to improve conditions as the project evolves (Scharnhorst 2004).

THE PLACE

Las Vegas already had a small contingent of light urbanization as early as the mid-1800s, when it was a railroad town (France 2011; see also chapter 13). The Wash itself was a very broad floodplain filled with beautiful honey mesquite trees. It was an ephemeral stream, receiving water only when it rained. Some groundwater was present as well. One of the things that Scharnhorst (2004) is most proud of in the restoration of the Wash is her work with one of the stakeholders there—one of the wastewater dischargers in the area—to create a demonstration wetlands project for wastewater treatment similar to those described in chapters 18, 19, and 20, and which also included a bird-viewing preserve. This particular organization even went as far as to produce its own short film, which included the following voice-over:

> Flying high over the desert of southern Nevada, your eyes are quickly drawn to the bright lights and high-rise buildings, beckoning eager visitors to the entertainment capital of the world. But from a bird's-eye view, the best destination resort in this valley is a unique quiet hideaway, offering cool, refreshing water and plentiful food and nourishment. A comfortable, protected place to build a home, raise a family, play with friends, eat a meal, and just get away from it all. The city of Henderson's bird-viewing preserve and water reclamation facility—a place to call home for the birds.

Who would ever guess you could do so much with wastewater effluent?

Las Vegas and its Wash are located in southern Nevada along the Colorado River. The Las Vegas valley itself is sandwiched between the Spring Mountains on the west and Lake Mead on the east, the latter being a national recreation area administered by the National Park Service. Lake Mead is an artificial reservoir formed by construction of the Hoover Dam on the Colorado River. The section of the Las Vegas

Wash developed into the Park (see chapter 13) is a linear stretch of seven miles, located upstream of Lake Mead and eight miles from the downtown Las Vegas Strip, at the lowest point in the valley.

The Las Vegas valley has a very interesting topography, hydrology, and geology (France 2011). The tallest peak is twelve thousand feet in elevation and all the water or snow that falls in the greater valley must run through the five major tributaries and then through the Wash before reaching Lake Mead. One interesting geologic feature that is contiguous to the Wash is the Rainbow Gardens, which include all of the major rock formation types that exist in the nearby Grand Canyon.

The area is, like the marshlands of southern Iraq and the Azraq wetlands in Jordan (see chapter 7), truly an oasis in the desert. Given the context of being located in a part of the Mojave Desert that gets only four inches of annual rainfall, the Wash is one of the few hydroriparian areas in the entire southwestern United States with a rich floral biodiversity. Indeed, over a hundred species of plants have been identified, seven of which are currently candidates by the U.S. Fish and Wildlife Service for listing as threatened or endangered species. Wildlife biodiversity in the Wash is also atypically high, with over two dozen threatened or endangered species, including the peregrine falcon, desert tortoise, several types of bats, and some species of birds that interestingly are also present in the Iraqi marshlands (Scharnhorst 2004).

The Las Vegas Wash is truly a unique place, but like any urban area there are attendant disbenefits associated with the presence of human development (Scharnhorst 2004). With construction of the Hoover Dam in the 1930s, the valley suddenly discovered that it had an abundant supply of electricity which would eventually lead to inexpensive air conditioning, which in turn would lead to the extreme urbanization of the valley floor. Indeed, the area today has the highest growth rate in the entire United States. Further, one of the first undertakings by the workers on the Hoover Dam and some of the early residents was to begin cutting down the mesquite trees for firewood. With the disappearance of the woody plants in the Wash, soils were no longer able to be held in place.

Following the increase in population, ephemeral streams became perennial as a result of wastewater being discharged to the lowest point in the valley. In consequence, stream channels began to become highly incised under this increased flow regime (Scharnhorst 2004). This resulted in the former broad floodplain becoming, by 1975, a raging river channel following the occasional high-intensity thunderstorms. And to give a perspective about how much water goes through this area at such times, the flow rate of the Colorado River that attracts rapid-running adventurers is generally about twenty-five thousand cubic feet per second, not too different from the amount of water that will actually go down the Las Vegas Wash during a hundred-year storm. The damage resulting from this increased discharge was enormous, with some reaches experiencing channel cutting in the order of thirty feet between 1978 and 1984 (France 2011).

The National Park Service used to have a road that went around Lake Mead that was built at grade with box culverts underneath. A particularly severe storm in 1975 washed the culverts out and the Park Service decided to improve the road by raising it up forty feet with a very long span bridge, thereby allowing the Las Vegas Wash to flow underneath. Today, the graffiti which had been scribbled on the bridge at that grade a quarter of a century ago is now more than thirty feet up in the air!

Given such water velocities, it is not just the soil that is lost but also the wetland plants. Prior to 1985, there were about 2,300 acres of wetlands in the Las Vegas Wash. By 2000, it was estimated that the total area or riparian habitat plus wetlands was only about seven hundred acres (Scharnhorst 2004). And prior to serious restoration efforts which began a few years later, the remaining wetlands had been reduced to an area of less than 10 percent of their original extent.

The entire region of the arid southwestern United States has received an enormous population boom that has deleteriously affected its desert ecosystems. So as has been the case for the Hula Swamp in Israel (chapter 6), the Azraq Oasis in Jordan (chapter 7), the Palmyra reed fields in Syria, and the southern marshes in Iraq (France 2007), the Las Vegas Wash is yet another glaring example of the destruction of that most incongruous and therefore rarest and most precious of all ecosystems: desert wetlands.

However, the saga of the Las Vegas Wash is also an encouraging story of environmental advocacy. Over time, people came to realize that this was truly an ecosystem that should be restored not just for its own sake, but also for its substantial educational and aesthetic values (France 2009). So it was in 1973 that the Las Vegas Wash Development Committee was formed, and in 1982 Scharnhorst and her colleagues began working on both the environmental assessment of the Wash as well as the first generation of a master plan of what would eventually become the Clark County Wetlands Park (see chapter 13).

PROTECTION EFFORTS

Another devastating storm in 1984 had interesting repercussions with respect to how people came to view the Wash (Scharnhorst 2004). There is a single water supply pipeline running from Lake Mead to the Las Vegas valley, and in that single storm the pipeline was exposed by the removal of over thirty-five feet of overlying sediment. Until that time, the thinking had been so isolated that the Las Vegas Valley Water District had stated that "the erosion in the Wash isn't our problem. We've got the money and the resources and we're going to bury that pipe two hundred feet below the Wash." The Regional Flood Control District in turn said that "it's not really our problem because the pipe is underneath the bottom of the valley and we just really handle the valley surface." The Clark County Parks and Recreation Department (CCPRD) said, "We think there is the potential for a great park, but it doesn't look like there will be anything left of the park for us to protect and enhance." And there were a number of other agency stakeholders who for a number of reasons wouldn't assume responsibility for the situation. So, after the 1984 flood another integrated planning study was drafted by another Wash taskforce made up of mostly agency individuals, and another stop-gap erosion mitigation project was put in place by the federal government. Finally, in about 1990 the CCPRD decided that they needed to take ownership of the problem and to really work to move toward restoring the Wash. Their strategy was to pass a state-wide bond initiative to raise the necessary funds, which if put before voters before 1990 probably wouldn't have passed because there was not yet a critical mass of laypeople and professionals in the valley who really cared much one way or the other about the Wash.

The Southwest Wetlands Consortium was formed (France 2011) and hired by the CCPRD to work on not just the Clark County Wetlands Park master plan (see chapter 13) but also a programmatic environmental impact statement to solve what had become quite a horrendous problem with respect to the Wash; in other words, the continued erosion was turning it from a treasure in the desert to simply a desert. Scharnhorst (2004) believes that there was not enough momentum in place for initiating such a restoration plan until the Wash had become so incised that it actually intercepted the shallow groundwater table in 1997. Before that time, nobody had really thought about exacerbating problems related to the erosion. Now the water plant operators at the Lake Mead intake, miles away from the Wash, suddenly found that they were beginning to detect perchlorate concentrations approaching those deemed dangerous in the Safe Drinking Water Act. The water managers began to become very concerned, realizing for the first time that they shared responsibility, benefits, and values with how the Wash functioned. The result was that for the first time, broad-scale support existed for undertaking restoration of the Wash. From then on, there was enough momentum in place to proceed with the work, even allowing for the plans to be retooled following several subsequent large and damaging storms in 1999.

ENVIRONMENTAL PROBLEMS

The original master plan had several goals based on stabilizing the natural resource of the Wash so that processes could be put in place to protect the valuable park.

Water is both the lifeblood of the Wash as well as the source of its potential ruin. In 1950, there were fifty thousand people living in the Las Vegas valley and producing less than a million gallons a day of wastewater flow to the Wash. Today, there are over 1.7 million residents in the valley with about 180 million gallons a day of wastewater being produced, most of which goes down the Wash. This amounts to a phenomenal growth rate of 60 percent per year, of not only people but also water being discharged into the Wash. What kind of ecosystem could be expected to handle that kind of abuse?

The wetlands of the Wash have three other sources of water besides the wastewater inputs: groundwater, urban runoff, and stormwater (Scharnhorst 2004). The wastewater effluent is of good quality in that it is tertiary treated to remove both phosphorus and nitrogen as the Colorado River is an extremely oligotrophic environment and thus very sensitive to nutrient pollution. Rather, it is the quantity of wastewater that's doing the more serious damage by contributing to the vertical incising of the receiving stream channel bed. Urban runoff from the overwatering of residential lawns, golf courses, and highway medians contains fertilizers, herbicides, and pesticides, but fortunately isn't very voluminous. The groundwater contains elevated concentrations of perchlorate and can be brackish. The biggest culprits and vandals to the environmental problems in the Wash are the stormwater surges (France 2001, 2011). In this respect, any high rainfall event can cause vertical erosion of up to twenty feet and lateral migration of the channel, which together can mobilize and send well over a hundred thousand tons of sediment downstream into Lake Mead.

The Las Vegas valley agencies have collectively accomplished a good deal in order to try to get water quality under control by implementing best management practices (Scharnhorst 2004). The perchlorate contamination resulting from chemical pollution stretching back half a century is being dealt with through installation of an interception system in the groundwater and subsequent treatment of the plume. The Nevada Division of Environmental Protection estimates that within the next ten years, all the gravel and sands of the Wash will have been flushed free of perchlorate to background concentrations.

So the major problem, therefore, remains the severe erosion in the Wash. For example, from the air it is possible to observe the massive sediment plumes entering Lake Mead which deposit from one to three hundred thousand tons of material a year.

RESTORATION EFFORTS

The restoration of the Las Vegas Wash provides an exemplary lesson about the value of taking early action (Scharnhorst 2004). Following the flood of 1975, experts estimated that four erosion control structures would be needed at a cost of about $200,000. By 1986, only five years later but subsequent to another very bad flood, estimates suggested the need for eleven erosion control structures (mostly aesthetically attractive, natural-looking installations) at a cost of $15 million. The erosion continued to worsen such that by 1991, fifteen structures were suggested to be required at a cost of $30 million. Today, the most current estimates are for $125 million to be spent to construct twenty-two difficult-to-build and highly engineered structures to be put in place to stabilize the Wash. The lesson here, of course, is that anytime one can obtain the financing, the resources, and the collaboration necessary to do something early on in terms of restoration—it doesn't matter if it is controlling invasive species, erosion, or whatever else is the issue—do it immediately (Scharnhorst 2004).

Las Vegas residents and environmental managers finally came to realize that if efforts were not made to immediately stabilize the Wash, there soon would be nothing left to save. Hydraulic and hydrologic modeling was used to map the fluvial and geomorphological processes in the Wash in order to understand sediment erosion and transport and thus aid in the design and construction of the stabilization measures. Stabilization is usually a three-pronged approach (Scharnhorst 2004): (1) downward channel bed degradation has to be stopped, (2) lateral channel migration must be halted, and (3) revegetation has to be recognized as being a major part of the solution.

Twenty-two erosion control structures are planned to create a stable elevation gradient of twenty feet per mile. The current grade of the Wash is between thirty-eight and eighty feet per mile, indicating the challenge involved with stabilizing and flattening out the grade so that erosion is halted. By 2004, eight weirs have been completed, four are in planning and design, and an additional ten are planned to be built over the next decade. Four miles of bank stabilization have now been completed, and twenty-three acres of former wetlands have been revegetated, all accomplished at a total cost of $25 million.

One of the things that Las Vegas is famous for is its demolition and reconstruction of casinos. The southern Nevada Water Authority has realized that each year, if they can stockpile and recycle the construction debris from some of this activity, they can save $10.00–15.00 a cubic yard on large rip-rap at the same time as keeping the material out of landfills. Therefore, much of the foundation for erosion control structures installed in the Wash came from those imploded buildings whose original presence in the city had done much to contribute the stormwater runoff which had caused the damage in the Wash in the first place (France 2011). This rip-rap was dumped into the channel bed, followed by revegetation through either natural colonization or deliberate planting by citizen volunteers.

ADAPTIVE MANAGEMENT

In April 2000, the Las Vegas Wash Committee approved a comprehensive adaptive management plan which would address monitoring channel stabilization, revegetation, and invasive plant management (France 2011). It was deemed important that the plan continue to maintain the momentum created over the decades of advocacy, be very adaptive to accommodate the dynamic system of the Wash, and include a large focus on biological and environmental monitoring and assessment which would be integrated with several of the other plans for the Wash. In this respect, adaptive management through monitoring is important to determine if the corrective measures have been effective—and, if not, to suggest ways in which to improve conditions (Scharnhorst 2004).

The revegetation program is an issue of extreme importance for the success of any wetland restoration project, be it either in Nevada or in southern Iraq. The erosion control structures have all been specifically designed to optimize the reintroduction of native riparian species (Scharnhorst 2004). But not all sites are equal, and the geomorphology and hydrology turned out to be the keys to the revegetation process. For example, greater success was soon found to occur if small cells of native riparian and emergent wetland species were intensively managed within what was really a nonnative matrix, rather than trying to loosely manage the entire area at any one time. And with respect to some of the invasive species, these were used to create some stand diversity in terms of canopy height or spacing as nesting habitat for migratory birds.

Regardless of the wetland restoration project anywhere in the world, invasives are a major problem due to their insidious presence. In the case of the Wash, there are three "most wanted weeds" that require constant management (Scharnhorst 2004). For one of these, for example—tamarisk, which comes from Eurasia—adaptive management has shown that in order to eliminate the plant, it is necessary to apply herbicide within fifteen minutes of their being cut or they will seal up and not succumb. If simply cut and left alone, this amazingly hardy invasive extrudes salt around it so that other plants cannot compete with it until it is ready to grow back. Further, tamarisks also pose a double problem in relation to the use of fire clearance as a management tool in that burning actually helps the plants propagate.

CONCLUSION

The restoration of the Las Vegas Wash and creation of the Clark County Wetlands Park were dependent on the organizational structure put in place from the very start. One of the major reasons for success is that over the years, there were over two dozen interested stakeholder groups made up of various agencies, regulators, nongovernmental organizations, and environmental permitting and regulatory agencies all intimately involved with the project. The bottom-line reason why the Las Vegas Wash is perhaps the most challenging and successfully implemented desert wetland restoration project in the world today is a direct consequence of how all these groups moved from being isolated individual organizations with their own agendas to a state of interdependence (France 2011). Successful environmental restoration projects have to find such a way in which to morph into a collaborative integrative community with an understanding that they hold shared responsibilities, shared benefits, and shared goals (Scharnhorst 2004). The only chance that the restoration efforts in the southern Iraqi marshlands have of succeeding is if such a feeling of confraternity develops beyond the self-interests of all the vested players working there currently and in the future.

ACKNOWLEDGMENT

Adapted from Scharnhorst (2004).

REFERENCES

France, R. L. 2001. (Stormwater) leaving Las Vegas. *Landscape Architecture*, August, 38–42.
———. 2011. *Designing new natures: People, ecology, and landscape repair*. Boca Raton, FL: Taylor & Francis.
———, ed. 2005. *Facilitating watershed management: Fostering awareness and stewardship*. Lanham, MD: Rowman & Littlefield.
———, ed. 2007. *Wetlands of mass destruction: Ancient presage for contemporary ecocide in southern Iraq*. Winnipeg, MB: Green Frigate Books.
Scharnhorst, V. 2004. Restoration of the Las Vegas Wash and other desert wetlands. Paper presented at the Mesopotamian Marshes and Modern Development: Practical Approaches for Sustaining Restored Ecological and Cultural Landscapes conference, Cambridge, MA, October.

Section II

Wetlands and Nature Reserves

Overview: Ecological and Cultural Context for the Restorative Redevelopment of the Iraqi Marshlands

Section 2 comprises eight chapters that deal with the ecocultural integrity of wetlands and nature reserves and their role in the restorative redevelopment of the Iraqi marshlands and other landscapes devastated by conflict or natural disasters. Together these chapters consider the management of large-scale wetland landscapes for people and wildlife; the innovative design, development, and management strategies for preserved or restored wetland parks or nature reserves; and how the most sustainable and therefore successful projects marry environmental protection to economic development. The most important lesson to take away from these chapters is that it is nonsensical to be concerned with only ecological restoration when dealing with the complexities involved in the restorative redevelopment of regional, and therefore by definition *ecocultural*, landscapes such as the Iraqi marshlands.

When considering ecosystems, chapter 9 by Steve Apfelbaum and James Ludwig reminds us that wetlands provide many goods and services to humans and that in order for landscape-scale restoration projects to be successful, a careful understanding of hydrology is essential. Chapter 10 outlines the planning and management approaches necessary to reconstruct waterfowl communities over large landscapes and how critically important it is to work in partnership with agricultural development.

In terms of the human element in managing or restoring wetlands and nature reserves, chapter 11 reviews two case studies where visitor centers, UNESCO designation, petrochemical industry funding, environmental education, public recreation, and produce marketing all play important roles. Chapter 12 demonstrates how scientific research can be supported by international volunteers and how this in turn can be used to help engage locals in environmental decision making and management. The integration of people in the planning and design involved in transforming a degraded wetland into the world's most visited desert wetland park is covered in chapter 13. And chapter 14 by Robert France and Evi Syariffudin introduces the many challenges involved in establishing ecotourism industries in two other wartorn regions of the world where indigenous people live suspended over shallow water.

Strategies for interlinking ecology and economics are described in chapter 15, where alternative futures modeling is used for predicting the impacts of various scenarios of tourism development on the natural resources of a threatened landscape. Finally, one of the world's most accomplished successes in conjoining nature conservation with sustainable development is reviewed in chapter 16, which concerns conjoining protected area management with regional livelihood sustenance.

Ecosystems

9 Hydrology and Wetland Restoration for Human Subsistence and Regional Biodiversity
The Challenge to Restore the Living Landscape of Iraq's Mesopotamian Marshes

Steven I. Apfelbaum and James P. Ludwig

CONTENTS

Introduction 132
Ecological Wetland Projects versus Living–Working Landscape-Scale Programs 132
Principles for Scaling Up 135
Examples at Scale 136
Red River of the North, Minnesota 136
    Understanding the Watershed Context for Wetland Restorations 136
Runoff Travel Time 139
Definition of Early, Middle, and Late Areas Relative to the Red River Main
Stem 139
General Hydrological Framework for Considering Wetland Restoration 140
    Biological Productivity and Diversity, and Hydrology Relationships 140
Diversity, Hydraulic Performance, and Outlet Design Characteristics Were
All Related 143
      Summary 144
Kankakee Sands, Rensselaer, Indiana 145
    The Restoration Process 145
      Summary 151
Conclusion 153
References 154

INTRODUCTION

Typical wetland restorations in developed western nations proceed from regulatory actions and the goals of conservation groups. Most restored wetlands are small, a few to several hundred acres, and not on a landscape scale. Some projects have aimed to restore historic hydrological conditions, but few achieve high-quality biological outcomes. Most projects contribute little to regional or global-scale biodiversity, water management, or other valuable functional outcomes. Rarely do these western projects provide much for human subsistence.

Few western watershed-scale restorations aim to support people as the key beneficiaries who will subsist on the biodiversity, productivity, and other attributes of restored wetland systems. Design of wetland restorations to provide healthy foodstuffs, water, flood damage reduction, fish, and wildlife for humans will require a restoration strategy very different than those driven by western-style regulatory actions or strict conservation goals.

ECOLOGICAL WETLAND PROJECTS VERSUS LIVING–WORKING LANDSCAPE-SCALE PROGRAMS

Restorations designed as conservation projects differ from restored living and working landscapes. Living landscape projects to support humans usually involve agricultural and animal husbandry. The planning and design process on small wetland projects reduces variability by constraining hydrology to a narrow operational range. Property boundaries define the limits of most smaller conservation projects to one owner's lands. Off-site impacts including flooding, surcharging shallow groundwater, and impacts on agricultural crops impose serious limits to the scale and hydraulic variability allowed in small wetland restorations. The typical project has few stakeholders. Often a deal is brokered between a conservation group, an agency, and the landowner, such as a farmer interested in waterfowl. For very large landscape-level restorations aimed to provide free goods and services to groups of people who will use the restored area for subsistence, the goals and impacts will be much larger and comprehensive. Goals will be driven by human needs rather than conservation concepts, although these are by no means opposed ideas. In fact, good landscape-scale restorations done for subsistence of people must be sustainable through time.

Large-scale living–working landscape restorations will have different kinds of variability than planners consider for a western-style wetland project. These restorations require user involvement from goal setting to design on through stewardship. For most western conservation projects, a strong biodiversity focus is typical (see chapter 10). Productivity, diversity, and system dynamics are typical measured endpoints; services and goods provided to resource-dependent human users are often considered to be the least important outcome. Reestablishing traditional relationships between humans and the ecological system from which they once obtained subsistence is not an important outcome in most western projects. However, in the struggle to restore Iraq's Mesopotamian marshes, reestablishing the traditional way of life, the marsh Arab culture and the goods produced by the ecological system for humans, must be foremost in the minds of participants. These will not be projects to restore rare threatened and endangered plant species, although most certainly these benefits will be a by-product of such projects.

Wetland restoration projects are task oriented. Because of the task and timeline focus, often plants, seeds, and money are the practical constraints that affect timing and project success. In contrast, living landscape programs are process oriented necessarily owing to the scale, the lack of funds to support detailed work, and the availability of key elements for detailed western-style restorations. For example, in Iraq there is no network of wetland seed and plant materials suppliers, but instead a poorly equipped agricultural infrastructure with few suppliers suited to support restoration of marshes of any type. User groups will be expected to provide labor as an important cultural investment. Often, it is the traditional extractive behaviors of inhabitants that determine the success of similar programs. The Mesopotamian marsh program must focus on changing those key factors that will initiate a chain of events and outcomes to benefit the users who will respond by traditional management and husbandry. This implies that the success of the program will depend more on understanding the culture of the marsh Arab peoples who live there than any elegant technical choices made by informed ecologists or landscape planners. Persons working in Iraq would be well advised to study the user peoples and their cultures before plunging into the intellectually satisfying tasks of elegant technical ecological restoration.

All wetland restoration projects and programs restore hydrology to varying degrees depending on the water supply, surcharge issues, substrate condition, and soil chemistry changes. Areas chosen for western conservation projects and biodiversity outcomes often restore isolated wetlands and relink fragmented habitat areas. Depending on budgets, individual projects may focus only on habitat restoration (e.g., restore the hydrology and hope the seed bank, invasion, and succession provide the biota), or may import seeds, plants, and even selected wildlife. Living–working landscape programs are designed from traditional human uses, reconnecting cultures to the land. In North America the closest parallels to the Iraq situation may be the nascent projects to restore traditional bison ranges based on native prairies on certain Native American and federal lands (see chapter 9). However, there are very few cross-cultural projects to be examined that could serve as models for the landscape-oriented Iraq Mesopotamian marsh restoration program.

In arid areas, water availability will constrain project outcomes (see chapters 7 and 8). Water rights and diversion for wetland restorations can present a significant impediment to success. Living–working landscape programs typically pit subsistence water-use against private, industrial, and urban users, unless a policy or political mandate predetermines that public-good benefits outweigh those of the individual property owners or governments. As we contemplate this immense task for Iraq, questions about Iraqi water law; the influence of other nations, particularly Turkey, on the water supplied to the upper reaches of the Tigris–Euphrates River systems; and the effects of other Iraqi users on supply and water quality emerge as significant technical unknowns. Restoration here must be considered in a political context (see chapter 1).

Programs to restore disturbed wetlands in arid regions often encounter highly modified substrates that were damaged fundamentally by changes that followed drainage (see chapter 6). The accumulation of high levels of salts or soil carbonates can destroy wetland soil structure, fracture aquacludes, or damage wetland seed banks. Where sodification or salinization is excessive, the soil structure will have collapsed (see chapter 21). This will increase vulnerability to subsequent wind and water erosion and may accelerate water losses to groundwater. Soil tensile strength is typically weakened where clay lattice structure collapses, especially in the salinized soda soils of arid places. All of these variables can affect the restoration of phreatic zones, especially where an adequate temporal water supply is not available, or other users or suppliers of river waters have caused changes in water quantity, quality, or timing of availability.

Changes in hydrology, chemistry, and water tables often favor exotic species that gravitate to stream courses and agricultural drains. Dried organic muck soils are often seriously depleted in organic matter through oxidative decomposition in heated climes. Nonwettable soil crusts often develop in former wetland soils that have been altered by dewatering and these physicochemical changes, especially in arid climates. Reflooded hydric soils often do not reestablish native species emerging from the seed bank as the dominants. Dense periphytic algal mats responding to the availability of nutrients in the water column may choke out key native plant species. Periphyton growth is often much faster in warm climates, especially in shallow waters of arid areas. Seed bank depletion from prolonged dewatering and heat occurs in substrates of arid regions. Some seeds may persist, but seed survival declines rapidly over time. Restoration success that depends on this seed bank is reduced.

In arid regions, rewetting dewatered wetlands can encourage rodent and waterfowl depredation of germinating seed banks and planted stock (see chapter 5). Rodent damage can also be severe through undermining water control structures (e.g., earthen irrigation flumes, levees, dikes, etc.) because of the increased localized water availability, moist soils, and more productive plant growth.

In many arid areas, montane snowloading far upstream provides hydrology for riparian systems and associated wetlands. Seasonal precipitation and seepage may control playa and other internally drained basins adjacent to riverine systems. Highly dynamic hydrology cycles depend on rainfall patterns and snowloading and can affect the extent, timing, and success of wetland restorations profoundly.

Ecotoxicological and biotoxic concerns are commonplace in the restoration of arid-area wetlands. Midsummer to fall wetland dewatering that coincides with waterfowl and shorebird migrations into wetlands also often coincides with avian botulism or cholera outbreaks. Wetland systems in arid areas often become anoxic (especially in highly altered systems), fish die-offs prevail, and subsistence uses of the wetlands can be damaged. This is especially problematic the first few years after restoration begins. Such outbreaks also occur where wetlands may be used for water quality enhancements receiving organic material loading rates that contribute to eutrophic and anaerobic conditions that support *Clostridium botulinum* types C & E, the principle botulism disease-causing bacterium (Simmers, Apfelbaum, and Bryniarski 1990; Ludwig and Bromley 1988). Exposure risks to wildlife from other contaminants, including organochlorines (Ludwig, Apfelbaum, and Giesy 1997), salts, selenium, boron, and other heavy metals, are common concerns in arid region wetland restorations—especially on rivers that originate in mineralized montane areas. Interestingly, many human parasites "piggyback" on outbreaks and spread to humans on programs that alter rivers and their hydrology such as occurred with the Egyptian Aswan High Dam experience. The Iraqi Mesopotamian marsh areas may be especially vulnerable at this moment in time owing to the excessive water pollution from the upgradient urban centers discharging increased volumes of untreated human sewage (pathogens) to the rivers as a result of infrastructure loss during the recent wars. Reconstruction of the extremely damaged Baghdad sewerage system should be a public health priority and may need to precede reflooding the Mesopotamian marsh systems (see chapters 18, 19, and 20).

PRINCIPLES FOR SCALING UP

Principles are important for wetland restorations at any scale. These become very important as project scale, regional aridity, and the percentage of urban, industrial, or agricultural land use and development change in the tributary watershed influence a restoration program. The following are understandings and principles used by Applied Ecological Services, Inc., in the restoration of other large wetland complexes that may be useful to consider for the Mesopotamian marshes.

1. Understanding the watershed context for wetland restorations is a critical factor in restoration success; existing (or possible) government policies related to the river may be critical.
2. Biological productivity and availability of larger (human) consumable foodstuffs often increase with some disturbance and nutrient loads in tributary watershed.
3. Biological diversity of native plant and animal species appears to be greatest in watersheds with the lowest agricultural conversion, lowest deforestation percentage, and lowest percentage of imperviousness or developed land.
4. Biological diversity of all species (including invasive species) may elevate with intermediate disturbance levels.
5. Biological diversity of all species declines with high hydraulic volatility and increased nutrient loading regardless of whether these changes are owing to impervious watershed conversion, row crop agriculture, industrial discharges, or domestic sewage.
6. On the annual hydrograph, a predictable seasonal high-water level and the trajectory to a seasonal low will be less predictable with an increased percentage of agricultural, urban, and industrial development.
7. For developed landscapes with the most land converted to annual crops and human development, land stewardship appears to be essential to maintain high levels of productivity and moderate diversity.
8. Significant incremental cost savings will accrue from seed bank and refugia analyses used to design an integrated stewardship program consistent with stakeholder participation and culture early in a project.
9. Cultural engagement is essential to any large-scale wetland restoration program. Success will be driven by the residents.

10. Cost thresholds and unit costs for restoration and stewardship will decline as larger scale ecosystem management strategies are used, rather than task-oriented and smaller scale interventions.

EXAMPLES AT SCALE

This section provides two North American project examples and some associated information we used to develop principles of large-scale restorations. An overview of each project with more detailed sources of additional information and data is cited in the references. These case studies are reasonably typical North American examples of large-scale restorations. However, these are also emblematic of what can be achieved when there is a common cultural context and a vigorous support infrastructure. It should not be assumed these techniques and results can be transplanted to Iraq easily. These case studies offer examples of the kinds of hydrological, biotic, and technical issues that apply in North America. We suspect the same biotic principles will apply in Iraq, but that the limiting factors associated with arid climate and a damaged marsh Arab culture will influence the achievable outcomes.

The Red River of the North in Minnesota is an example provided to illustrate basic relationships between wetland hydrology, land use, and the potential influences on the achievable biodiversity and productivity in wetland restorations in large riverine basins. The Kankakee Sands wetland restoration in Indiana illustrates the pragmatic design process and steps used to restore a 7,300 acre drained wetland basin.

RED RIVER OF THE NORTH, MINNESOTA

UNDERSTANDING THE WATERSHED CONTEXT FOR WETLAND RESTORATIONS

The Red River basin upstream from Winnipeg and glacial Lake Winnipeg has a drainage area of about 45,000 square miles. This includes 17,806 square miles (4,611,833 ha) in Minnesota, 20,820 square miles (5,392,472 ha) in North Dakota, 573 square miles (148,409 ha) in South Dakota, and the balance (about 5,800 square miles [1,502,225 ha]) in Manitoba. Similar to the Tigris–Euphrates system in Iraq, the Red River of the North descends from a set of upper subwatersheds to a very flat plain that once supported vast tracts of flood plain marshes and forested wetlands. Between European settlement and 1970, the majority of these wetlands (85 percent) were drained for annual crop agriculture; the flood-attenuating natural wetlands disappeared largely from the system. The result has been a recent history of devastating floods. The project summarized here is the attempt by Minnesota agencies, farmers, and the public at large to understand how to ameliorate these forces to improve all aspects of this large riverine system. Unlike most riverine wetland restoration studied in North America, this is a rural watershed ($\geq$ 90 percent in agriculture) physically similar to the Iraqi system, but without a montane origin or in a similar arid landscape.

The Red River basin in Minnesota is shown on the shaded relief map in Figure 9.1. To allow for direct comparison, the horizontal and vertical scales are consistent for all cross-sections.

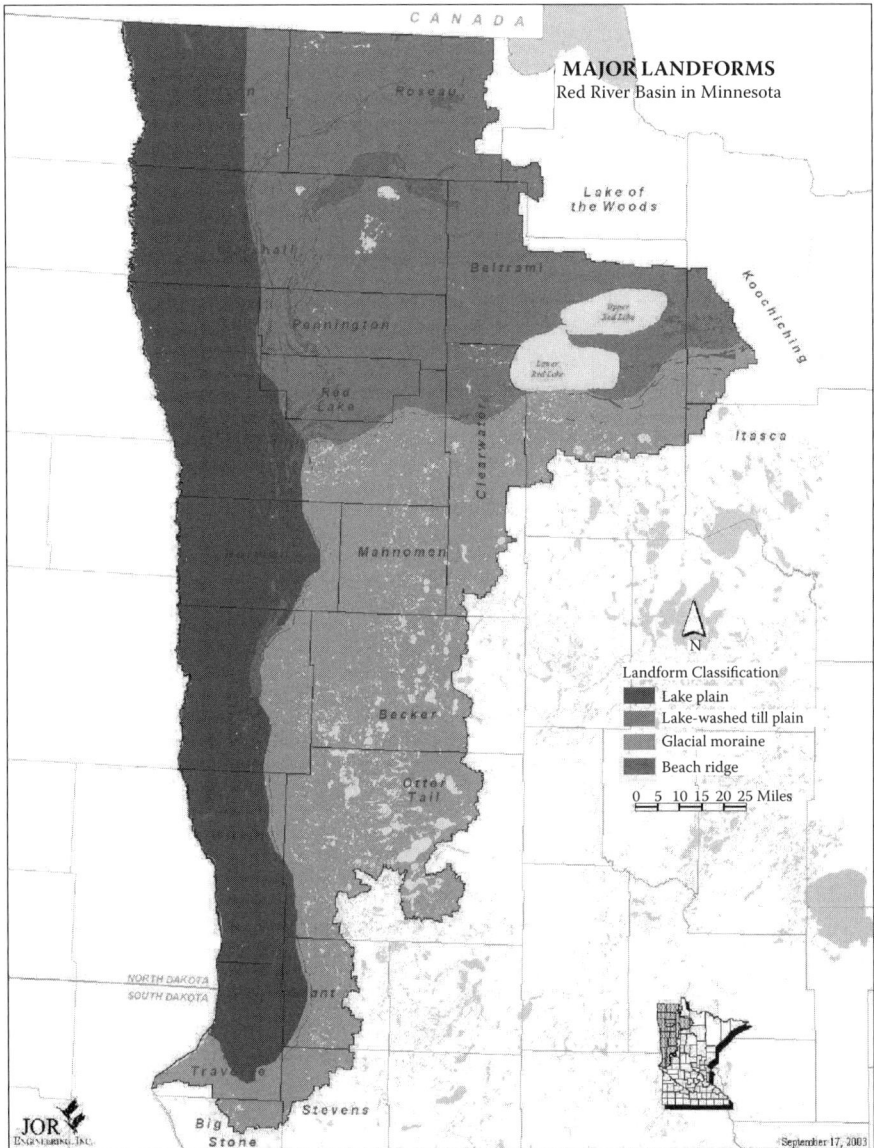

FIGURE 9.1 Major landforms of the Red River basin in Minnesota.

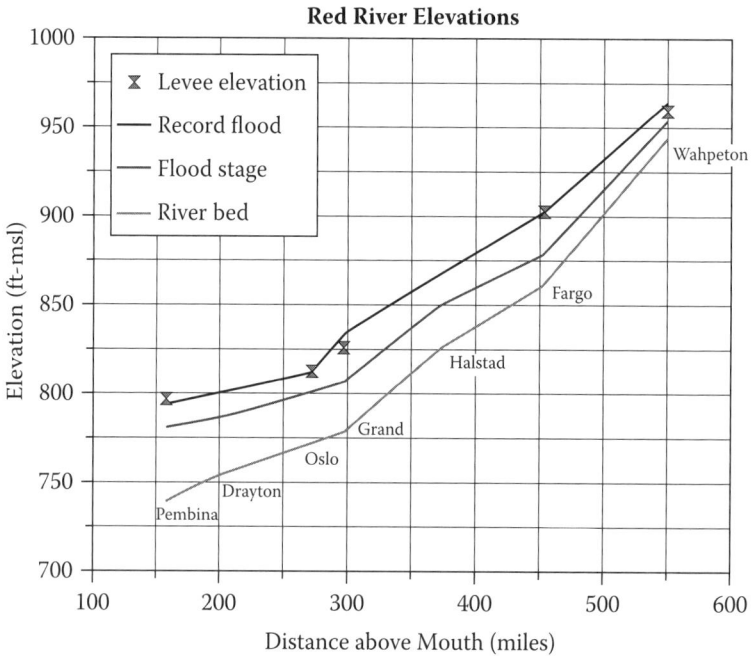

FIGURE 9.2 Profiles of the Red River (data from U.S. Army Corps of Engineers and Federal Emergency Management Agency 2003).

The Red River basin has four general landform regions characterized by elevation, topography, soils, and stream characteristics, as indicated in Figure 9.2. The following descriptions of these landforms are from the highest to the lowest in elevation.

1. *Glacial moraine*: This headwaters region is characterized by rolling hills, lakes, depressional wetlands, and variable soils associated with glacial ground moraine.
2. *Lake-washed till plain*: This region of the basin is characterized by land of gradual slope with large areas of nondepressional wetlands and poorly developed stream networks. Surface soils include large areas of peat lands.
3. *Beach ridge areas*: The beach ridge areas of glacial Lake Agassiz are characterized by sandy soils, multiple levels of beach ridges, relatively steep slopes, and incised rivers with relatively narrow floodplains. Wetland areas often exist on the upstream side of beach ridges.
4. *Glacial lake plain*: The lowest elevations are within the lake plain of glacial Lake Agassiz. The land within the glacial lake plain region is extremely flat with very low surface and river channel gradients. This flat area originally included large areas of wetlands. The soils are dominated by relatively impervious lacustrine silts and clays.

The highest land in the Minnesota portion of the Red River basin is located in Clearwater County in the Wild Rice River watershed at an elevation of 2,010 feet above sea level. The lowest land, located in Kittson County along the Red River near the Canadian border, is 750 feet above sea level.

The most flood-prone areas generally are those with the least slope and those downstream from areas of steep slopes.

The gradient of the Red River ranges from a little over 1.0 foot per mile north of Breckenridge, to about 0.5 foot per mile in the vicinity of Grand Forks, to about 0.2 foot per mile at the Canadian border. Profiles of the Red River from Wahpeton to the Canadian border are shown on Figure 9.1.

RUNOFF TRAVEL TIME

Another way to help understand the hydrology and delivery of water to restorations in large riverine basins is to identify early, middle, and late tributary areas relative to the river main stem. Computed runoff travel times (Figure 9.3) based on topography and land use show the relative travel time of runoff (the darker the shading, the longer the travel time).

The map only depicts time in transit. Therefore, the detention effect of wetlands, lakes, and other storage is not shown.

DEFINITION OF EARLY, MIDDLE, AND LATE AREAS RELATIVE TO THE RED RIVER MAIN STEM

Figure 9.4 identifies early-, middle-, and late-runoff areas within the Red River basin relative to the main stem. This generalized map was based on the evaluations of historical flood hydrographs, knowledge of more recent floods, and computed runoff travel times. The design of hydrological restoration in Iraq's Mesopotamian marshes will require data on seasonal water flow and the timing of flood flows.

The Red River of the North is one of only a few major rivers in North America that flow north. This increases its spring flood potential, because snow in the southern headwaters of the basin often melts before snow in the northern areas, causing peak flows from downstream tributaries to coincide with the flood crest on the Red River. The northward flow of the river also results in more ice jam problems than most southern-flowing rivers experience. In addition, the Red River is located within the broad, flat bottom of glacial Lake Agassiz, which has only a slight northward slope. As a result, the main stem and tributary rivers in the glacial lake plain area of the basin frequently overflow onto broad floodplains. A significant body of knowledge and data on the hydrology and flood hydraulics relationship is available for the Red River basin (see references). However, hydrological relationships with biological diversity or productivity remained unexplored. A series of studies were designed and implemented to test and understand these relationships and to provide an applied basis for the design and implementation of wetland restorations in the Red River basin. The following summarizes some of the general findings from these investigations (Table 9.1).

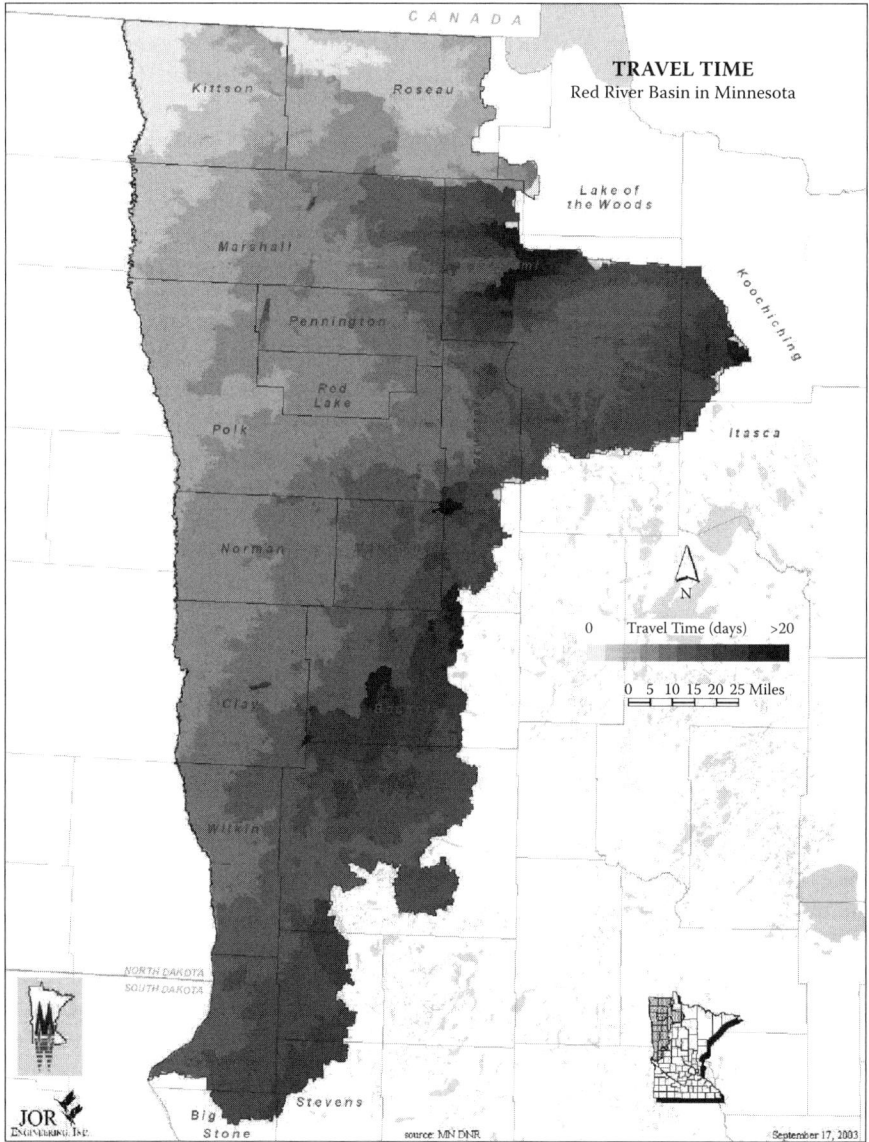

FIGURE 9.3   Computed runoff travel time to the U.S.–Canadian border.

GENERAL HYDROLOGICAL FRAMEWORK FOR CONSIDERING WETLAND RESTORATION

BIOLOGICAL PRODUCTIVITY AND DIVERSITY, AND HYDROLOGY RELATIONSHIPS

Two years of study of biodiversity and hydrological features in twenty-eight restored and natural wetlands provided a basis for understanding the potential for achieving

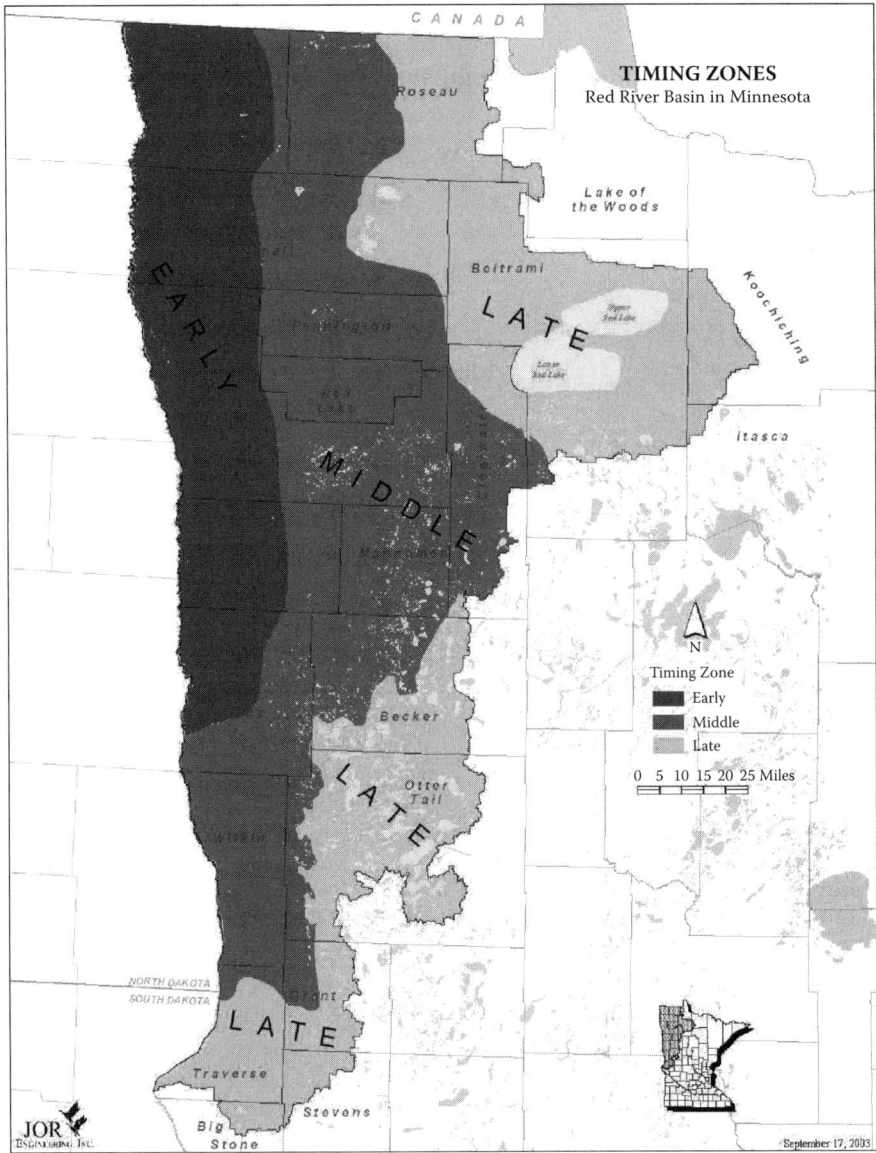

FIGURE 9.4 Timing zones, Red River basin in Minnesota.

diverse and productive restorations (Apfelbaum, Eppich, and Solstad under review; Technical and Science Advisory Committee [TSAC] 2004). This wetland study was preceded by several years of additional research in more focused topics (TSAC 1998a, 1998b, 1998c, 1998d, 1998e, 1998f, 1998g, 2004).

Through the research we documented strong inverse relationships where wetland restorations in watersheds with increasing percentages in row crop agricultural lands had consistently lower bird, macroinvertebrate, and native plant diversity. Correlated

TABLE 9.1

General Design and Siting Criteria for Wetland Restorations in the Red River Basin

Wetland Setting or Design Criteria	Wetland Quality Expectation		
	Low	Intermediate	High
Watershed to wetland area ratio	> 30:1*	5:1–40:1*	< 10:1
Percentage of watershed in agricultural production	> 50%*	20–70%*	< 20%
Potential for additional wetland restorations within watershed	Limited	Possible	Likely
Condition of the watershed (water quality surrogate)	Highly modified with limited application of BMPs	Variable	Minimal modification or intensive use of BMPs, including restoration
History of restoration site	Intensive agricultural production for several decades	Variable	Minimally Disturbed
Upland buffers around wetland	Typically absent	Typically absent	Present
Restoration of tributary watershed to native vegetation	Typically absent	Limited to buffer around wetland	Integral part of project, large percentage of the watershed
Prevalence of additional factors which may adversely impact water quality, including road and urban runoff, feedlots, and other industry	Factors present	Limited presence	Not present
Growing season bounce maximum* (10-year, 24-hour design storm event)	≤ 3 ft	≤ 2 ft	≤ 1 ft

* This study did not examine the relationship between bounce and nesting success of overwater nesting birds.

with an increasing percentage of agricultural lands were increasing amplitude, frequency, and duration of bounce events which were also inversely linked to declining biodiversity in birds, macroinvertebrates, and plants.

Recognition of wetland biodiversity as a measure of wetland quality and the relationships with tributary watershed conditions were found to influence wetland restoration potential strongly. While the lowest quality wetlands were located in watersheds with the highest levels of row crop agricultural land use and impervious landscapes, these same wetlands typically had the most productive biota measured by numbers of individual macroinvertebrates and birds. We found the converse also true: the highest quality wetlands had more species of birds, macroinvertebrates, and plants; higher diversity; but lower abundances of the species present.

DIVERSITY, HYDRAULIC PERFORMANCE, AND OUTLET DESIGN CHARACTERISTICS WERE ALL RELATED

In many instances, watershed and site characteristics had a much greater influence on wetland quality and the frequency and duration of bounce events than the outlet design. In settings where the watershed ratio is low and upland perennial vegetation can be planted to buffer the watershed, wetland restorations may be higher quality. Operable outlets may be used in this setting primarily for vegetation management, and possibly for controlled pre-spring drawdown during wet climatic cycles. Outlet design wetlands with large watershed ratios tended to be two-stage, with relatively higher outflow capacity than for the low-watershed-ratio wetlands. Higher and more frequent bounce is more probable and natural seasonal drawdown less probable for these wetlands, due to their larger watershed ratios.

High-quality wetlands were believed difficult to achieve in watersheds with larger watershed ratios and where land use is primarily agricultural. Wetlands in watersheds with higher watershed ratios showed declines in biodiversity.

Bounce relationships examined the contributors to bounce maximum amplitude, duration, and number of bounce events in > 0.5 foot, > 1.0 foot, > 2.0 foot, and > 3.0 foot categories. This analysis showed very strong multilinear relationships between bounce metrics in wetlands and land use within watersheds High-quality wetlands were found to have lower amplitude and lower frequency bounce events, and medium- and low-quality wetlands to have durations generally meeting a ten-day bounce drawdown and less than 2.0 foot bounce amplitude. Where high-quality wetland biodiversity is a desired outcome in restorations, lower bounce levels, frequencies, and durations during the growing season are encouraged.

An interesting discovery of the research was the effects of wetland water level drawdown rates on biodiversity. We found a polynomic relationship where slowest and fastest drawdown rates were routinely correlated with the lowest biodiversity. Intermediate rates correlated with the highest biodiversity for all biotic groups measured. Example rates are provided in Table 9.2.

A basin-wide tributary water management framework is needed to ensure a coordinated approach to achieve outcomes that will be effective at the local, watershed, and basin levels. This framework should encourage the implementation of measures that reduce local, watershed, and main stem flood damage potential. Development of a basin-wide framework requires an understanding of critical factors that affect flooding, including runoff timing and volume. When implementing individual restoration projects, it is necessary to know how water from any given area will arrive.

TABLE 9.2

Hydrologic Comparison of Outlet Types

	Natural	One Stage (Ditch Plug)	Two Stage/ Operable
Average number of summer bounce events > 0.5 ft	3.7	10.1	9.6
Average summer bounce height (ft)	0.98	0.87	0.99
Average bounce duration (days)	72.7	43.6	37.7
Average spring bounce (ft)	1.00	1.36	1.52
Average yearly fluctuation (ft)	1.63	2.18	2.11
Measured Water Levels			
Total range of measured water levels (ft)	2.30	2.50	2.99
Seasonal drawdown (ft/day)	0.015	0.020	0.021

SUMMARY

This study provided a strategic understanding and a framework for large-scale wetland restorations. We also became aware of a keen need for an adaptive approach, and sideboards to define response–nonresponse approaches, to restoration as follows:

1. The study of nature can show the most successful way to restore ecological systems.
2. Strategically locating start-ups is very important, especially in larger projects.
3. The design process for projects is very important. For larger projects in particular, defining adaptive management triggers, designs to allow stochastic events to play out (e.g., fire, insect infestation, hydrological cycles, and blowdown in forests), is very important.
4. In the design of wetland restoration projects, matching the desired outcomes of a project to the watershed and site characteristics determines if restorations fall within three broad wetland quality categories for deeper and shallow depressional systems.
5. Site selection for wetland restoration and the strategy for buffering hydrology, hydraulics, and water quality from agricultural and developed watersheds will closely influence the levels of wetland biodiversity and quality achievable in a wetland restoration.
6. If higher biodiversity wetland restorations are desired in locations with higher watershed ratios, additional measures will be required. A reduction in watershed ratio will provide associated reductions in bounce, and increased water quality and biodiversity. This can be done by increasing the size of the proposed wetland, increasing upstream storage, perhaps restoring other dispersed wetlands or otherwise impounding water, and diverting water around the subject wetland restoration site. Buffering, planting more perennial vegetation in the watershed, and active vegetation management

can improve the quality and reduce the quantity of water entering the wetland restoration site, which may help offset a higher watershed ratio and support a higher quality wetland restoration.

These studies in the Red River basin of Minnesota have provided important relationships between floodwater performance and the biological diversity of wetlands. The biological diversity and productivity of the historic and restored Mesopotamian marshes may respond similarly to the same variables. These studies demonstrate the important link between watershed and wetland land uses, acreages, and the outcomes of wetland restoration. To restore a living landscape for human sustenance, floodwater uses in wetlands, water system turnover rates and processes will greatly affect the health of human foodstuffs and their reliable production.

KANKAKEE SANDS, RENSSELAER, INDIANA

The Nature Conservancy purchased 7,300 acres of former drained wetland and glacial lake basin in northwestern Indiana. AES was retained to design, permit, and build the first two thousand acres of this restoration project. This land was purchased because once restored, it could serve as a conservation link to nature preserves to the north, east, and south.

THE RESTORATION PROCESS

A long history of land drainage and farming and the presence of significant infrastructure such as center pivot irrigation systems, over two hundred miles of legal drainage ditches (Figure 9.3), thousands of miles of buried drain tile (Figure 9.4), and annually maintained surface drainage ditching on this very low-relief landscape greatly affected the restoration design, engineering, and implementation program.

We initially hired new aerial photography and 6' aerial topographic maps for the entire project area and a slightly enlarged area beyond the property (Figure 9.5). On-site ground control and horizontal and vertical surveys of surrounding infrastructure, such as roads, drainageways that crossed property lines, tile inverts shared by adjacent property owners, enabled the surveying the locations of the house foundation and septic, and road surface elevations, center lines and road edges, and cross-sections of the drainage ditch systems. Because of the adjacent private properties needed to maintain the agricultural land use and the drainage infrastructure, we walked all ditches, found and located shared tiles, and surveyed in property boundaries.

Many ditches were legal drains that could not be modified by the restoration efforts (Figure 9.6). Mapping of the ditch system and legal and regulatory constraints associated with the modification of each was an important step in the process. We conferred with the local county surveyor and engineer to understand the regulatory constraints on our use and modification of any ditches.

Shallow surface ditches maintained annually by each property owner were shallow, 1–2 feet in depth and 2–6 feet wide (Figure 9.7). These ditches had no legal

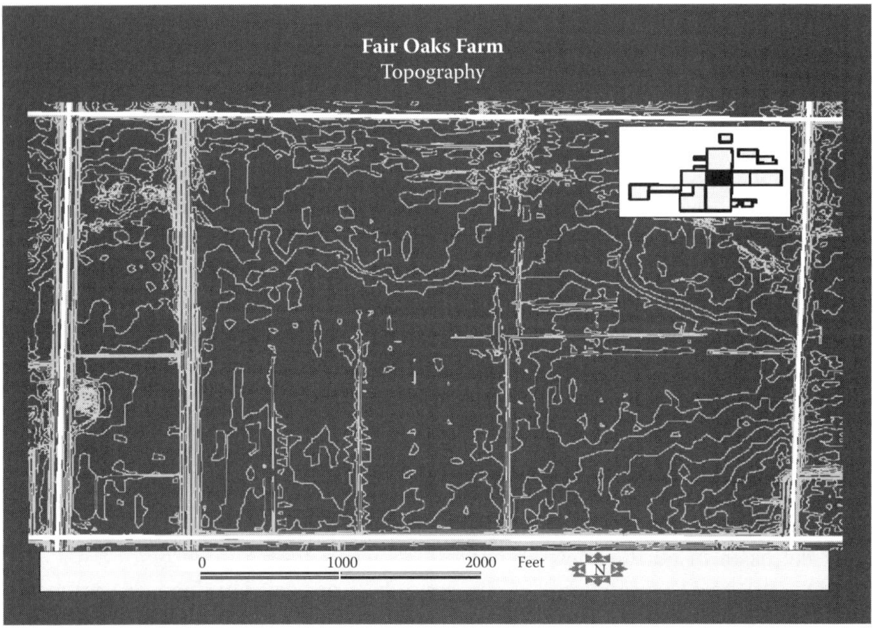

FIGURE 9.5   Example of Fair Oaks Farm topography in a typical square mile. The image shows 6 inch contour intervals.

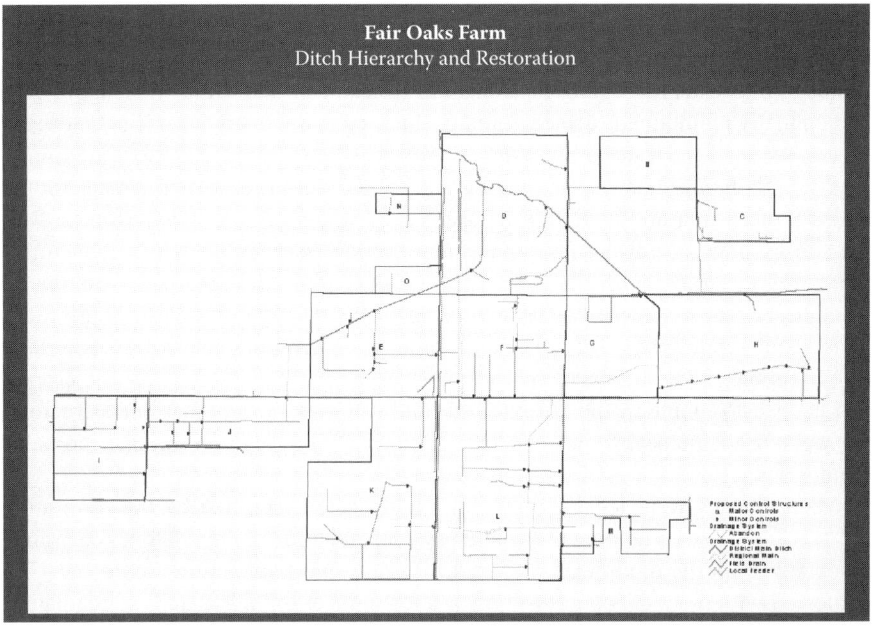

FIGURE 9.6   The mapped ditch systems, including legal drains and farmer-maintained ditches.

or regulatory standard, and we were allowed to readily modify or eliminate these ditches. Center pivot irrigation units such as in Figure 9.7 watered nearly five hundred acres; these were abandoned and sold.

Tile lines were surveyed by finding their outlets into the surrounding ditches followed by review of historic installation records and by surveying invert systems and inlets to develop overall distributional maps (Figure 9.8). Invert elevation surveying was also conducted.

Because of the heavy investment in drainage infrastructure and the intensive agricultural uses of the land, we had significant concerns whether a native plant seed or propagule bank was present in the soil. We randomly stratified each major soil type and collected eight replicates at each of 2,300 sample locations, where up to 1 kg of substrates was collected with two inch diameter soil probes to a depth of 60 cm (Figure 9.9). These soil samples were immediately iced and removed to the AES greenhouse, where one-half of the sample was put into cold storage, and the remaining ones were spread in greenhouse flats and placed in a temperature-, humidity-, and moisture-controlled propagation greenhouse. Each flat was cycled through three wet and dry cycles; all germinating seedlings were identified and enumerated to understand the diversity and abundance of the seed bank plant response. These data were very useful in saving significant money by taking advantage of the presence of a significant native plant seed bank in some soil types. These data and quantitative plant compositional data from reference natural areas within fifteen miles of the project site contributed to our planning for the site. Both

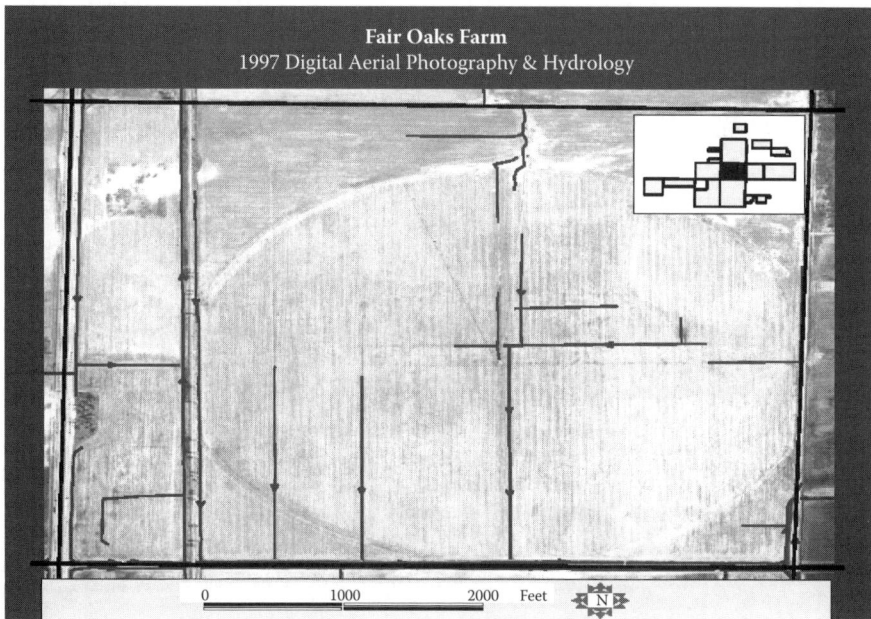

FIGURE 9.7   Surface shallow ditches maintained annually by farmers.

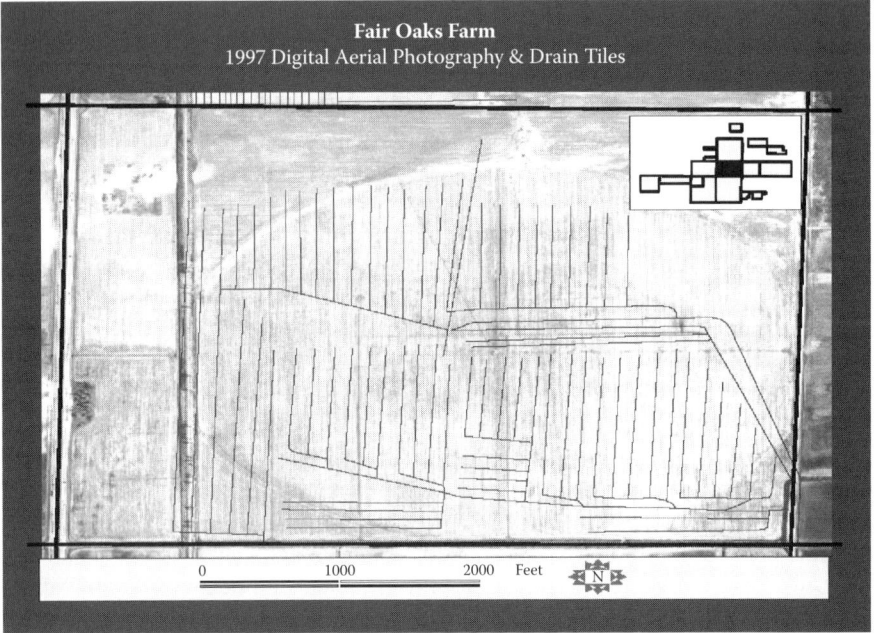

FIGURE 9.8   Example of tile mapping showing the current active tile system. Two dysfunctional historic tile networks were found beneath this active system.

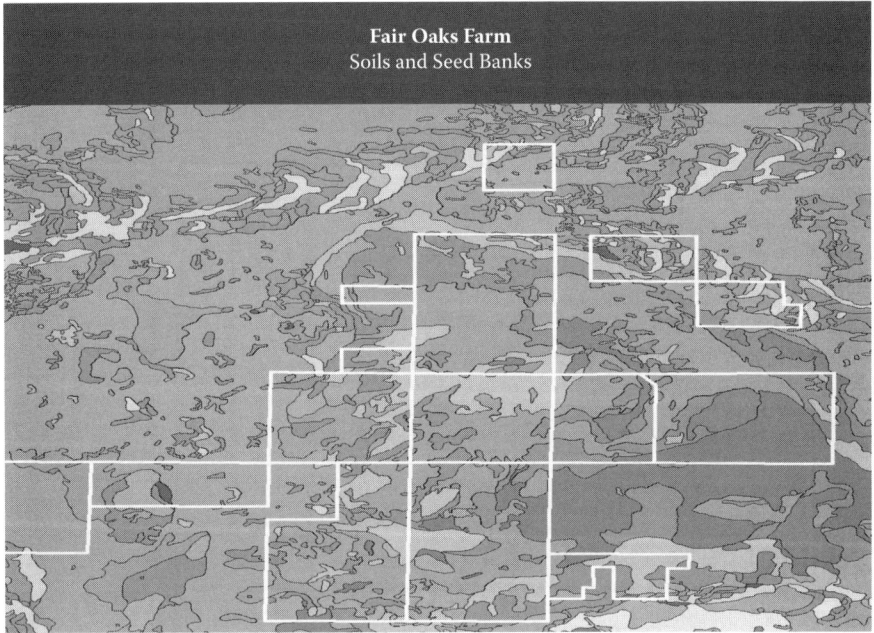

FIGURE 9.9   Mapped soil types were useful along with seed bank analysis to determine seed bank response.

data sets were used to design planting specifications and the restoration program for the Kankakee Sands project.

The restoration plans consisted of detailed hydrological, seeding and planting, vegetation management, and soil preparation specifications for the entire 7,300 acres (Figure 9.10). A total of 5,200 acres of several key wetland types, 1,900 acres of various prairies, and about 200 acres of savanna were targeted in the restoration plans.

A phasing program plan was prepared commensurate with financial resources and labor and equipment availability to implement the program (Figure 9.11). The initial stages of work involved collecting local genetic seed stocks from about a thirty-five mile radius for approximately three hundred native plant species. Seeds were dried and stored, and some percentage of supplies was direct seeded into a 180 acre nursery established on-site and designed to provide enough pure live native plant seed after year 3 to restore approximately one thousand acres of land. A small fraction of seed was used for producing about two hundred thousand plant plugs grown in the AES greenhouses that were also planted in the nursery (Figure 9.12). A first-year planting of plugged New England aster (*Aster novae angliae*) in the nursery during the first year produced over thirty pounds of pure live seed per acre.

Disablement of tiles was done by simply backfilling ditches that were not encumbered by regulatory or legal protection. Using earthmoving equipment, filled-in

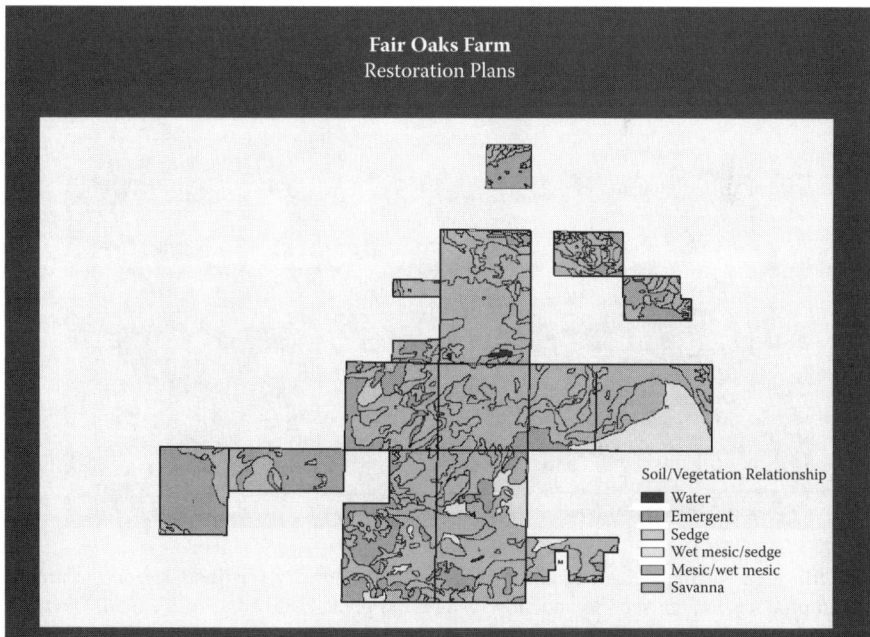

FIGURE 9.10 Restoration plans were completed by taking soils and their seed banks, determining hydrological restoration potentials, and using reference natural areas for design definition.

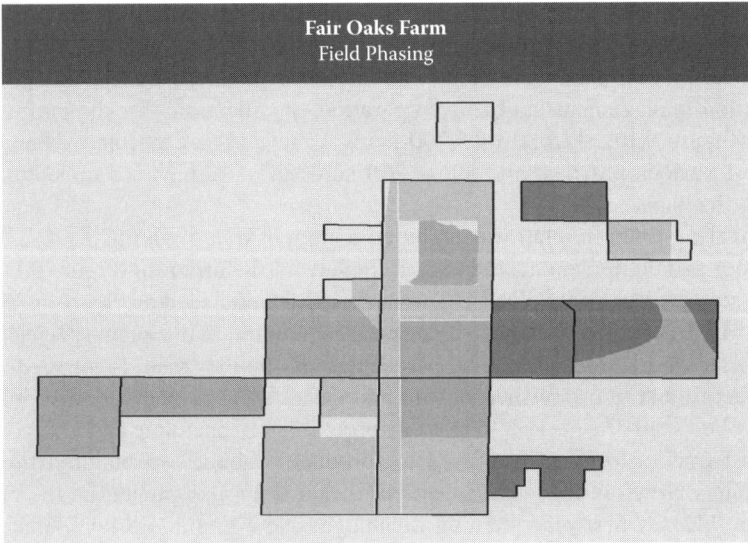

FIGURE 9.11 A phasing plan was prepared that proposed to restore the best agricultural lands last to provide a revenue stream to finance the restoration of the poorer soil areas.

FIGURE 9.12 An approximate 200 acre native plant nursery was installed on the project site to produce enough seed to annually restore and plant 1,000 acres of wetland. Over one hundred native species were planted. In the nursery, the first-year plantings of New England aster (*Aster novae-angliae*) produced > 100 lb of pure live seed per acre, valued at ~US$100–200 dollars per pound.

ditches immediately surcharged water to the land surface and up into tiles, and prevented their ability to self-scour and clean influent sediments. Owing to the very flat gradients on the ditches and tiles, this strategy resulted in a very inexpensive way to disable the tiles (Figure 9.13).

All soils that were to be planted were initially disked lightly to corrugate the soil surface and then seeded using several techniques. The most efficient seeding technique used an auger drive fertilizer spreader with two hundred foot long booms (Figure 9.14). Using this equipment and a flowable formulation of the diversity of harvested native seeds and carrier agents (e.g., cracked corn cobs, and sand), we perfected a method to broadcast seed approximately one acre every forty-eight seconds, allowing for the planting of one thousand acres in up to two days' time. After being broadcast, seeded soils were rolled with a cultipacker pulled behind a tractor. Projects such as this can benefit endangered migratory species such as the whooping crane (*Grus americana*) (Figure 9.15).

SUMMARY

The restoration of large wetlands, particularly in highly modified riverine watershed systems, always requires significant forethought to match the potential for hydrological restoration with biological outcomes. Modifications to soils (e.g., structure, chemistry, and organic matter) during dewatered periods, such as over the years of agricultural use, can affect the potential for recovering native vegetation systems greatly (see chapter 5). In addition, legal and physical constraints imposed by retained hydrological and built infrastructure (e.g., legal drainage ditches, culvert

FIGURE 9.13   Restoration of hydrology required backfilling ditches and disabling the tiles. Immediately after this was conducted, hydrology returned to the soil's surface, establishing saturated soils and standing water in surface topographic depressions.

FIGURE 9.14   Native wetland plant seeding was conducted with innovative low-cost techniques such as using agricultural fertilizer spreaders that broadcast seeds over the lightly disked land surface at a rate of > 1 acre per minute and a cost of less than US$6.00 per acre.

FIGURE 9.15   This restoration project is anticipated to have far-reaching benefits, including providing a migratory resting area to support the recovery of whooping cranes (*Grus americana*).

inverts, roads, and other infrastructure elevations) represent an important design consideration. Biological changes such as seed bank depletion and conversion to agronomic and noxious weedy species can contribute significant cost and reduced levels of restoration performance, and can lengthen the restoration period.

Innovative strategies for assessing seed banks, modifying hydrology, and planting and managing the land are important for success. For example, by understanding the seed bank through least plot analysis, hundreds of thousands of dollars were saved by eliminating significant seed purchases and reducing the cost of wild harvesting those native plants projected to appear on their own from the seed banks.

Agricultural herbicide residues and salinity from highway runoff were found to affect the seed bank and planted native species growth and development greatly. In locations with residual toxicity, soil amendments to chelate (change the chemical form) helped restoration success. In arid regions, addressing salinization and toxicity will be more important than in Indiana, where regular precipitation has helped reduce this carryover effect (see chapters 17 and 20).

CONCLUSION

Restoration of the Iraqi mesomarsh ecosystems will be a very complex undertaking informed by at least the following principles and needs.

- The users—the "marsh Arabs"—must be a part of the planning and implementation from the beginning to the end of restoration efforts. These peoples will be instrumental in the construction, use, and protection or exploitation of the restored ecosystems. Any technical restoration program that ignores the key role of these peoples and their cultures is much more likely to fail than a program that places these people front and center. We ought not to model ecological restorations on the ill-planned U.S. military activities in postwar Iraq. The abject failures of U.S. policy since the end of formal military combat should support the strategy of placing Iraqis in charge of restoration programs, including prioritization of local projects from the outset.
- Data on the hydrological cycles of the rivers are essential. Understanding the hydrological fluctuations and water qualities endemic to these systems is fundamental to restoration.
- Data on the survival and availability of plants in the seed bank of previously drained areas are very important. Prerestoration efforts may include restoring native plant nurseries as a prerequisite to successful restoration.
- Water quality data for these river systems will be required. Restoration successes may be influenced strongly by waste disposal practices upstream, especially in Baghdad for human pathogens, or owing to toxic residuals from decades of war. Such toxic agents as heavy metals, organophosphates, organochlorines, oil, polycyclic aromatic hydrocarbons (PAHs), substrate chemical stocks, and depleted uranium from military weapons are certain to be common contaminants in Iraqi watersheds after combat. These will

find their way into surface waters and the rivers, and should be inventoried. Their specific threats and pathways to the human population must be understood before money is allocated. We ought not to repeat the schistosomiasis debacle of the Aswan High Dam project in Egypt or the Minimata mercury disaster in Japan out of ignorance borne of preprocess oversights in the interest of saving money.

- A wide variety of preprocess baseline data on soils and preprocess site conditions should be allowed for in each site-specific project. For example, the potential for soil sodification or collapse following drainage and rewetting should be anticipated to be crucial at some sites, but insignificant at other sites. Site-specific data are essential.
- Long-term management of these restored ecosystems will require plans for monitoring and adaptive management consistent with the climate, soils, and hydrological cycles.
- The capabilities, needs, and cultures of the resident peoples will affect management strongly. Another demonstration of blatant American hubris will simply poison all efforts to restore the Iraqi mesomarshes. We must listen to the people who live there and follow their leadership. These restoration efforts will not have the infrastructural support we take for granted in the West. Adaptive management and a sense of humility will be as essential as anything of a technical or science-driven nature that North Americans and Europeans bring to this process.

REFERENCES

Apfelbaum, S. I., J. D. Eppich, and J. Solstad. Under review. Hydrology and biodiversity relationships in Minnesota's Red River basin wetlands. *Ecological Applications*.

Ehrenfeld, J. G., and J. P. Schneider. 1991. Wetlands and suburbanization: Effects on hydrology, water quality and plant community composition. *Journal of Applied Ecology* 28(2): 467–90.

Gleason, H. A. 1952. *The new Britton and Brown illustrated flora of the northeastern United States and adjacent Canada*, 3 vols. New York: Macmillan.

Goff, F. G., G. A. Dawson, and J. J. Rochow. 1982. Site examination for threatened and endangered plant species. *Environmental Management* 6(4):307–16.

Hershfield, David M. 1961. Rainfall frequency atlas of the United States for durations from 30 minutes to 24 hours and return periods from 1 to 100 years. Technical paper no. 40, prepared for Engineering Division, Soil Conservation Service, U.S. Department of Agriculture. Washington, DC: Cooperative Studies Section, Hydrologic Services Division, U.S. Weather Bureau.

Kuehnast, Earl L., Donald G. Baker, and James A. Zandlo. 1982. Climate of Minnesota, part XIII: Duration and depth of snow cover. Technical bulletin no. 333-1982. Minneapolis: Agricultural Experiment Station, University of Minnesota.

Ludwig, J. P., and D. D. Bromley. 1988. Observations on the 1965 and 1966 mortalities of alewives and ring-billed gulls in Saginaw Bay, Lake Huron ecosystems. *Jack Pine Warbler* 66:2–19.

Ludwig, J. P., S. I. Apfelbaum, and J. P. Giesy. 1997. Ecotoxicological effects of watershed contamination. In *Proceedings of 1996 USEPA Symposium assessing the cumulative impacts of watershed development on aquatic ecosystems and water quality*. Washington, DC: Environmental Protection Agency.

Maclay, R. W., T. C. Winter, and L. E. Bidwell. 1972. Water resources of the Red River of the north drainage basin in Minnesota. Water-Resources Investigations 1-72. Washington, DC: U.S. Geological Survey.

McCune, B., and M. J. Mefford. 1999. *PC-ORD: Multivariate analysis of ecological data*, v. 4. Gleneden Beach, OR: MjM Software Design.

Miller, John F. 1964. Two- to ten-day precipitation for return periods of 2 to 100 years in the contiguous United States. Technical Paper no. 49, prepared for Engineering Division, Soil Conservation Service, U.S. Department of Agriculture. Washington, DC. Washington, DC: Cooperative Studies Section, Office of Hydrology, U.S. Weather Bureau.

Ownbey, G. B., and T. Morley. 1991. *Vascular plants of Minnesota A checklist and atlas*. Minneapolis: University of Minnesota.

Pennak, Robert M. 1989. *Fresh-water invertebrates of the United States protozoa to mollusca*, 3rd ed. New York: John Wiley.

Red River Basin Flood Damage Reduction Work Group. 2001. *A user's guide to natural resource efforts in the Red River basin*. Minneapolis, MN: Red River Basin Flood Damage Reduction Work Group.

SPSS. 2002. *SYSTAT*, v. 10. Chicago: SPSS.

Simmers, J. W., S. I. Apfelbaum, and L. F. Bryniarski. 1990. Assessment of avian botulism control pilot project at the Dike 14 confined dredged material disposal facility, Cleveland, Ohio. Miscellaneous paper El 90-23. Cleveland, OH: U.S. Army Engineer Waterway's Experiment Station.

Technical and Science Advisory Committee (TSAC). 1998a. An overview of the impacts of water level dynamics ("bounce") on wetlands. Technical Report no. 1. St. Paul, MN: Red River Flood Damage Working Group, Minnesota Department of Natural Resources.

———. 1998b. An overview of the impacts of water level dynamics ("bounce") on wetlands. Technical Paper no. 1. St. Paul, MN: Red River Basin Flood Damage Reduction Work Group Technical and Scientific Advisory Committee.

———. 1998c. Basin strategy: Hydrologic analysis. Technical Paper no. 10. St. Paul, MN: Red River Basin Flood Damage Reduction Work Group Technical and Scientific Advisory Committee.

———. 1998d. Integration of flood damage reduction and natural resource enhancement in the Red River basin. Technical Paper no. 8. St. Paul, MN: Red River Basin Flood Damage Reduction Work Group Technical and Scientific Advisory Committee.

———. 1998e. Siting and design of impoundments for flood control in the Red River basin. Technical Paper no. 4. St. Paul, MN: Red River Basin Flood Damage Reduction Work Group Technical and Scientific Advisory Committee.

———. 1998f. The effectiveness of agricultural best management practices for runoff management in the Red River basin of Minnesota. Technical Paper no. 3. St. Paul, MN: Red River Basin Flood Damage Reduction Work Group Technical and Scientific Advisory Committee.

———. 1998g. Watershed modeling of various flood damage reduction strategies. Technical Paper no. 6. St. Paul, MN: Red River Basin Flood Damage Reduction Work Group Technical and Scientific Advisory Committee.

———. 2004. Wetland hydrology and the effect of water level bounce on natural resource wetland diversity. Technical Paper no. 12. St. Paul, MN: Red River Basin Flood Damage Reduction Work Group Technical and Scientific Advisory Committee.

U.S. Army Corps of Engineers. 1988. *Timing analysis*, vol. 1 of *Technical resource service, Red River of the North*. St. Paul, MN: U.S. Army Corps of Engineers, St. Paul District.

U.S. Army Corps of Engineers and Federal Emergency Management Agency, Region V and Region VIII. 2003. *Regional Red River flood assessment report, Wahpeton, North Dakota/Breckenridge, Minnesota to Emerson, Manitoba*. St. Paul, MN: U.S. Army Corps of Engineers, St. Paul District

U.S. Weather Bureau. 2003. Evaporation maps for the United States. Technical Paper no. 37. Washington, DC: U.S. Department of Commerce.

———. 2006. Frequency of maximum water equivalent of March snow cover in North Central United States. Technical Paper no. 50. Washington, DC: U.S. Department of Commerce.

Van der Valk, A. G, L. Squires, and C. H. Welling. 1994. Assessing the impacts of an increase in water level on wetland vegetation. *Ecological Applications* 4(3):525–34.

Washburn, C. F. 2001. Timing of managed seasonal drawdowns to promote the establishment of wetland plant species on the Tule Lake National Wildlife Refuge, California. PhD diss., University of Washington. UMI publication no. AAT 9983562.

Zar, H. J. 1984. *Biostatistical analysis*, 2nd ed. Englewood Cliffs, NJ: Prentice Hall.

10 Rebuilding Wetlands and Waterfowl Resources in North and South America

Importance of Capacity Building and Landscape Perspective*

CONTENTS

Introduction 157
Regional Management Prioritization 159
Management Approach 160
Case Studies 163
    The Pantanal 164
    Delta Marsh 165
    Ring Water Basin 165
    Chesapeake Bay 168
Conclusion 169
Acknowledgment 169
References 169

INTRODUCTION

The marshes of southern Iraq comprised one of the most ecologically important wetland ecosystems in the entire world (Patrow 2001; Evans 2002; France 2009). It has been estimated that as many as 3 million birds from over one hundred waterfowl species used the marshes during their annual migration to and from Africa and central Russia. Pictures by Nik Wheeler in Young (1977) show skies completely filled with thousands of birds in flight, and he and Young sustained themselves during their trips there in the 1970s by hunting coots and wild ducks whenever possible (Wheeler 2007). There is, of course, a long tradition in relying upon the birdlife in

---

* Adapted by Robert L. France from Young, D., and B. Batt. 2004. Rebuilding wetlands and waterfowl resources. Paper presented at the Mesopotamian Marshes and Modern Development: Practical Approaches to Sustaining Restored Ecological and Cultural Landscapes conference, Cambridge, MA, October.

the marshlands, as contested to by the many ancient carvings indicating waterfowl being captured in nets by Mesopotamians.

Ducks Unlimited is a nonprofit organization based in North America and is possibly the world's leading practitioner of wetland conservation. Its operating mission is to conserve, restore, and manage wetlands and associated habitats for North American waterfowl. Though the focus is on waterfowl, there are a wide variety of benefits accruing from its work to other wildlife in addition to people, including hunters (Ducks Unlimited 2008). The organization was founded in 1937 and today has more than 1 million supporters worldwide, including forty-five thousand volunteers and a staff of nearly one thousand in North America. Over fourteen thousand projects have been accomplished with more than 11 million acres of wetlands being conserved. The annual operating budget is over $200 million, and during its history Ducks Unlimited has invested more than $2 billion in projects related to wetland conservation.

Don Young and Bruce Batt (2004) describe Ducks Unlimited's operating philosophy as very strongly science based, with probably the highest concentration of wetland scientists gathered under one roof anywhere in the world. Ducks Unlimited depends on a significant grassroots constituency with virtually every community in North America touched by its volunteers or staff. Other managing strategies include being very responsive to local landowners' needs in a nonconfrontational manner, being solution oriented, working with a wide array of partners (e.g., landowners, other nongovernmental organizations [NGOs], government agencies, and the private sector), and being highly efficient in its delivery of its work (e.g., over 86 percent of the money brought in the door goes back into its conservation mission). Offices exist in the United States, Canada, and Mexico, with a Latin American and Caribbean program administered from the United States. Ducks Unlimited is also affiliated with like-minded organizations in both Australia and New Zealand (Ducks Unlimited 2008).

Unlike many other wetland organizations, the focus of much of Ducks Unlimited's conservation work is conducted on a broad landscape basis rather than on discrete hundred-acre units due to the realization that comprehensive ecological restoration and protection need to be tackled on a much larger scale (Young and Batt 2004). And for this reason alone, the corpus of work, established over more than half a century of field experience, has direct relevance to the situation in southern Iraq with respect to the sustainable development of the once-extensive marshlands there. Within this landscape framework, Ducks Unlimited's approaches to conservation (Young and Batt 2004) include the following:

- Restoring wetlands and other important wildlife habitats that have disappeared
- Enhancing habitats that have been degraded
- Protecting endangered habitats
- Managing wildlands for wildlife
- Influencing wildlife-friendly legislation at local, national, and international levels
- Encouraging the next generation of wetland leaders through curricula development and the establishment of interpretive centers for wetland education

Partner networks are the key to Ducks Unlimited's continuing success in wetland conservation. In particular, because the vast majority of wildlands in North America are in private ownership, it becomes essential to bring these individuals on board in order to establish landscape-wide protection and restoration endeavors that will be sustainable through time. In this regard, there is an obvious parallel to the situation in southern Iraq, where land abutting the marshes is held in a complex mosaic of diverse ownership of pastoral and agrarian uses.

The guiding document by which Ducks Unlimited delivers its work is its International Conservation Plan (Ducks Unlimited 2008), which provides a framework that helps direct on a day-to-day basis where to work in response to ecological threats and needs. Additionally, Ducks Unlimited operates within the guidelines advanced in the North American Bird Conservation Initiative, a plan constructed by ornithological expertise from across the continent in an attempt to highlight key regional landscape typologies and areas where work should be directed. From this, Ducks Unlimited created its own North American landscape tableau, and in each region they provide a general description, identify its importance to waterfowl, list other wildlife values, investigate existing conservation programs, establish goals, recognize assumptions, and formulate strategies for action (Young and Batt 2004). This strategy enables Ducks Unlimited to establish priorities for its work based on population levels, priority species, habitat conservation needs, and emerging threats and opportunities.

REGIONAL MANAGEMENT PRIORITIZATION

Even given Ducks Unlimited's generous financial resources (in the range of tens of millions of dollars to be spent annually), funds are nonetheless finite and it's important to make sure the organization can deliver its conservation work in the most important areas. And so, when Ducks Unlimited makes choices about where it is going to work (and there always are choices), all of it clearly can't get done and therefore it considers carefully where to best direct its efforts (Young and Batt 2004).

As a result, different areas are assigned varying levels of priority based on a suite of criteria. The number-one priority areas in North America are the boreal forest of Canada; the prairie pothole region spanning both Canada and the United States; the central valley of California which, comparable to Iraq, has "lost" over 90 percent of its wetlands; and the Gulf Coast marshes. The number-two priority areas include the Pacific Northwest, the Great Lakes, and the great plains of the central part of the continent. Number-three and -four priority areas are more broadly distributed. The important thing to carry away from these four different priority areas is that Ducks Unlimited is adopting an approach whereby they're delivering conservation in virtually every province and state in Canada, the United States, and Mexico. And within each of these different priority areas, Ducks Unlimited is adopting an adaptive management approach allowing progress in its conservation work to be measured at a spatial scale of preparedness to enable the modification of management strategies based upon knowledge gained along the way (Young and Batt 2004).

MANAGEMENT APPROACH

The success in Ducks Unlimited's management comes about through marrying general ecological science with important technology tools, particularly geographic information systems (GIS) and remote sensing, that enable it to deliver its work. The approach helps Ducks Unlimited to target its conservation areas to specifically track the delivery of its conservation work, including monitoring compliance, gauging the effectiveness of its programs, determining emerging problems at the landscape scale, and quantifying all accomplishments. By walking and not just talking conservation, Ducks Unlimited, through its ongoing quantitative self-assessments of projects, is able to build what it refers to as a "competitive discriminator" in the conservation marketplace by attracting partners for its restoration work (Young and Batt 2004). And this tool has, in turn, proven helpful for marketing conservation needs in terms of fostering an ability to communicate to the public at large about the vital ecological significance of wetlands.

An example from the prairies in the center of North America, which again also happen to be one of Ducks Unlimited's number-one priority areas for conservation delivery, can help to illustrate the management approach. Although many often think on a day-to-day basis of far-off exotic places as being of particular significance, the prairie pothole region is one of the world's most significant wetland regions in terms of supporting waterfowl (Young and Batt 2004). Rainforest areas of South America and the Iraqi marshes are also of ecological significance, but the prairie pothole region of Canada and the United States represents the most important waterfowl habitat for grassland- and wetland-associated species in the entire world. It is, therefore, just as important for North Americans to make sure that this is not taken for granted as it is for Iraqis in relation to their own marshland ecosystem. Likewise, the challenge for the worldwide wetland conservation community (and also for Ducks Unlimited as an organization) is to recognize that these pothole ecosystems are areas that are under tremendous threat. This pothole area, despite some of the environmental conservation protection measures and restoration efforts in place, is, like the Iraqi marshes, still under tremendous siege.

The biggest threat to the prairie pothole region today, as it was in the central California basin throughout most of the last century, as well as to the Iraqi marshes in pre-Saddam days, is abutting agricultural development (Figure 10.1). On a day-to-day basis, these pothole areas are being converted from grasslands and wetlands into soybean fields. So how does Ducks Unlimited effectively deal with these problems in a prairie setting given the international significance of this area as constituting one of the most significant ecological settings in the entire world?

The prairie pothole region is characterized by the juxtaposition of abundant and diverse wetlands on a landscape covered with grasslands. Waterfowl and a variety of other wetland-dependent species, including shore birds, are very much attracted to such a complex land mosaic (Forman 1998). In these settings, birds are attracted to the water (Figure 10.2), but what allows them to be successful is the abundance of grasslands (Young and Batt 2004). When over 40 percent of the landscape is incredibly productive grasslands, this must form the basis of what guides the management

FIGURE 10.1   Dioramas at the Fort Whyte Nature Centre in Winnipeg, Manitoba, a cement mine that has been reclaimed for wetland interpretation. One diorama shows how a pothole in the original tall-grass prairie might have looked, and the other illustrates the landscape changes associated with modern agriculture.

work. But where does one choose to work when you're dealing with a scale on the magnitude of hundreds of thousands of square miles?

One of the ways that Ducks Unlimited tackles the challenge—and an approach that will be absolutely de rigueur for undertaking such management efforts in the immense marshlands of southern Iraq—is by using overlaying layers of GIS-interpreted data in landscape assessment. One might, therefore, map the locations

FIGURE 10.2 Waterfowl resting at a restored water body at the Fort Whyte Nature Centre in Manitoba during their fall migration.

of all existing wetlands. And through good data collection, the known distribution of waterfowl can be mapped on that particular landscape, indicating variable densities. Another layer of information Ducks Unlimited uses is to predict the densities of birds over vast, unsurveyed regions based upon the geographic characteristics of the sampled area, including wetlands and grasslands. These distributions of data are then overlain to indicate the highest priority areas and thus to provide information to enable a management prescription within which Ducks Unlimited will undertake its varied conservation work. The approach enables a decision matrix to be created allowing determination to be made of where intervention can be most effective—a prescription, therefore, of where management should be delivered (Young and Batt 2004).

Some areas leap out as the optimal setting for management such as protecting existing habitat because it represents the best of the best in supporting waterfowl. Other areas are indicated where the management organization would probably choose not to work, due to, for example, either irreparably degraded wetland systems or an absence of any grassland cover. In other areas, the grassland cover might be reasonably good and the management team might want to go back and undertake some wetland restoration. For another area, active management might be suggested in a landscape characterized by an abundance of wetlands but little grassland. This is an oversimplification, but basically what developing such a framework has done is that, by using good data, it has allowed Ducks Unlimited to be able to deliver a product—a very effective conservation product—in a complex landscape which, like the Iraqi marshlands, spans international borders and is truly immense in its scope (Young and Batt 2004).

The Iraqi marshlands comprise a wetland and agricultural landscape inhabited by wildlife and humans (Young 1977; France 2009, 2010). Ducks Unlimited has developed considerable managerial experience in such areas where it has had to become engaged with local farmers and other stakeholders in, for example, rotational grazing schemes. This becomes a very important dimension because in order to deliver its management product, Ducks Unlimited has had to work in concert with the land owners. Much of these efforts are guided by good social science due to the recognition that if the conservation work cannot be undertaken without good socioeconomic underpinnings, the program delivery will not be very effective. So the objective in such circumstances is to keep multigenerational ranchers on the landscape by providing them with an economic incentive to continue a ranching lifestyle that is compatible with Ducks Unlimited's own conservation needs (Young and Batt 2004). So, in such a case, Ducks Unlimited works with ranchers to develop rotational grazing schemes, allowing grasslands to be kept intact in order to meet waterfowl objectives rather than converting the land into soybeans. Additionally, Ducks Unlimited promotes conservation easements for grasslands to preserve these ecosystems. Ducks Unlimited is also engaged in working with the agricultural and the academic communities to develop different strains of winter wheat crops that will provide good cover, reduce input costs for the rancher and the farmer, and also provide wildlife benefits. And, finally, Ducks Unlimited is also heavily involved in direct wetland restoration work in degraded landscapes.

CASE STUDIES

Ducks Unlimited is involved in over fourteen thousand restoration and management projects. The few examples reviewed in this chapter pertain to those dealing with larger landscapes, on a scale similar to or even exceeding in size the predrainage area of the marshes of southern Iraq. Each of these projects is characterized by a high degree of ecological and institutional complexity and is based on the integration of expertise and data assembled over a long time frame, and thus might be considered to offer a sustainable solution which is, of course, to be considered to be the ultimate goal in the forecasted restoration efforts in Iraq. In this respect, the work is directly relevant to the situation in southern Iraq (Young and Batt 2004).

THE PANTANAL

Based on Ducks Unlimited's demonstrated capability in using GIS analysis for large-scale wetland planning, the three countries involved in the Pantanal wetlands—Brazil, Bolivia, and Paraguay—requested the organization to come and apply that technology with (not for) them in their countries. The important point in this example is to demonstrate that capacity building—which all agree is the key to eventual success in the sustainable development of the Iraqi marshlands—is really the most effective way to undertake conservation (Young and Batt 2004).

The Pantanal, at 485,000 square kilometers in size, is the largest wetland in the world (see also chapter 12). This enormous size means that it is obviously too much to tackle in its entirety given that the technology to be utilized had yet to be proven in the region and also the people needed to be trained to use it. As a result, Ducks Unlimited decided to develop a pilot project in an area shared by all three countries. The ultimate objective was to develop a tool for the whole Pantanal by creating a GIS database that would help formulate a conservation planning program that could be delivered to the entire river basin.

Specifically, the objective of the pilot project was to develop the international capacity for people in the countries to create a GIS tool so that they would have ownership in terms of understanding how it operates and being able to use it independently. The end product was that over two dozen individuals from the three countries associated with the Pantanal received training and actually developed the pilot project themselves. What began as a training exercise for the locals with assistance from eleven different partner organizations produced a product, which was the pilot study. This pilot study was the paper product. But the most important result of the enterprise was the fact that people were trained and developed the capacity to carry on in the future on their own.

The strategy in starting with pilot projects is a well-established one in watershed management (France 2005, 2006) and is the obvious first step on the pathway to undertaking the complicated restorative redevelopment work in the Iraqi marshlands. In the Pantanal, for example, the pilot project has been completed, and funding secured and the wherewithal obtained to next target the entire wetland. And indeed that is exactly the process that is currently underway there. The locals who have now been trained in this capacity have gone ahead and are managing their wetland resources on their own, which is just the way it should be done in Iraq (Young and Bratt 2004).

Some of these local initiatives include a highway-planning project being undertaken in Brazil (the issue of roads and wetlands is of great importance, as described in chapter 17), a hydrological study and precipitation analysis developed by individuals from all three countries, the creation of a program concerned with planning and managing a state park in Brazil, and another project that deals with fire monitoring (a major variable affecting landscape evolution in this region). So the Pantanal story is a very successful example of capacity building and a valuable approach that is important to carry to other areas, including Iraq.

Delta Marsh

Ducks Unlimited has been involved with other organizations in a major wetlands ecology research project in Manitoba, Canada, called the Marsh Ecology Research Program (MERP). This is a long-term (twelve-year), multidisciplinary study involving a series of replicated experiments dealing with the ecosystem, specifically examining the wet–dry cycle that is characteristic of wetlands in the northern prairies as well as elsewhere. It is critical that this type of careful science be conducted because, as described in chapter 6 for the Hula Swamp in Israel (an ecosystem with many parallels to the marshes in southern Iraq), the eventual success of restoration efforts may very well be predicated on a careful understanding of wetland soil aeration during dry periods. Because all scientists recognize the value of trying to do multidimensional, multidisciplinary, experimental work at the ecosystem level, Ducks Unlimited actually set out to engage in this research (Young and Batt 2004).

The work was undertaken in the Delta Marsh, a large and ecologically very significant wetland in southern Manitoba (Figure 10.3). Experimental marshes were established in which water levels were manipulated, and ensuing biological changes were tracked. The objectives were to understand ecological processes: basically to develop information, test ideas that already existed, and move the science of wetlands conservation further ahead. But Ducks Unlimited also wanted to take that information and provide recommendations for the management of wetlands. And as a side benefit, because many people are involved in doing this kind of work, Ducks Unlimited wanted to engage students in the process so that they would receive training and go on to become wetland scientists, or at least have empathy for wetland science. Such scientific education (see chapter 12 for one such example) will be essential for creating a cadre of experts in the sustainable wetland management in Iraq. Following the research the results were published in various places, including the book *Prairie Wetland Ecology*. As the restoration work unfolds in Iraq, there are probably techniques and methods for dealing with this scale of wetlands that may be helpful based on the work undertaken in the Phragmities marshes of southern Manitoba (on an aside, as described in France 2011). It is interesting to note, based on current geological theories, that Manitoba is intimately tied to the destruction and creation of the original marshlands of Iraq seven thousand years ago as a result of the cataclysmic draining of glacial Lake Agassiz, the largest freshwater body in history, and consequent rise in global ocean levels which led to the breaching of the Strait of Hormuz and creation of the Persian/Arabian Gulf.

Ring Water Basin

The Ring Water Basin is located in Nebraska and is an important area for waterfowl with up to 15 million birds staging there each spring. In fact, 30 to 50 percent of

FIGURE 10.3 Images of Delta Marsh, Manitoba, a vast wetland complex whose location between agricultural flatlands and a large open-water body is reminiscent of the Iraqi marshlands. On an interesting and tangential note, recent geological evidence points to events in Manitoba being linked to the destruction of the original Mesopotamian marshes, now underneath the Persian/Arabian Gulf (see France 2012).   **(Continued)**

FIGURE 10.3 (*Continued*) Images of Delta Marsh, Manitoba, a vast wetland complex whose location between agricultural flatlands and a large open-water body is reminiscent of the Iraqi marshlands. On an interesting and tangential note, recent geological evidence points to events in Manitoba being linked to the destruction of the original Mesopotamian marshes, now underneath the Persian/Arabian Gulf (see France 2012).

North America's pintail and mallard populations converge there at this time (Young and Batt 2004). Wetland loss in this area has been enormous, to a degree similar to that in Iraq (i.e., over 95 percent of the historic wetlands are now gone). This is a very modified landscape with some of the best corn and bean agriculture in the world. As in present-day Iraq, what happens in Nebraska is that all the birds showing up in the spring are now concentrated in the few remaining patches of wetlands. In Nebraska, this has had the result of enormous die-offs of waterfowl which are likely to increase in the future. Mortality is due to avian cholera, and it's anticipated that in the right combination of weather, wetland scarcity, and concentration of birds, many hundreds of thousands, maybe millions, could die here in a single year. The management issue then becomes how to restore wetlands and to distribute the birds farther apart so that they're not caught in these huge and potentially deadly concentrations. These are the kinds of issues that are by no means endemic to Nebraska but rather are characteristic of many wetland systems around the world (Young and Batt 2004).

In order to address the problem, Ducks Unlimited purchased land in Nebraska with the aim of restoring the wetlands by taking a former cornfield and converting it into a more desirable kind of habitat for waterfowl and other environmental interests. The landscape is a complex mosaic of wetland basins that required a major planning exercise based on obtaining considerable biological information as well as establishing a close network with the local community. A Global Positioning System (GPS) survey of land contours was needed to develop a detailed topographic information base from which to plan future restoration efforts. And because the view of many people was that the area should be a cornfield and not wetlands, very careful coordination was needed with the local community. Engineering plans were developed, and invasive trees removed, and because some of these wetlands had been so silted-in due to poor farming practices, bulldozers were actually required for much of the restoration work. Project success was based on involving local volunteers. Future plans are to expand upon the success of the pilot project with a goal to restore a major wetland complex in the greater river basin for the primary purpose of redistributing the birds. By spreading out the birds, the chances of disease transmission will be reduced as will pressures for food availability during the breeding and pre-breeding seasons.

CHESAPEAKE BAY

Chesapeake Bay is a large estuary with strong waterfowl interests given that during winters, a very high proportion of North America's mallards reside there. The environmental management issue here is the loss of food for waterfowl due to the aquatic vascular plants diminishing tremendously as a result of estuarine eutrophication and other causes. Collectively, this has resulted in a marked decrease over the last four decades in the use of the area by ducks (Young and Batt 2004). This is of concern to Ducks Unlimited as well as other organizations with interests beyond waterfowl alone because these diving ducks are operating very much like the classic canaries in the mine through alerting us to the fact that all estuarine life is threatened.

Using GIS analysis again, Ducks Unlimited and their partners targeted the watershed, specifically addressing those counties and communities whose various land uses were contributing to the problems. Because Chesapeake Bay is tightly coupled to its watershed, the restoration approach therefore addressed the integrity of riparian forests throughout the entire system whose presence was deemed essential to reduce the transport of eutrophying chemicals. This is really the standard strategy in dealing with agricultural pollution (U.S. Department of Agriculture 1998; Mitchell 2002; France 2006) and will almost certainly be an issue of concern in southern Iraq. Another strategy was to restore wetlands in order to capture runoff water and thus interrupt the transport of nutrients, an approach that has met with success in many other locations (e.g., France 2003, 2006).

CONCLUSION

These case studies represent some of the issues that Ducks Unlimited has been addressing over the years in its management of wetlands on the landscape scale, and that might be expected to be useful in the restorative redevelopment of the Iraqi marshlands. Young and Batt (2004) firmly believe that technology transfer in terms of capacity building will really be the key element of the restoration process of southern Iraq. In this regard, local communities will need to be engaged in the process in order to ensure that restoration will have the greatest likelihood of success (France 2008; see also chapter 2). Additionally, information from long-term monitoring is expected to also be crucial for the development of an institutional program of adaptive resource management, which might be profitably based on implementing some of the methodologies described above.

ACKNOWLEDGMENT

Adapted from Young and Batt (2004).

REFERENCES

Ducks Unlimited. 2008. [Home page]. www.ducksunlimited.org.

Evans, M. I. 2002. The ecosystem. In *The Iraqi marshlands: A human and environmental study*, ed. E. Nicholson and P. Clark. London: Politico's.

Forman, R. T. T. 1998. *Land mosaics: The ecology of landscape and regions*. Cambridge: Cambridge University Press.

France, R. L. 2003. *Wetland design: Principles and practices for landscape architects and land-use planners*. New York: Norton.

———. 2005. *Facilitating watershed management: Fostering awareness and stewardship*. Lanham, MD: Rowman & Littlefield.

———. 2006. *Introduction to watershed development: Understanding and managing the impacts of sprawl*. Lanham, MD: Rowman & Littlefield.

———. 2008. *Handbook of regenerative landscape design*. Boca Raton, FL: CRC Press.

———. 2011. *Restoring the Iraqi marshlands: Potentials, practices*. Sussex, UK: Sussex Academic Press.

———. 2012. *Back to the Garden: Searching for Eden in the Mesopotamian marshes*. Cambridge, MA: Harvard University Press.

Mitchell, F. 2002. Shoreland buffers: Protecting water quality and biological diversity (New Hampshire). In *Handbook of water sensitive planning and design*, ed. R. L. France. Boca Raton, FL: CRC Press.

Patrow, H. 2001. *The Mesopotamian marshlands: Demise of an ecosystem.* Nairobi: United Nations Environmental Programme.

U.S. Department of Agriculture. 1998. *Stream corridor restoration: Principles, processes, and practices.* Washington, DC: U.S. Department of Agriculture.

Wheeler, N. 2007. Witness to a lost landscape: The marshes in the mid-seventies. In *Wetlands of mass destruction: Ancient presage for contemporary ecocide in southern Iraq*, ed. R. L. France. Winnipeg, MB: Green Frigate Books.

Young, D., and B. Batt. 2004. Rebuilding wetlands and waterfowl resources. Paper presented at the Mesopotamian Marshes and Modern Development: Practical Approaches to Sustaining Restored Ecological and Cultural Landscapes conference, Cambridge, MA, October.

Young, G. 1977. *Return to the marshes: Life with the marsh Arabs of Iraq.* London: Collins Books.

People

11 Preservation, Rehabilitation, and Management of Heritage Wetlands in Mexico[*]

CONTENTS

Introduction 173
Pantanos de Centla 174
    History and Description 174
    Environmental Threats 175
    Park Management 175
    Nature Center 180
Xochimilco 184
    History and Description 184
    Destruction 191
    Cultural Significance 191
    Environmental Restoration and Regeneration of Natural Capital 192
    Ecological and Recreational Parks 197
Lessons for the Iraqi Marshlands 206
    Pantanos de Centla 206
    Xochimilco 206
References 209

INTRODUCTION

Parallels for the environmental situation in southern Iraq have been sought for and voiced by several scientists within the international community. Usually these case studies have been proposed based on what is perceived to be either a hydrological (e.g., the Mekong Delta in southern Vietnam) or restoration scale (e.g., the Everglades in Florida) similar to that in Iraq. Indeed, Iraqi scientists have even been taken on tours of the latter site to better prepare them for undertaking their important work upon returning home. Restorative redevelopment, however, such as that taking place within Iraq, is really a form of what might be called "regenerative landscape design" (*sensu* France 2008), and therefore has as much to do about rebuilding human communities as it does about rebuilding ecological ones (Nicholson and Clark 2002; France 2007). In this regard, I would like to suggest that there are historical and

---

[*] Authored by Robert L. France

cultural similarities to, and planning and managerial lessons that can be gained from, several prominent case studies in Mexico that concern the preservation, rehabilitation, and management of culturally and ecologically important wetlands.

PANTANOS DE CENTLA

HISTORY AND DESCRIPTION

The Pantanos de Centla wetlands are located on the Gulf of Mexico coast in the province of Tabasco (Figure 11.1) and have a similarly rich and ancient history of inhabitation as the Iraqi marshlands (Plata 2002). The wetlands are located nearby several prominent Olmec sites, one of the oldest civilizations in the New World, which in 1200 BCE were contemporaneous with those in Mesopotamia. For the most part, the Olmecs were a wetland-based civilization living, as they did until recently in the Iraqi marshlands, in reed huts (Figure 11.2). PEMEX, the Mexican oil company, destroyed La Venta (Greenpeace 1997), the oldest city in North America, whose famous basalt busts were rescued and moved to an archeological park in the modern city of Villahermosa (Figure 11.3). Also nearby is the late-classical Mayan-period (700 CE) city of Colmalcalco, which contains the oldest brickwork in the New World (Figure 11.4) and is located in very flat and ephemerally flooded forest wetlands.

The Pantanos de Centla Park consists of a matrix of over 300,000 ha of wetlands and seasonally flooded forests and is one of the most important sites for biodiversity in Latin America. The region contains 15 percent of all Mexican species, including

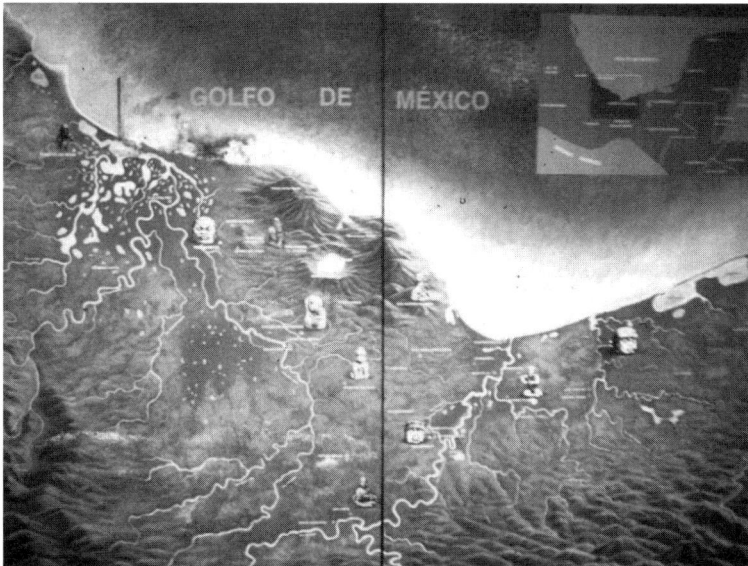

FIGURE 11.1   Schematic map showing historical Olmec cities along the Gulf of Mexico. The Pantanos de Centla are situated at the eastern (right) side of the map.

FIGURE 11.2   Idealized Olmec village, circa 1200 BCE.

most notably manatees, jaguars, osprey, garfish, crocodiles, tapirs, peccaries, toucans, howler monkeys, and iguanas (Plata 2002; World Wildlife Fund 2006).

As a result of this rich biodiversity and deep history, the region is sometimes called Mexico's "Eden," a moniker also used of course in reference to the Iraqi marshlands.

ENVIRONMENTAL THREATS

Because soils in the region are the most productive in the country, the Pantanos de Centla region faces intense pressure for agricultural development. Deforestation for cattle farming is widespread and has consumed 15 percent of the total area of the wetlands. Road building, which can cause substantial environmental impacts itself (chapter 17), has enabled access to inner regions of wetland and allowed for the development of oil exploration facilities, which in turn have produced industrial pollution. Poaching from the more than two thousand fishermen in the region who rely upon the wetlands as an important economic activity is an ongoing problem. And finally, as in Iraq, there are plans for the construction of several large dams and water diversions which promise to produce dire ecological consequences to the long-term ecological integrity of the area should they be implemented.

PARK MANAGEMENT

The story of the management of Pantanos de Centla Park also offers constructive lessons for what might one day transpire in Iraq; in this case, however, these lessons take the form of what not to do. Park managers have created a set of designated critical "core zones," but there is essentially no enforcement of bylaws due to there being only three vigilance personnel on the payroll. With only a single enforcement official per 100,000 ha of wetland park, the result is that new settlements, oil drilling,

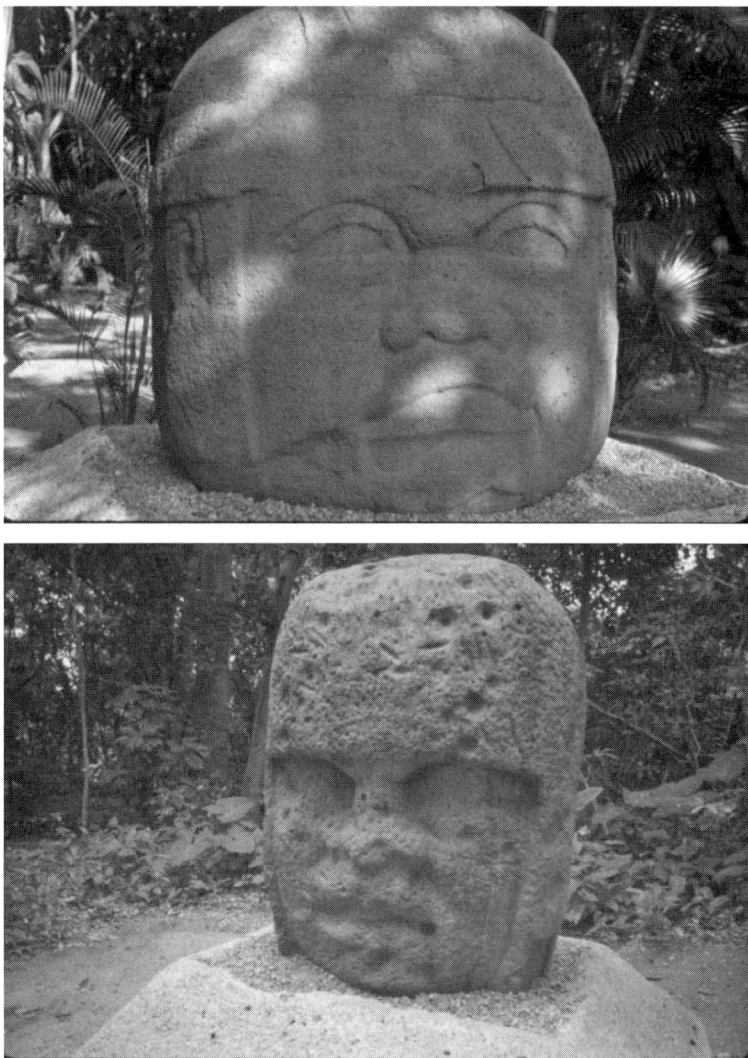

FIGURE 11.3 Characteristic Olmec basalt busts from La Venta, now on display in Villahermosa.

and deforestation all continue within the so-called protected core zones. Also, these core areas seem to have been drawn on a map based solely on ecological attributes and as a result bear no relation to the hundreds of people who have been living there for decades, making their livelihoods by exploiting wetland resources. There is also confusion concerning federal and local authority and who ultimately has the responsibility for managing and enforcing the park bylaws.

Parks Watch International has taken a polarized viewpoint of stating that absolutely no humans should be allowed to reside within the core zones, and has called for rezoning the reserve to create clarity in the management of overlapping uses

FIGURE 11.4   Late-classical Mayan city of Colmalcalco.

(Parks Watch 2006). Once again, this is the old chestnut in environmental management based on a desire to separate humans from nature in order to protect the latter.
In contrast to this, there have been fledgling attempts made at establishing an ecotourism industry in order to empower locals and engage them in sustaining the park
rather than exploiting it (see also chapters 5, 12, 15, and 16 for similar attempts to
work with locals in conservation efforts).

Today it is possible to hire a variety of tourist boats (Figure 11.5) and go out into
the large area of the park to see lakes, lagoons, mangrove islets, and the mixture of
freshwater and saltwater wetlands (Figure 11.6). And in contrast to the wishes of
those who would like to remove people from the park, what is particularly interesting

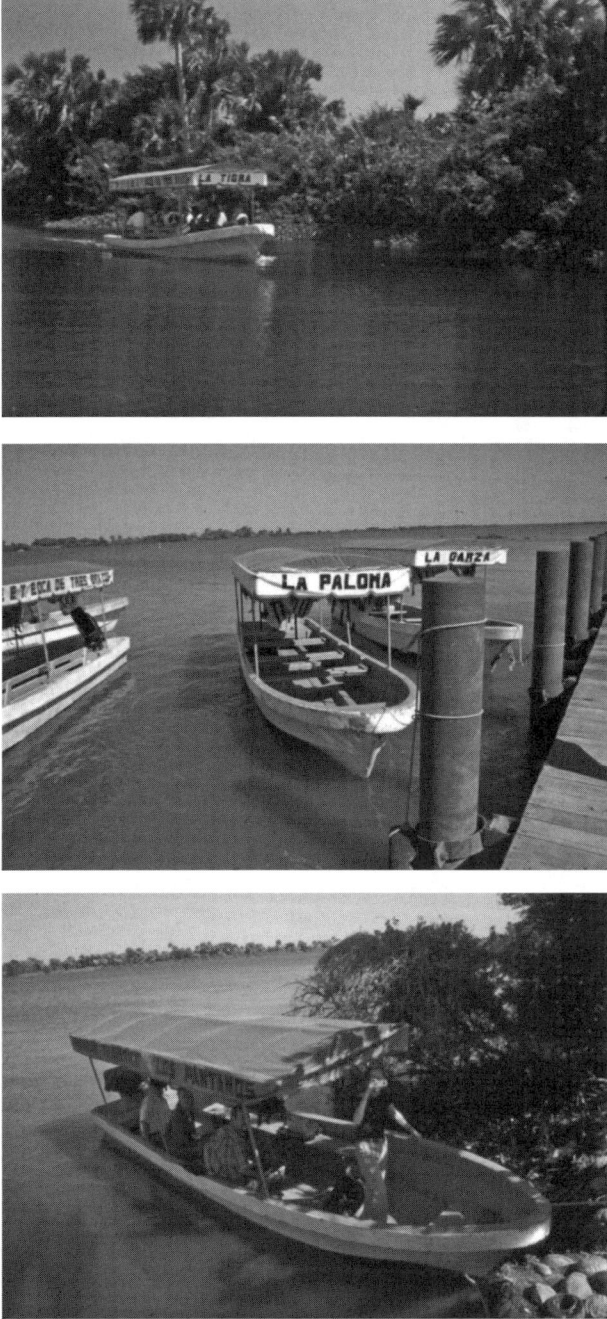

FIGURE 11.5   Tourist boats for hire for transport into the wetlands reserve.

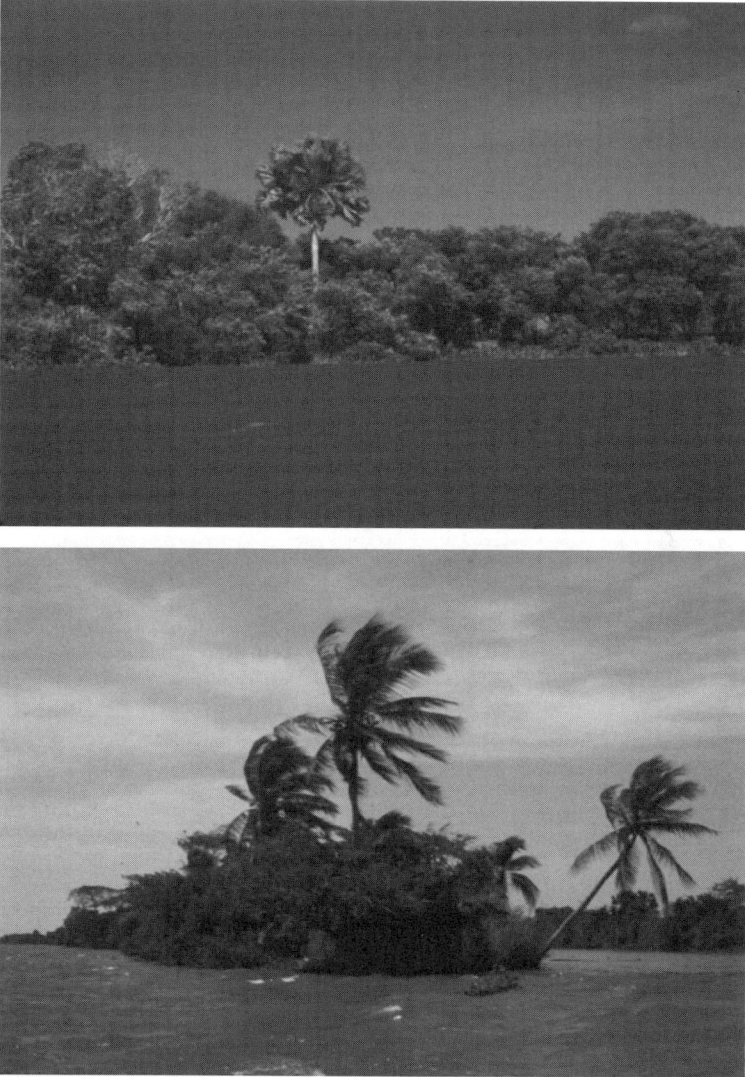

FIGURE 11.6   Forested islands in the lagoon.

about such tours is the ability to be able to enter into a cultural landscape and discover a whole array of people living within the wetlands, fishing, hunting, cattle ranching, or selling gasoline to the tourist launches (Figure 11.7). And similar to the marshlands of southern Iraq are the bovines seen along the shores (Figure 11.8). These glimpses of indigenous wetland living are perhaps the most attractive attribute of being a tourist in the Pantanos de Centla Park but must be balanced of course with the recognition that such development must be tightly regulated in order to preserve the ecological integrity of the region.

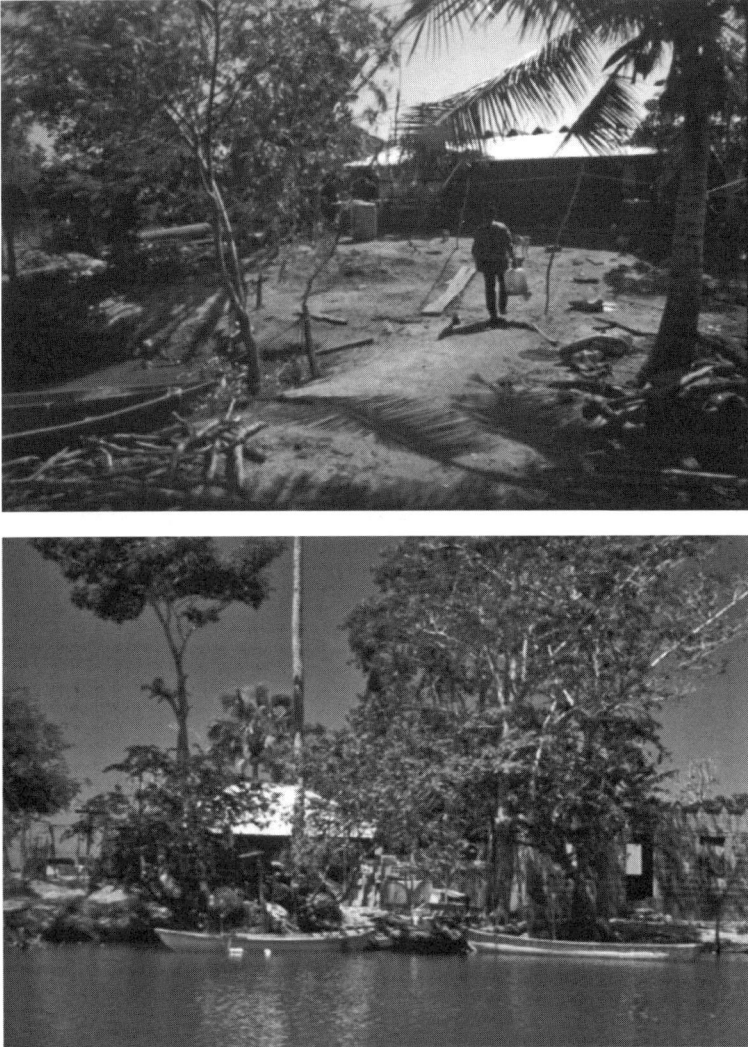

FIGURE 11.7   Wetland living in the Pantanos de Centla.   *(Continued)*

NATURE CENTER

Another important attribute of the Pantanos de Centla Park is its new nature center, which uses attractive and functional architecture to bring people out over the water (Figure 11.9). There are also houses for visiting scientists and workers as well as what must certainly be one of the tallest observation towers located at any wetland interpretation center (Figure 11.10). The boardwalks are very interesting in that they are supported by recycled oil pipes (Figure 11.11) provided by PEMEX, which was a major supporter of the project and donated millions of dollars to its construction. It is important to have such walkways due to the presence of crocodiles who, somewhat

FIGURE 11.7   *(Continued)* Wetland living in the Pantanos de Centla.

alarmingly, seem to frequent the visitor center (Figure 11.11). Radiating out from the visitor center is the beginnings of a network of boardwalks that go out into the wet-land (Figure 11.12) to provide that important immersion experience (France 2003) and that are planned to be extended in the future. There is an attractive and functional interpretive center that pays some attention to their use of alternative decentralized wastewater management (Figure 11.13), such as that described in chapters 19 and 20. A series of dioramas instructs visitors about the local biodiversity (Figure 11.14), and panels contain information about the multifaceted role of the park in terms of envi-ronmental study, restoration, protection, education, and sustainability (Figure 11.15).

FIGURE 11.8 Grazing livestock.

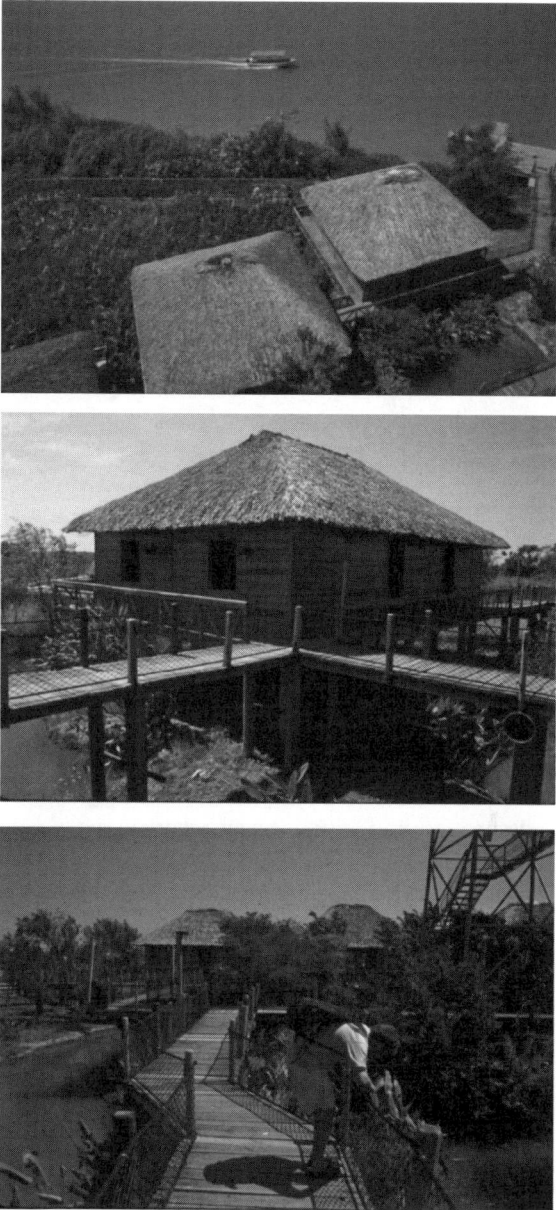

FIGURE 11.9   Pantanos de Centla visitor center.

FIGURE 11.10   Visitor accommodation and observation tower.

All in all, this nature center in Mexico is the equal to many such wetland interpretation centers within the United States (Kusler 1998) or indeed anywhere else in the world (France 2003).

XOCHIMILCO

HISTORY AND DESCRIPTION

In their own right, the ecological rehabilitation and economic revitalization of the Xochimilco wetlands and residents at the edge of Mexico City represent one of the most interesting landscape restoration projects anywhere in the world. And furthermore, in the present context, Xochimilco offers many parallels to the Iraqi marshlands and thus may be able to offer insight to that situation.

Lake Texcoco was a large, shallow body of water in the Central Valley of Mexico that once covered an area of more than 1,000 km². The lake and wetlands supported a regional population of about two hundred thousand, including what was probably the largest city in the world at its time and what would later give rise to Mexico City. Up to 2 million people lived in the entire Central Valley during the time of Teotihuacán (700 CE), the Toltecs (1000 CE), and the Aztecs (1500 CE). Teotihuacán is the largest archeological site in the New World with its own Mesopotamian ziggurat equivalents in the form of the famous Mesoamerican pyramids of the Sun and Moon, the largest such structures in the world.

"Xochimilco" means the place where flowers bloom in the native Nahuatl language. The first Spanish accounts rhapsodize about the location being "a magical landscape that resembles the stuff of dreams" wherein the city seemed to float on islands of flowers and vegetables, with intoxicating scents, canoes gliding by on extensive networks of canals through the marshes, flights of singing birds, and the like (Schjetnan 1995, 1996).

Early dikes were built soon after the conquest in order to protect Xochimilco from salt intrusions originating from the brackish lakes to the north caused by

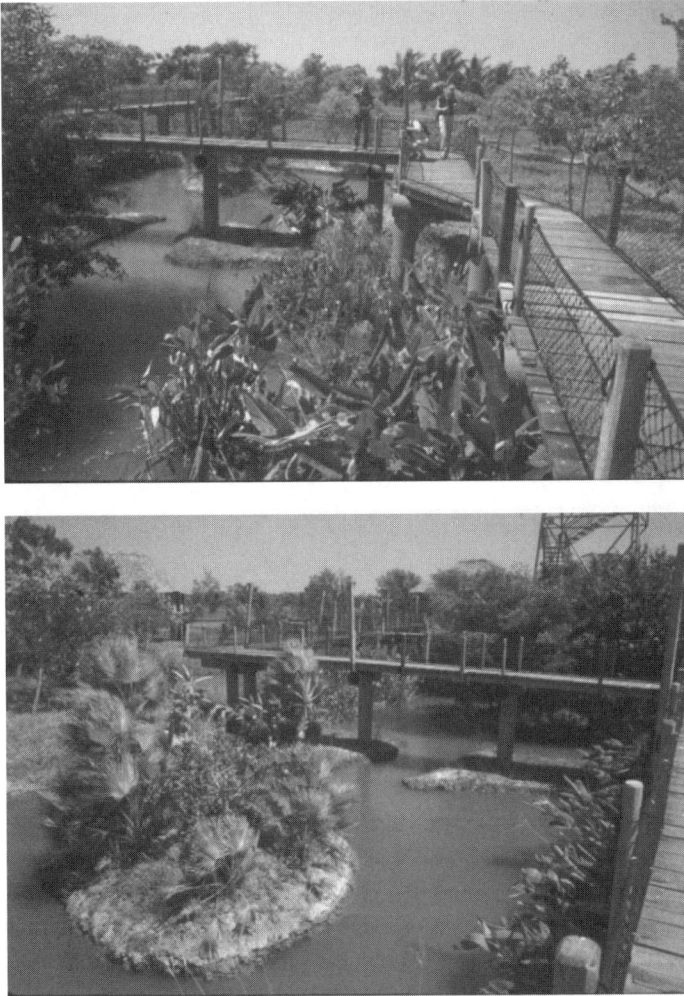

FIGURE 11.11   Walkways and crocodiles at the visitor center.   *(Continued)*

extensive evapotranspiration (ET), a characteristic of such hot and dry climates (see chapter 18). This was a lacustrine and wetland-based civilization much like Sumeria (Pournelle 2003) and different from most of the rest of ancient cultures, which were predominantly maritime based.

One particularly interesting feature of the peoples in the Central Valley from ancient to recent times is the *chinampas,* raised beds of matted organic matter and sediment from wattle built in shallow water and bordered by anchoring trees. This represents the birth of hydroponic cultivation, which has been called the greatest technological invention of the pre-Columbian New World (Coe 1964; Armillas 1971; Outerbridge 1987; Duffetel 1992). Chinampas began in 1000 bce and were expanded by the Aztecs until they covered an area of more than 120 km^2. Drawings from later

FIGURE 11.11   (*Continued*) Walkways and crocodiles at the visitor center.

FIGURE 11.12   Newly constructed boardwalks leading into the extensive wetlands.

(Continued)

FIGURE 11.12   (*Continued*) Newly constructed boardwalks leading into the extensive wetlands.

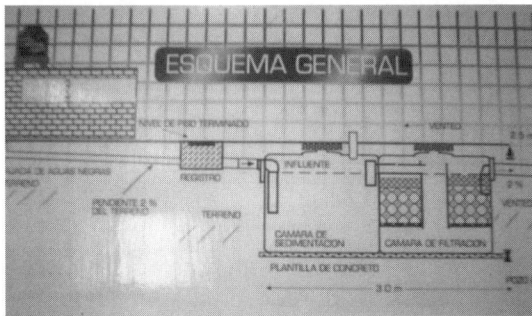

FIGURE 11.13   Lecture hall and interpretive signs inside the visitor center.

FIGURE 11.14 Biodiversity dioramas inside the visitor center.

Mayan codices show people heading out on boats into the chinampas. Estimates suggest that at one time there were over a hundred thousand watercraft plying the shallow lake and wetlands.

Because chinampas allowed for three crops to be grown per year with no need for irrigation, they could support the large population of the Central Valley and were really one of the most intensive and productive farming systems ever devised. Therefore, whereas Mesopotamians had to rely upon irrigation (which they invented) for producing their crops grown on dryland amidst the wetlands, the ancient Mesoamericans did them one better by actually farming right inside their wetlands.

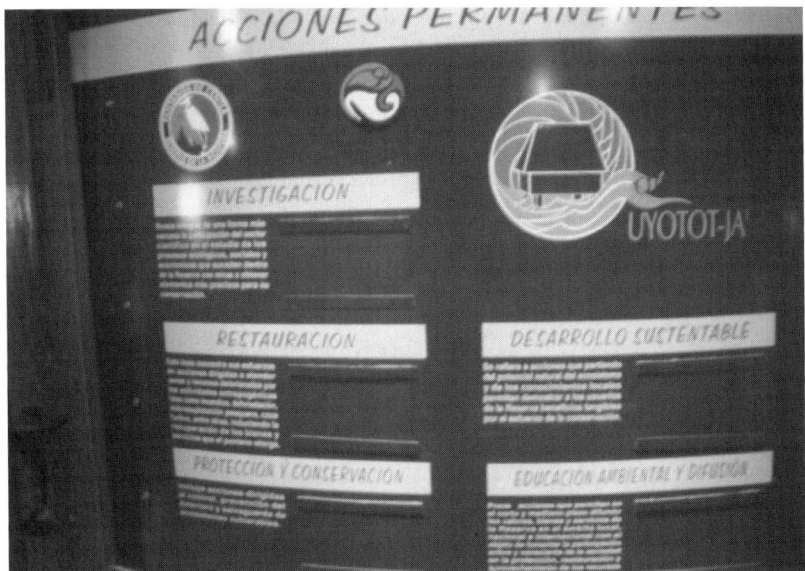

FIGURE 11.15   Interpretive sign explaining the role of the bioreserve park.

DESTRUCTION

Soon after the Conquest at the beginning of the sixteenth century, the Spaniards drained the lake as a move to reduce flooding to their new, increasingly terrestrial city (Wirth 1997; Schjetnan 1996). The flooding had increased no doubt as a consequence of the widespread overgrazing by the Spaniards' cattle. The ludicrously aggrandizing Ba'athist Party names for their massive drainage ditches in Iraq find their Spanish equivalents here in the Grande Madre Canal, one of the largest to be built outside of China up until that time. It was constructed by sixty thousand Mexican slaves, of whom ten thousand died from "accidents" and another ten thousand from sickness. Further drainage for cattle ranching took place in the nineteenth century as Mexico City grew into the world's largest metropolis. Eventually this led to the near complete destruction of one of the world's most historically significant lake–wetland complexes and the death of a unique water culture.

Today Xochimilco provides 20 percent of the drinking water for Mexico City but receives a raw deal in exchange, in that for about three decades it was the receptacle for raw sewage discharged from the city. Due to eutrophication of the few remaining canals, water hyacinths invaded and clogged them to such an extent that most waterways had been abandoned by the early 1960s. And salinization of agricultural soils became widespread due to overirrigation (see chapter 21), some with the use of contaminated water.

CULTURAL SIGNIFICANCE

Xochimilco's lake and wetlands are as important in Mexican history and cultural identity as are the Mesopotamian marshlands in Iraq. Prosaic writing has described

Xochimilco as "a landscape deeply embedded in our collective consciousness ... form[ing] part of our symbols and myths as a mystical landscape like a Paradise Lost" (Hiriart 1992), or as "a cultural symbol of the city ... and a center of profound community life and a living vestige of the Prehispanic world in the Valley of Mexico" (Schjetnan 1995). It is no surprise, then, that in 1987 UNESCO declared the area a World Heritage Site in order to protect what they referred to as "the living cradle of civilization," a term often used, of course, with reference to Mesopotamia (e.g., Kramer 1967). As a result, many Mexican artists have painted Xochimilco over the years, infusing it with an almost unreal, Edenic romanticism.

Environmental Restoration and Regeneration of Natural Capital

The "rescue" or "resurrection" of Xochimilco (*sensu* Elizondo 1996; Schjetnan 1996) involved four general goals:

- To provide botanical, historical-cultural, recreational, and economic improvements and regeneration
- To attend to and update the antiquated sewage disposal system
- To encourage agricultural sustenance
- To provide a forum for ecological education

Specifically, the objectives were to restore the lake and wetland system, maintain and increase the valley's natural water supply (which was excessive a few times during the year and insufficient for much of the rest of the year), reactivate the traditional hydroponic agricultural systems of chinampas and assist in marketing of the produce grown, redefine and enforce zoning bylaws to regulate development, enlarge the available green space, and produce historically appropriate architecture whenever appropriate to help foster a sense of place.

The project has been extremely successful. Over 3,000 ha of chinampas were protected from further degradation and another 1,200 ha were restored, including some 200 km of canals that were cleaned and made navigable for the first time in decades. More than 1,000 ha of land were desalinized and made productive for agriculture. A new wastewater treatment plant was built that services twenty thousand residents. Four large detention basins were constructed for stormwater management, and twenty small dams built to control floods. Finally, forty archeological sites from 700 to 1500 were excavated which yielded an amazing fifteen *tons* of artifacts.

Although tens of millions of dollars in outside funding were raised to accomplish these and other tasks, it is important to note that the project also developed its own "restoration economy" (*sensu* Cunningham 2002; and see chapter 3). Nurseries were built to grow the 20 million trees per year needed for the restoration tasks, and three thousand jobs were created to undertake this work. Up to seven thousand people are now employed in the agricultural activities in the area, aided by a new $7 million credit line created for farmers which now supports two hundred families and over a thousand people. Agricultural improvements such as greenhouses, irrigation systems, access roads, and the like have contributed to greatly improving productivity.

Today Xochimilco is a suburb of the sprawling megalopolis of Mexico City and is easily reached by a simple and inexpensive train ride and thus accessible to tourism, which has played a dramatic role in "restoring the natural capital" (*sensu* Aronson, Milton, and Blignaut 2007) of the area. Up to twenty thousand people now regularly visit Xochimilco on a weekend day to ride about the network of restored canals on *trajinera* boats that are often operated by former farmers whose lands were returned back into wetlands (Figure 11.16). Xochimilco has therefore become the Venice of the New World and is recognized as being a great place to visit, especially to celebrate one's birthday as a floating party accompanied by boats selling music

FIGURE 11.16   Boats for hire in Xochimilco.

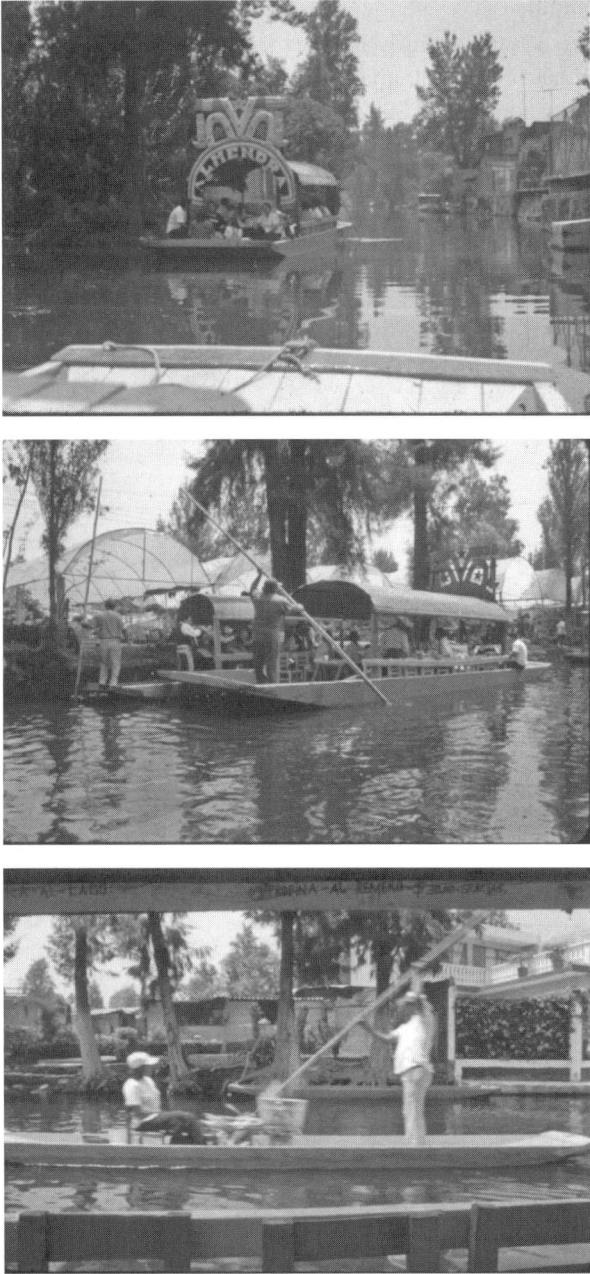

FIGURE 11.17   The "movable feast" of canal life.

and food (Figure 11.17). Real estate values have soared, with affluent individuals leaving the city to build expensive canal-side homes in their new Mexican Venice (Figure 11.18).

Farther afield away from most of the partying tourists, it is possible to move along wonderfully pastoral canals (Figure 11.19) through a landscape filled with floating mats of young plants, mostly flowers where in the past it had been vegetables (Figure 11.20). These flowers are being grown for the nearby newly constructed flower market that has two thousand stalls over an area of 11 ha and is the largest such operation in Latin America (Figure 11.21). It is this element that elevates this

FIGURE 11.18   Villas alongside the restored canals.

FIGURE 11.19   Boating along the restored canals.

case study from being yet another romanticized ecological restoration project to a much more complex example of "regenerative landscape design" (and also a much more useful model for the Iraqi marshlands) (France 2008). For as one celebrant of the Xochimilco project aptly stated, "[This is] not just a sentimental evocation of the past, but a space geared to the present and future, because it deals with modern problems and aspirations of the community." And, as the television commercials say, "And there is more!"

FIGURE 11.20   Canalside growing of flowers.   (*Continued*)

Ecological and Recreational Parks

A 215 ha ecology park was built in 1993 at a cost of $400 million (Figure 11.22); at one time, it employed a hundred landscapers to manage the 1 million trees, and more than three hundred endemic species were installed. Today, the park struggles to be financially self-reliant through admission fees, guided tours, a small shop, and an accredited water chemistry laboratory.

As well as being an ecological reserve for research, the park has become a major tourist attraction with half a million visitors per year and, interestingly, is managed not by an ecologist but by an anthropologist. The attractive visitor center complex, including a rooftop observation platform, a terrace, and a plaza with an artistic water tower, has won numerous design awards (Trulove 2002; Schjetnan 1996; Meier 1996;

FIGURE 11.20 (*Continued*) Canalside growing of flowers.

and see Figure 11.23). Park elements include flower pagoda walkways and ornamental and recreational ponds, the latter with paddle boats, boardwalks, sun shelters, and restored chinampas (Figure 11.24).

The park is also a functional landscape that receives 300 m^3 of tertiary treated wastewater (from the new treatment plant) that is supplied to the major lake through what must be the world's most beautiful discharge outlets for treated wastewater (Figure 11.25). This water is combined with stormwater runoff that is itself treated and harvested from swales located in the median strip of the abutting highway (Figure 11.26) and all supplied in the morning to reduce ET losses (i.e., based on a water budget strategy, as described in chapter 18).

FIGURE 11.21   Flower market.

FIGURE 11.22   Parque Ecologico de Xochimilco.

FIGURE 11.23   Viewing terrace and water tower.

FIGURE 11.24   Amenities at the Parque Ecologico de Xochimilco.   (*Continued*)

FIGURE 11.24   (*Continued*) Amenities at the Parque Ecologico de Xochimilco.

FIGURE 11.24   (*Continued*) Amenities at the Parque Ecologico de Xochimilco.

Over a raised walkway across the busy highway from the ecological park is the sister recreational park that is governed locally. Here it is possible to play a variety of sports, have picnics, take carnival-type rides, paddle boats about small lakes, feed birds, and construct religious shrines (Figure 11.27).

Entrance admission for the recreation park is a fraction of the cost for the accompanying ecology park, and this has unfortunately resulted in class segregation. Structural elements in both parks, but particularly within the recreation park, are now in serious need of repair and thus of a much better program of long-term sustainable, economically viable maintenance.

FIGURE 11.25 Outfall pipes for treated wastewater.

FIGURE 11.26   Treatment swale for stormwater runoff from the divided highway.

LESSONS FOR THE IRAQI MARSHLANDS

There are a number of lessons from these two Mexican wetland projects that are applicable to the restorative redevelopment of the Iraqi marshlands.

PANTANOS DE CENTLA

- It is important to put in place rigidly defined and strictly enforced zoning to avoid allowing confusing gray zones to evolve which can lead to mismanagement (see chapters 15 and 16).
- An active ecotourism program and construction of a high-visibility visitor center are useful to generate worldwide interest which might prove useful for further fundraising efforts (see chapter 16).
- Though at first pass it seems dangerous, it may also prove profitable to develop some sort of cautious relationship with oil exploration companies to support visitor center infrastructure projects.
- There is a cardinal need to engage locals in the tourism industry in order to diminish poaching and overfishing (there are many examples of this in African wildlife refuges where former poachers operate now as game wardens; see also chapter 16).

XOCHIMILCO

- A very useful strategy is to capitalize on the romance, nostalgia, and history of the location to leverage World Heritage Site status and visibility.
- A high-visibility visitor center designed by renowned architects and landscape architects can grab worldwide attention and interest (such as that

FIGURE 11.27 Site features at the recreation park across the highway from the ecological park. *(Continued)*

FIGURE 11.27   (*Continued*) Site features at the recreation park across the highway from the ecological park.

at the Dana Nature Reserve in Jordan, as described in chapter 16, and in contrast to the small and scruffy center at the Hula Swamp in Israel, as described in chapter 6).

- There is a need to work with neighboring farmers to compensate them for land returned to wetlands or to create a new market for their products (see chapter 10).
- Harvest "waste" water; in other words, don't waste it (see chapters 18, 19, and 20).
- Ecological education can be very important as a change agent (see chapters 12, 13, and 16).
- Don't forget or be snobby about recreation along with ecological re-creation (such actions will certainly be important in building sustainable communities at the edge of the Iraqi marshes).
- Restoration is a growth industry that employs many (see chapter 3).
- Establishment of a sustainable funding base is required to ensure ongoing maintenance of the ecological refuge or park.

REFERENCES

Aronson, J., S. J. Milton, and J. N. Blignaut. 2007. *Restoring natural capital: Science, business, and practice*. Washington, DC: Island.

Armillas, P. 1971. Gardens on swamps. *Science* 174:653–61.

Coe, M. 1964. The chinampas of Mexico. *Scientific American* 211:90–98.

Cunningham, S. 2002. *The restoration economy*. San Francisco: Berrett-Koehler.

Duffetel, D. 1992. A dream garden: A brief history of the chinampas and three dreams. *Arte de Mexico* (20; special issue).

Elizondo, J. G. 1996. *Mexico City: Images of a rescue: Xochimilco, centro histórico*. Berkeley: Graduate School of Design, University of California.

France, R. L. 2003. *Wetland design: Principles and practices for landscape architects and land-use planners*. New York: Norton.

———. 2007. *Wetlands of mass destruction: Ancient presage for contemporary ecocide in southern Iraq*. Winnipeg, MB: Green Frigate Books.

———. 2008. *Handbook of regenerative landscape design*. Boca Raton, FL: CRC Press.

Greenpeace. 1997. La Venta, the cradle of the Olmec culture converted by PEMEX into petroleum coated wetlands. http://archive.greenpeace.org.

Hiriart, H. 1992. The sleeping garden: Impressions of Xochimilco. *Arte de Mexico* (20; special issue).

Kramer, S. N. 1967. *Cradle of civilization*. New York: Time-Life.

Kusler, J. A. 1998. *Guidebook for creating wetland interpretation sites including wetlands and ecotourism*. Windham, ME: Association of State Wetland Managers.

Meier, R. 1996. Xochimilco. *Garten + Landschaft* 6:11–15.

Outerbridge, T. 1987. The disappearing chinampas of Xochimilco. *The Ecologist* 17:76–83.

Nicholson, E., and P. Clark. 2002. *The Iraqi marshlands: A human and environmental study*. London: Politico's.

Parks Watch. 2006. Pantanos de Centla Biosphere Reserve. www.parkswatch.org.

Plata, F. 2002. *Historias de Pantanos*. Mexico City: Espacios Naturales y Desarrollo Sustentable A.C.

Pournelle, J. R. 2003. Marshland of cities: Deltaic landscapes and the evolution of early Mesopotamian civilization. PhD diss., University of California, San Diego.

Schjetnan, M. 1995. The ecological park of Xochimilco.

———. 1996. The "resurrection" of Xochimilco. In *Mexico City: Images of a rescue: Xochimilco, centro histórico*. Berkeley: Graduate School of Design, University of California.

Trulove, J. G. 2002. *Ten landscapes: Mario Schjetman*. Beverly, MA: Rockport.

Wirth, C. J. 1997. The governmental response to environmental degradation in the Xochimilco ecological zone of Mexico City. http://lasa.international.pitt.edu/LASA97/wirth.pdf.

World Wildlife Fund. 2006. Pantanos de Centla (NT0148). www.worldwildlife.org/wildworld/ profiles/terrestrial/nt/nt0148_full.html.

12 Involving People in Science and Sustainability in the Pantanal Wetlands of Brazil*

CONTENTS

Introduction 211
Engaging People 212
The Earth Watch Model 213
Case Study: The Pantanal Wetlands 215
Acknowledgment 217
References 217

INTRODUCTION

There will come a time (hopefully not too far in the future) when the unrest in Iraq is over and a large number of people from around the world whose collective imagination has been captured by reading the compelling accounts of British adventurers of the twentieth century (Maxwell 1966; Thesiger 1964; Young 1977) or whose collective consciousness has moved them to want to help to repair the damage (see chapters 1 and 2; and France 2007a, 2007b), will make them eager to visit the marshes in a constructive, nonconsumptive way, possibly through contributing to the scientific study of the ecosystem as support for ongoing restoration or management actions. The basis of this chapter is to support the thesis that science can help foster a dialogue that can create constituencies for change by engaging an international populace. And one of the leaders in developing this strategy is the Earth Watch Institute, which has spent thirty years based on constructing field research experiences for the general public.

The mission of Earth Watch is to create a constituency for a sustainable environment. Mark Chandler (2004) believes that such a mind-set not only involves the natural world but also must consider a human dimension of that environment from the iconic and spiritual viewpoint. Earth Watch regards the world from a viewpoint

---

* Adapted by Robert L. France from Chandler, M. 2004. Role of volunteers in sustainability research in the Pantanal wetlands. Paper presented at the Mesopotamian Marshes and Modern Development: Practical Approaches for Sustaining Restored Ecological and Cultural Landscapes conference, Cambridge, MA, October.

211

of desiring a sustainable global environment and generating action and policy toward that end (Earth Watch 2008). But in order to get there, informed and implemented decisions are needed.

There are five different sectors that could contribute to those decisions (Chandler 2004): having better information through research is clearly one of them, broad stakeholder participation in the process is another, and an educated and motivated public is a third. And these are the three sectors that Earth Watch focuses on. Clearly, socioeconomic factors and politics (Chandler's fourth and fifth sectors) are also involved, but these are not directly involved through Earth Watch activities.

This chapter will briefly touch upon engagement of people in general, focus on the Earth Watch model of how to engage people through science in particular, and, finally, examine the case study example of the Pantanal wetlands of Brazil that were also discussed in chapter 10.

ENGAGING PEOPLE

One of the key pieces beyond simply understanding the natural dynamics and the planning process concerns the need to engage people (Chandler 2004). In particular, what are the stakeholders required to engage in during the process to make land use decisions? One of the lessons that Earth Watch has learned is that though it might be easy to initially capture people's interest in environmental work, it is often very difficult to maintain that level of engagement.

The second lesson is that it's not an all-or-nothing relationship. In other words, there are different grades of engagement. For example, on one end of the spectrum of engaging a community or individuals is the passive approach, where you just provide them with information by giving a presentation and informing the public about what is going to happen. Or, on the other end of the spectrum, the engagement can take the form of self-mobilization and empowerment, where during that dialogue the public are actually able to inform the ultimate decisions in the activities being undertaken. In these two extremes, and everywhere in between, there are different ingredients of how to facilitate that engagement ranging from answering questions to consulting for solutions. So one of the key aspects in building this form of relationship is to set clear expectations with the stakeholders as to the level of engagement that they're expecting and what your role as an agency will be in facilitating that engagement (Chandler 2004).

Clearly important in developing a relationship of engagement is the listening aspect, which is sometimes forgotten. The challenge is to transform concerned citizens from being passive recipients of information to becoming actively engaged participants in the decision-making process. And one of the key approaches that Earth Watch has developed in this regard is to use experiential techniques (Chandler 2004), including the planning of the meetings, workshops, and so on, such that individuals actually find themselves participating in some activity that helps shape the outcome and understanding of the environmental issue of concern.

The other key aspect of engagement that Earth Watch is involved with is in dealing with real-life, real-world situations. This is similar to the Habitat for Humanity model where concerned citizens are actually involved in the construction of houses, as opposed to an experience that is canned like a laboratory experiment or something that you just go through where there are certain pre-scripted outcomes of that activity. Developing a form of engagement where the outcomes are not known is important in terms of modulating what sorts of experiences are generated (Chandler 2004). Earth Watch believes that informed stakeholders are the key to a better decision-making process concerning protecting the world's environmental and cultural heritage (Earth Watch 2008). For this to happen, it is necessary for people to not only understand but also "buy into" and care about the outcomes of their and others' actions. With this motivation comes a feeling of empowerment where individuals who have actually performed something together feel that they have more sort of community that is dedicated to actually carrying out certain outcomes (see also chapter 2 for how ecological restoration is one way of building such a confraternity of action). This networking component is very important in terms of enabling people to share personal information and to develop capacity building in terms of the training aspects learned from involvement in any activity. Earth Watch has often seen that when people engage in these activities, they decide to make dramatic career choices toward doing something they've always wanted to do but have never previously had the sort of empowerment or ideas necessary to actualize such long-held aspirations.

Earth Watch engages a local as well as an international or regional component in their sponsored projects. Clearly with respect to the Iraqi marshes, for example, involving Turks or others from outside of the immediate sphere will be important in terms of these engagement pieces. Also, the engagement of leaders—people who have multiplier effects within the community—from the political and corporate world, as well as educated students, will be equally important. And one other group to target for fostering widespread engagement in environmental management is those key people within communities who share information and help build capacity, such as storytellers and artists (France 2003; France and Fletcher 2005).

THE EARTH WATCH MODEL

The mission of Earth Watch is to engage people worldwide in scientific field research in education in order to promote the understanding that's actually necessary for a sustainable environment (Earth Watch 2008). The key piece is the scientific field research, so that individuals participating in peer-reviewed field research projects are those who are supported. Earth Watch has been doing this sort of work for over three decades and in any year supports about 130 different individual field research projects with 260 researchers in close to fifty countries. In order to do this work, Earth Watch solicits about four thousand paying volunteers annually to participate on these projects and to support them.

In terms of the use of volunteers in the scientific field research projects, one of the important aspects is in providing a labor force necessary to accomplish the goals. In other words, whether it's replanting or eradicating exotics, there's often a labor-intensive component to many restoration projects that can be satisfied by volunteer efforts (see chapter 2). And there is an educational component similar to museums or national parks using volunteers to help inform the public. Earth Watch therefore looks at using previously untrained but highly motivated and usually knowledgeable volunteers to support the field research, often through their participation in active, community-based monitoring programs (Chandler 2004).

The funding aspect is something that Earth Watch has developed strongly with volunteers contributing about $2,000 toward their participation. Half of this amount goes to the organization of the research projects, and half goes to the researcher for arranging accommodation. This represents an important source of funding for both the research as well as the local economy. And in the end, the researcher has a team of interested and dedicated participants to help with running the project.

Through its activities, Earth Watch has become involved in developing tourist economies in some places where previously such did not exist, thereby enabling large numbers of tourists to visit such locations. Especially, at some future date, in a place like Iraq, which may have a perception problem, such an experience may be a critical first step toward establishing international tourism. There are four constituencies that Earth Watch uses to lay the groundwork for enabling such a tourism industry to become established (Chandler 2004): the scientific constituency, the academic community, the volunteerism network of concerned individuals, and the educational component. Pioneer tourists visit somewhere like the marshes of southern Iraq and help develop the infrastructure, and lay out some of the groundwork, the training, and so on that is then followed through with a sort of larger scale market approach. In this way, Earth Watch volunteers and volunteerism in general initiate a fledgling tourism economy in a local area which over time can mature to a viable local economy.

Clear benefits accrue to Earth Watch volunteers, including everything from forming a new awareness about the environmental issues to building specific skills through participation in the projects. There's an opportunity to share cultures in a new environment among the individuals who participate, leading to the development of both a local and a global perspective. Earth Watch uses teams of between eight and a dozen volunteers who work in the field for about two weeks at a time. Depending upon the composition of the team, a good deal of exchange and personal growth occurs among team members through their combined work efforts. In several different kinds of projects, alumni networks have developed, allowing volunteers to continue the experience through shared information (Chandler 2004).

Typically, Earth Watch supports three to five teams a year on a research project, some of which are based on ecological restoration. This has become an issue of increasing interest amongst Earth Watch's own volunteer network (supporting the thesis developed in chapter 2 about the attraction of engaging in such activities) as well as being a subject of widespread interest among many thousands of potential volunteers wishing to participate in the postwar rebuilding of Iraq (France 2007a; see also chapter 2).

Some of the key aspects in terms of getting the most out of engaging volunteers actively in field research are establishing realistic expectations and preparing people accordingly in order to create a more beneficial experience (Chandler 2004). Earth Watch has also developed mechanisms for engaging schoolchildren (see chapter 16 for similar efforts by the Royal Society for the Conservation of Nature in Jordan). Their "Life in the Field" web-based reports allow teacher-volunteers participating, for example, on projects in the Pantanal wetlands of Brazil to be able to communicate to their classrooms back in São Paulo or Boston. This process extends and multiplies the participation experience through back-and-forth questions and answers on a daily basis between students and their teachers because not everyone can participate in the field.

Earth Watch does have a number of projects around the world that deal directly with wetlands (Earth Watch 2008) such as the rehabilitation of acid-rain damaged headwaters in the Czech Republic, the measurement of glacier surges and sediment transfer and flooding in Iceland, the community-based restoration of mangroves in Kenya, and the restoration of biodiversity to coastal wet grasslands in Estonia. One such example that chief researcher David Harper from the University of Leicester in England has been working on with Earth Watch for fifteen years concerns studying the wetlands around several lakes in Kenya. Over this time period, seven hundred Earth Watch volunteers from all over the world, from Kenyans to Australians, have participated. An important offshoot of the project is its generation of local capacity, such as an entire environmental nongovernmental organization (NGO) as well as many university theses and peer-reviewed publications. The project has both also been awarded a Ramsar Prize and been designated as a Ramsar site of international wetland importance. This case study is a nice testament of what can be achieved when you engage volunteers from around the world in the scientific field research process (Chandler 2004).

One of the principal questions for scientists is, can volunteers actually contribute to collecting meaningful data? And the answer is, it depends on the task, the specific nature of the project, and the training and preparation of the volunteers. Earth Watch volunteers have undertaken a vast variety of tasks in many water-related projects such as biological, biogeochemical, and physical sampling, and for landscape-related projects such as vegetation/habitat mapping and sampling. Volunteers have also proven useful with respect to the social sciences, whether it's archaeological research, architectural surveys, mapping cultural landscapes, or interviews and attitudinal surveys. Clearly there are cultural and language issues that need to be dealt with depending upon the sorts of volunteers available or the people being engaged with. But there are tremendous options, and in terms of being able to conduct rigorous science, many stakeholder groups can be engaged toward meaningfully contributing to the data collection. And again, this gets back to the idea that in this Earth Watch model, when people are engaged, it's not simply in a canned version of how science is conducted (Chandler 2004). Volunteers actively participate in the collection of data that will be used through engaging in a real-life situation that adds value to the experience for all.

CASE STUDY: THE PANTANAL WETLANDS

In addition to individual research projects, Earth Watch supports five to ten projects under what are called "conservation research initiatives," where a field director on the ground searches for and cultivates important research needs that can be supported through the volunteer model. Emphasis is placed on disseminating the results to multiple stakeholders as rapidly as the research results are gathered (Chandler 2004).

One of these conservation research initiatives takes place in the Pantanal, which at a size of over 200,000 km^2 is the world's largest freshwater wetland ecosystem (see also chapter 10). The principal threats to the Pantanal are threefold:

- Drainage projects, which would channelize and remove a significant amount of water from the watershed
- Cattle ranching, which has become intensified, resulting in the fragmentation of the landscape surrounding the wetlands
- Agriculture, with the area now becoming the world's largest producer of soy, and the consequent pollution due to uncontrolled release of sediments and nutrients

Earth Watch formed a partnership with Conservation International and ran a series of workshops which identified a number of wildlife corridors and critical habitats that needed protection. Work began in a single location, and efforts have now been expanded to five other areas throughout the whole Pantanal in order to corroborate the original baseline information gathering. But the key lesson in this example that is very applicable to the evolving situation in Iraq is Earth Watch's use of partners to identify the research priorities and how the research will be subsequently used in the management plan (Chandler 2004).

The Pantanal conservation research initiative has been running since 2000 as nine multidisciplinary, simultaneous projects in which the researchers share information and learn from one another. Volunteers have repeatedly returned to participate in different projects involving a whole range of issues related to wetland ecology, fisheries and wildlife management, and wildlife disease control. In terms of capacity building, a large number of young Brazilian scientists have been supported in this initiative with a program of mentoring and support from experienced American researchers. Annually, Earth Watch places over two hundred volunteers in the Pantanal, half of whom are paying people who have selected the location and one of its projects as the destination for their working holiday, and half of whom— teachers, corporate members, and people from the local community—whose transportation and participation are paid through fundraising by Earth Watch (Chandler 2004). From this fundraising, Earth Watch is able to provide about a quarter of a million dollars in grants to the scientists to help them do the field research, as well as contributing directly to the local economy through food and accommodations. In this respect, there is a direct financial gain through this model as well (see chapter 16 for another approach linking ecology and economy).

One of the characteristic aspects of all natural systems, in particular wetlands, is their integration. Earth Watch capitalized on this idea in the Pantanal by supporting multiple projects that all feed into one another, in a way reminiscent of Cunningham's (see chapter 3) clarion call for the need to integrate across sectors and think more broadly about how we approach ecological restoration. Money is used to target those individuals (teachers, students, conservation professionals, anyone involved in environmental management, and corporate employees) who have a greater effect, or "multiplier effect," in terms of getting things done (Chandler 2004). Corporations, for example, are solicited to not only support local engagement but also have their own employees take part. As a result, employees from the Rio Mining Company, Shell Oil, HSBC Bank, and Starbucks have all participated on the Pantanal projects. Earth Watch believes that it is very important for corporate employees to understand the relationship between science, the environment, and their own corporate activities.

The goal in all these efforts is to foster a dialogue and build a constituency to be able to manage the Pantanal in a different way than is currently happening (Chandler 2004). Managing watersheds involves managing people, and focusing on communication is critically important to fostering awareness and stewardship (France 2005, 2006). It may be some time before volunteers are able to work on restoring the environment of Iraq, but there are certainly opportunities at the present time to begin operations to bridge the gap between ongoing scientific field research and the many vested and varied stakeholder interests. The Earth Watch model is available to be able to make that happen, both from a capacity-building and from an empowerment and information-sharing perspective. The benefits of adapting this model include building viable connections through direct engagement in problem solving and management; developing greater contextual understanding of environmental problems and potential solutions; sharing experiences across different sectors to build a community of mutual concerns and possible objectives; creating an opportunity for dialogue between scientists, local environmental managers, and community members; enabling new ideas to emerge through cross-fertilization; and, finally, expanding the belief horizons of all participants.

ACKNOWLEDGMENT

Adapted from Chandler (2004).

REFERENCES

Chandler, M. 2004. Role of volunteers in sustainability research in the Pantanal wetlands. Paper presented at the Mesopotamian Marshes and Modern Development: Practical Approaches for Sustaining Restored Ecological and Cultural Landscapes conference, Cambridge, MA, October.

Earth Watch. 2008. www.earthwatch.org.

France, R. L. 2003. *Deep immersion: The experience of water*. Winnipeg, MB: Green Frigate Books.

———. 2005. *Facilitating watershed management: Fostering awareness and stewardship*. Lanham, MD: Rowman & Littlefield.

———. 2006. *Introduction to watershed development: Understanding and managing the impacts of sprawl*. Lanham, MD: Rowman & Littlefield.

———. 2007a. A tale of two wetlands: The view from abroad. In *Healing natures, repairing relationships: New perspectives in restoring ecological spaces and consciousness*, ed. R. L. France. Winnipeg, MB: Green Frigate Books.

———. 2007b. *Wetlands of mass destruction: Ancient presage for contemporary ecocide in southern Iraq*. Winnipeg, MB: Green Frigate Books.

France, R. L., and D. F. Fletcher. 2005. Watermarks: Imprinting water(shed) awareness through environmental literature and art. In *Facilitating watershed management: Fostering awareness and stewardship*, ed. R. L. France. Lanham, MD: Rowman & Littlefield.

Maxwell, G. 1966. *People of the reeds*. New York: Pyramid.
Thesiger, W. 1964. *The marsh Arabs*. New York: Penguin.
Young, G. 1977. *Return to the marshes: Life with the marsh Arabs of Iraq*. London: Collins.

13 Planning and Development of a Desert Wetland Park in the United States*

CONTENTS

Introduction 219
People, Paper, Places 219
Clark County Wetlands Park: The Place 222
Integrating People, Planning, and Design 222
The "New" Place 224
Implementation 225
Acknowledgment 226
References 226

INTRODUCTION

Deserts are some of the most romanticized and mythologized of all landscapes, and as a result they, and especially their associated oases, are becoming popular tourist destinations (e.g., chapter 7). Significantly, tourism, if approached sensitively with respect to both the physical and social environments, has the potential to play an important role in the sustainable development of desert communities (chapter 16). The transformation of the environmentally degraded Las Vegas Wash (chapter 8) into the award-winning Clark County Wetlands Park (France 1999, 2011) is one of the most successful examples of how tourism can be a positive agent for social and ecological change in arid regions, and one that may offer lessons for plans to convert a portion of the restored Iraqi marshlands into a national park at some future date.

PEOPLE, PAPER, PLACES

Restoring, preserving, and protecting special landscapes in ecosystems are not just about the environment. Becky Zimmerman (2004) believes that careful integration of

---

* Adapted by Robert L. France from Zimmerman, B. 2004. Clark County Wetlands Park: A restored desert wetland. Paper presented at the Mesopotamian Marshes and Modern Development: Practical Approaches for Sustaining Restored Ecological and Cultural Landscapes conference, Cambridge, MA, October.

environmental conditions, social equity, economic impacts, and thoughtful aesthetics can create legacies in terms of sustainable places of timeless beauty, enduring quality, and untold values for future generations. Successful restoration and place-making efforts must begin with a mission or vision that is held passionately over the long period of time that it often takes to accomplish these types of projects. An example of such a mission statement might be to create beautiful, imaginative, and environmentally sensitive places where people meet, children play, and the curious explore. Another mission statement might be to preserve unique landscapes, memorialize historic sites, protect the land, and meet the recreational needs of the community. Design Workshop has capitalized on such a strategy to create exceptional parks and open spaces for people and nature around the world. Of key importance is the need for cultivating a collaborative environment involving people, paper, and places.

Restoring wetlands or any other type of environment begins with an idea or a dream that is researched, reality tested, and then used to generate alternative approaches in order to reach the desired outcome (see chapter 4). The final vision is a result of the process of testing those initial ideas against scientific, cultural, and economic realities. In other words, restoring wetlands requires people. Environments do not restore themselves; it takes *people* to believe in what an environment should become or be returned to (Zimmerman 2004; France 2007a). Environments are rarely successfully restored without plans, without detailed studies, and without imaginative solutions, all of which are captured on *paper*. Restored environments create landscapes which, most importantly, become valued *places* (France 2008).

The focus of this chapter is the Clark County Wetlands Park. But first, it is important to briefly review several other examples from Design Workshop of the importance of people and culture in restoring ecosystems, special habitats, and refuges (for both wildlife and people). The eight thousand acre Walnut Creek National Wildlife Refuge, now called the Neal Smith National Wildlife Refuge, in Iowa was created at a time when the U.S. Department of the Interior was directing its funding toward projects that served more of the public good. This was a significant departure from the business model of one of its departments, the U.S. Fish and Wildlife Service, whose traditional role had been in creating wildlife refuges solely for the preservation and conservation of wildlife habitat with little or no public access opportunities. As a result, this new orientation brought about new challenges and benefits (Zimmerman 2004). The newly designed refuges would now have to encompass education, interpretation, and explanation of the complex natural and cultural forces at work on this site over time (such as previous agricultural use). But this focus would also foster some great opportunities in terms of community partnerships that would create volunteers to help the refuge staff, and to also generate funding in order to expand that supplied from federal revenues.

The design approach to the Walnut Creek Refuge was both pragmatic and sensitive. The team first completed a market study to determine if this site would have the demand to be a publicly visited facility. In other words, a market study was used to determine if an environment should be restored in the first place. This is, of course, a very novel concept in the arena of environmental restoration (Zimmerman 2004). This was followed by an intensive period of research supported by the design team, wildlife biologists, botanists, and educators. From the start, it was deemed important for the design to capture opportunities for interactive learning about the restored environment in order to help ensure that future generations would become stewards of the repaired land (see France 2005 for more about the importance of education in fostering stewardship).

A series of collaborative charrettes was completed to help generate the physical form for the project's intended structures, based on exploring the materials and development patterns of the Native Americans and the early European settlers in this region of Iowa. So the result is a place that immerses visitors in the landscape by both preserving and containing views, through creating places that give a sense of what this place had originally been like. Also, the work of the early "prairie school" of designers provided the vernacular form for the visitor center and for the three outdoor classrooms which resonate with the broad horizontal character of this landscape.

The project also uses natural processes to support the land (Zimmerman 2004), by using constructed wetlands to filter waste as described in chapters 18, 19, and 20, while the solar orientation of the buildings with the earth–berm design provides some natural cooling and heating for the visitor facility complex. Eroded stream banks were reshaped, regraded, and reformed in order to accept the extensive amount of reseeding that would occur on the site. Native seed was collected by hand from the few remaining original tall-grass plant species that were still found on the surrounding landscape. Today, volunteers still continue to collect that seed.

There are now more than two hundred indigenous plant species established back on the site. Significantly, the few remnants of the original and rare tall-grass ecosystem not only have survived but also now have actually flourished. With a thriving population of dry-land buffalo (occupying a similar ecological and anthropological role to the water buffalo of the Iraqi marshlands), the restored landscape has become a refuge for both wildlife and people. And this is what is most important in this project by Design Workshop in terms of addressing the original premise: plans without people are just plans. One can create places to celebrate species, but a great deal of social benefit (or natural capital *sensu* Aronson, Milton, and Blignaut 2007) is lost if opportunities to interpret and provide education about that environment are not simultaneously advanced (Zimmerman 2004; France 2006). In the end, places without people are just spaces. It's often the ability to enjoy a place that really creates the long-lasting benefit.

Places are also composed of the palimpsest of layered history, nowhere thicker than in southern Iraq (France 2007b). As in Iraq, sometimes the historical aspects of an environmental landscape have been wiped clean. It is necessary to celebrate these landscapes of memory by telling the story of the site's ecological and cultural history. By allowing people to see and interact with educational displays, history becomes part of their future (Zimmerman 2004). Educational interpretation is therefore a critical vehicle to allow people to connect to and envision their role in shaping the places in which they inhabit (France 2005). And at Clark County Wetlands Park, historic events and people's dedication to direct environmental destiny have made a difference and have produced what may be *the* signature success story in desert wetland restoration.

CLARK COUNTY WETLANDS PARK: THE PLACE

Only eight miles beyond the twenty-four-hour lights of the Las Vegas Strip, the Las Vegas Wash has been an often overlooked treasure—so much so, in fact, that it was almost destroyed by the years of development expansion in southern Nevada (chapter 8). People have been part of this land for a thousand years, with several native Indian groups occupying the area, even simultaneously at times. The Las Vegas springs quickly developed into a stop along the Spanish trail and later became a place for early Mormon settlement.

The city of Las Vegas has always been an influence on regional water quality and water use activities (France 1999). This is particularly the case at the present time given that protection of the unique wash environment involves wastewater treatment, water rights, return flow credits, and salinity control to the Colorado River (France 2011). And as urban development continues in the valley, the natural resource values of the Las Vegas Wash also continue to increase.

Until recently, the timeline here was really a story of key decisions and delayed action, the consequence of which was that in 1984 there was a significant storm event that resulted in severe erosion (chapter 8). At the same time the Wash became the city's dumping ground as well as a location for illegal target practice and a camping location for vagrants. But there is another timeline which highlights the decades of advocacy (Zimmerman 2004) such as the Wash Development Committee's formation in 1973, the Wash environmental assessment conducted in 1981, leading to the first Clark County Wetlands Park Master Plan in 1982, comprehensive and integrated planning activities in 1987, formation of the Wash Task Force in 1988, initiation of the Wash erosion mitigation project in 1989, voting by Nevada residents to approve a $13.3 million bond for Wash restoration activities in 1992, generation of a new master plan in 1995, followed by a programmatic environmental impact statement for the planned park in 1996, formation of the Wash Coordination Committee in 1998, and finally construction of the Nature Preserve beginning in 2000.

An important key to the success of the Las Vegas Wash–Clark County Wetlands Park story has been the ability to change the public's perception of the place from being one of fear and loathing to an acceptance by all that it was actually a community asset (France 1999; Zimmerman 2004). For example, in the early years of transformation, up to twenty tons of trash were collected from this site and shipped off in big trailers. Protective fencing was erected around key places, and security patrols put into effect. And the fines for illegal dumping were increased and enforced.

INTEGRATING PEOPLE, PLANNING, AND DESIGN

Project success is predicated upon developing a set of clearly stated goals in which to direct the mission of the master plan (France 2006). In this case, the goals were focused on balancing public access and environmental sensitivity, creating a community sense of pride and ownership, developing a critical educational venue, and conserving and restoring a degraded landscape. The integrated planning and design process for the Las Vegas Wash needed to be comprehensive, congruent, and compelling, and it really wasn't going to work if any one of these points were left out (Zimmerman 2004).

The issues and constraints to successful implementation of the planning goals were many:

- How to stop the loss of wetlands and even try to gain back some of this habitat
- How to turn the place from a dumping ground into a recreational asset
- How to rehabilitate the Wash so that area residents could use it and take pride in it
- How to create a synergy of edges where the Park and urbanization interface
- How to improve water quality, particularly within the realm of financial constraints

The multiyear planning process combined innovative techniques with logical conclusions to reach a master plan that had large public buy-in and multi-agency support. The key elements of this planning process (Zimmerman 2004) included public participation, inventory and analysis, programming, alternative plans, refinement, and documentation.

The public participation component of the plan was early on deemed to be critical to its success. Consequently, there was a great outreach in sending out newsletters to the community and hosting multiple meetings and workshops, purposely going to where the people were instead of trying to get the community to come to the planners.

Effective planning depends on good base information (chapter 4). In the case of the Clark County Wetlands Park, the first part of the planning process was dedicated and devoted to gathering and analyzing information, including information on geology and soils, surface and subsurface hydrology, visual impacts, landownership, infrastructure conditions, wetlands delineation, and archaeological sites, and conducting a survey of most threatened species (France 2011).

At public workshops, people were actively engaged to help determine what the appropriate and inappropriate uses for this new park would be, given the underlying caveat that all uses had to be compatible with the environment which was going to be restored. This was no easy task because these public meetings were attended by everyone from the Audubon Society, who wanted to eliminate all trails, except for maybe one, in the entire 3,200 acres; to the ATV (all-terrain vehicle) clubs, who felt like every time a place was improved, their members got pushed further and further away from the city; and all kinds of special interest groups in between. But in listening to and working with these various groups, Zimmerman (2004) and colleagues were confident that they could come up with some viable alternatives for the future park.

Three alternatives with purposely very distinct characteristics were generated so that they could be compared and contrasted (see chapter 4 for an overview of this process and chapter 15 for another case study). One alternative was termed "Conservation" and was basically the Audubon Society version of a single trail fed from a very limited number of trailheads. This alternative, therefore, had a small amount of public access and was really about the environment. Another alternative was termed "Recreation" and was focused on determining what types of recreation, with the exception of active recreation locations like ballparks and soccer fields, could be accommodated on the site. And the third alternative, "Full Development," was the most controversial because it was based on generating money. This alternative was directed toward creating the biggest impact and the biggest change on the landscape, as well as restoring the environment. Because in both cases a considerable amount of funding would be needed, the planning team looked at locations where there could actually be a land trade. One such location that was proposed to be sold for development in order to generate revenue was located adjacent to an existing development.

As so often happens, and what the planning team had expected based on the way that they had orchestrated it (Zimmerman 2004), no single of these alternatives was the one that was selected in total, but bits and pieces from each of them were what registered with the public. The final master plan that went through this public process also identified concerns that were raised through a simultaneous environmental impact statement and expanded the program to integrate the very unique site conditions. The resulting documents that were created (Southwest Wetlands Consortium 1995a, 1995b) had even more importance than Zimmerman and her colleagues had realized. This is because both documents, though produced in 1995, are still very much in use today. In this respect, the documents became the holder of the original vision as well as very utilitarian in that they were not only concerned with the environmental conditions of the park, but also focused on all the implementation steps needed to get to that stage, including both funding and operations and maintenance (Zimmerman 2004).

THE "NEW" PLACE

The programming and land use of the Clark County Wetlands Park are focused on integrating the facilities into the surroundings, providing a place of reflection and respite from the commotion of the incredible nearby city, and balancing the public uses with natural habitat enhancement (Zimmerman 2004). The goal of the programming has been to ensure a land use mix that can be enjoyed by future generations.

Existing conditions considered important for creating a framework for site design were related to views, vegetation, and hydraulic conditions. And water and its movement, logically, became the main theme in the guiding spirit for the overall design. The planting palette was based on the use of natives to withstand the tough climatic conditions of the region in terms of sun, wind, and heat.

Trying to create a sense of place in this vast desert landscape posed a tremendous challenge (Zimmerman 2004). The design objectives, therefore, focused on enhancing an environment that has a timeless spirit, at the same time as reflecting the characteristics of this unique landscape (i.e., wetlands in the middle of the Mojave Desert at the edge of the fastest growing city in the country). So throughout the Park are entrance points and gateways that are accentuated and expressive of each site's qualities and provide an opportunity to identify the protected landscape at key locations for assisting visitors in orientating to the place (Southwest Wetlands Consortium 1995a). This orientation is important because once the landscape of tall grasses is entered, it is possible to quickly lose all sense of direction (which, of course, can be regarded as either an advantage or not).

A clear hierarchy of trails was established to reduce the impacts of numerous pedestrians and eliminate vehicular trails that had once crisscrossed the site (France 2011). Trail notes inform visitors about rest stops, interpretive education facilities, and scenic windows or vistas into and out of the Park. Existing trails are being improved and new trails constructed to lead visitors to specific areas, link to other recreational opportunities on adjacent federal lands, and provide an educational storyline about the various habitats throughout the Wash. A system of interpretive pavilions and graphic signage helps give visitors a better sense of the history and human impacts on this environment. By offering educational opportunities throughout the Park, the Las Vegas Wash has become the region's largest classroom (Zimmerman 2004).

Although residents of the Las Vegas Valley are always searching for active recreation (there seem to never be enough ball fields and such), the emphasis in the Clark County Wetlands Park was on creating passive recreation opportunities such as bird watching, picnicking, horseback riding, photography, mountain biking, and so on for all ages and abilities, always keeping in mind the need to be compatible with the overriding conservation and restoration goals. Because most of the water in the Wash is treated effluent (France 1999, 2011; chapter 8), trails were actually sited to discourage interaction with the water, except for research purposes.

IMPLEMENTATION

So how are all these great ideas and good intentions actually used get something built? In order to protect nature surrounded by urbanization, management was directed to the following actions (Zimmerman 2004):

- Control of the perimeter of the Park
- Strong advocacy for the designed trail system
- Elimination of off-road vehicle use
- Reintroduction of native plant species
- Use of real data to inform design, construction, and management
- Concentration of visitor and interpretive areas within the park
- Portions of the Park left undeveloped

The first major implementation in the Master Plan was the creation of a 120-acre, $4.3 million nature preserve. And the challenge became how to design features in such a dynamic environment—one in which, for example, three 100-year flood events occurred within two years. Because the entire site can be inundated, careful attention to landscape grading became essential for building new structures outside of the pathways of water. And the sense of place was created through visual orientation and a connectedness to adjacent habitat in distant mountains. And because this site was designed to be a place for people within nature, a circular trail system was constructed based on differing travel times to allow walkers to interact with as much wildlife and plants as they desire.

In conclusion, there were several prominent keys to the success of the Clark County Wetlands Park (Zimmerman 2004). One was the ability to create and sustain momentum within the community. Another was the ability to leverage finances and labor from concerned and committed partners. One of the biggest challenges has also been that dollars for new construction projects tended to be easier to obtain than funding for continual operations and maintenance. Yet once that significant initial investment has been made to create such a place, it is essential that it must be cared for. Many restored environments have, unfortunately, been financially orphaned and physically abandoned.

What is the bridge to the future of the successful restoration of landscapes like the Las Vegas Wash? Zimmerman (2004) is adamant that such projects will only succeed in the long term if people have a sense of ownership in the transformation of the place (see also chapter 2). The lesson for the Iraqi marshlands and elsewhere is that people must be empowered to serve as stewards of that repaired land, and that they must remain passionate about the outcome of the restoration.

ACKNOWLEDGMENT

Adapted from Zimmerman (2004).

REFERENCES

Aronson, J., S. J. Milton, and J. N. Blignaut. 2007. *Restoring natural capital: Science, business, and practice*. Washington, DC: Island Press.

France, R. L. 1999. (Stormwater) leaving Las Vegas. *Landscape Architecture Magazine* 8:38–42.

———. ed. 2005. *Facilitating watershed management: Fostering awareness and stewardship*. Lanham, MD: Rowman & Littlefield.

———. 2006. *Introduction to watershed development: Understanding and managing the impacts of sprawl*. Lanham, MD: Rowman & Littlefield.

———. ed. 2007a. *Healing natures, repairing relationships: New perspectives in restoring ecological spaces and consciousness*. Winnipeg, MB: Green Frigate Books.

———. ed. 2007b. *Wetlands of mass destruction: Ancient presage for contemporary ecocide in southern Iraq*. Winnipeg, MB: Green Frigate Books.

———. ed. 2008. *Handbook of regenerative landscape design*. Boca Raton, FL: CRC Press.

———. 2011. *Designing new natures: People, ecology, and landscape repair*. New York: Routledge.

Southwest Wetlands Consortium. 1995a. *Clark County Wetlands Park master plan*. Las Vegas, NV: Southwest Wetlands Consortium.

———. 1995b. *Clark County Wetlands Park planning process*. Las Vegas, NV: Southwest Wetlands Consortium.

Zimmerman, B. 2004. Clark County Wetlands Park: A restored desert wetland. Paper presented at the Mesopotamian Marshes and Modern Development: Practical Approaches for Sustaining Restored Ecological and Cultural Landscapes conference, Cambridge, MA, October.

14 Living over Water
Introduction to Ecotourism Development Concerns for the Tonle Sap, Cambodia, and Lake Titicaca, Peru

Robert L. France and Evi Syariffudin

CONTENTS

Introduction..227
Tonle Sap Great Lake...228
Lake Titicaca...272
Conclusion 293
References..293

INTRODUCTION

Wetlands and shallow-water lakes have always served as convenient refugia for peoples fleeing persecution. The relief image in the British Museum of cowering Akkadians hiding from Assyrian warmongering despots amidst the tall marsh grasses in southern Mesopotamia (Young 1977; France 2007) is a scene oft repeated throughout history in many places around the world besides Saddam's modern Iraq (nowhere more famously perhaps than the first settlements in the Venetian lagoon; France 2008, 2010). Indeed, two particular examples—the Tonle Sap Great Lake in Cambodia and Lake Titicaca in Peru—bear striking resemblance to the situation in Iraq (e.g., overwater living, heritage water body of national importance and worldwide ecological uniqueness, surrounding archaeological ruins of global historical significance, burgeoning tourism industries, and rampant development) and can perhaps offer insight into what future concerns might be worthy of attention once societal peace and environmental restoration are achieved in Iraq. Ecotourism is recognized as a means to conserve the environment, educate and support local populations, and create an awareness of fragile and precious ecosystems in the rest of the world (Liu and Syariffudin 2003). However, the major challenge is to initiate such a program in a region such as Iraq (and also Cambodia and Peru) where a strong foundation of environmental protection has not been established. The present chapter thus serves to briefly introduce these two case studies in a hope of encouraging future, detailed research on the interesting parallels between them and what could evolve in southern Iraq.

TONLE SAP GREAT LAKE

Due to its unique hydrological phenomena, the Tonle Sap Great Lake in Cambodia is "a real natural wonder of the world" (Anonymous 2004). During the dry season it is 150 km in length and on average 20 km wide, occupying a surface area of 2,500 km^2. In May, the "wonder of nature" begins when meltwater from the Himalayas flows down the Mekong River coincident with the spring monsoons. So great is this volume of water that the 120 km Tonle Sap River, which normally drains the lake into the Mekong at Phnom Penh, begins to back up and reverse its flow. The resulting inundation of the Great Lake causes it to swell to 250 km in length and 100 km wide, with an area of over 13,000 km^2, a change in volume from 5 to 80 billion m^3 and representing a fourfold increase in area, making it the largest lake in Southeast Asia (Figure 14.1). Twenty percent of the Mekong River's floodwaters are absorbed by the Tonle Sap (Asian Development Bank [ADB] 2005b, 2005c). In this regard, as well as in providing an extremely important fish-breeding ground, the lake's importance transcends Cambodia (Anonymous 2004). Dry-season water depths between 1 and 2 m become 10 m at this time, causing major flooding of the forests, sometimes to distances of over 50 km from their dry-season shorelines. Trees begin to drop their leaves, and the submerged forest becomes prime habitat for fish and birds (Figure 14.2). In November, the flow reverts back to normal, leaving behind thousands of square kilometers of fertile sediment which become important for rice planting (Figure 14.3). These floodplains supply 12 percent of the total rice production in Cambodia (ADB 2005b).

This environment supports one of the largest freshwater fisheries in the world. In fact, the fishery ranks first in the world for productivity and fourth for total catch (ADB 2005c). Indeed, at the peak of the fish migration, an unbelievable 34 tons, or about 3 million individual fish, are caught each hour in bag nets (ADB 2006). Cambodia has one of the world's highest per capita consumption rates of fish, about half of which comes from the Tonle Sap. Over 15 percent of the country's population depends on the lake for their livelihood. During the wet season, large regions (some up to 300 km in size) of the inundated forests are cordoned off with bamboo and netting fences to entrap and contain fishes (Figure 14.4). Over two hundred species of fish inhabit the lake, including the rare Mekong River Catfish, which, at 3 m in length and 300 kg, is the world's largest freshwater fish (Poole 2005).

The Tonle Sap Lake was, like the Mesopotamian marshes, situated at the heart of one of history's greatest civilizations. The Khmer Empire and its predecessors settled around the lake, most particularly in the area to the northwest near the modern-day city of Siem Reap, and from 100 BCE to 1450 CE built some of history's most spectacular structures (Figure 14.5). Angkor Wat is one example, which though the most famous, is really only one of a vast complex of temples in what is the world's

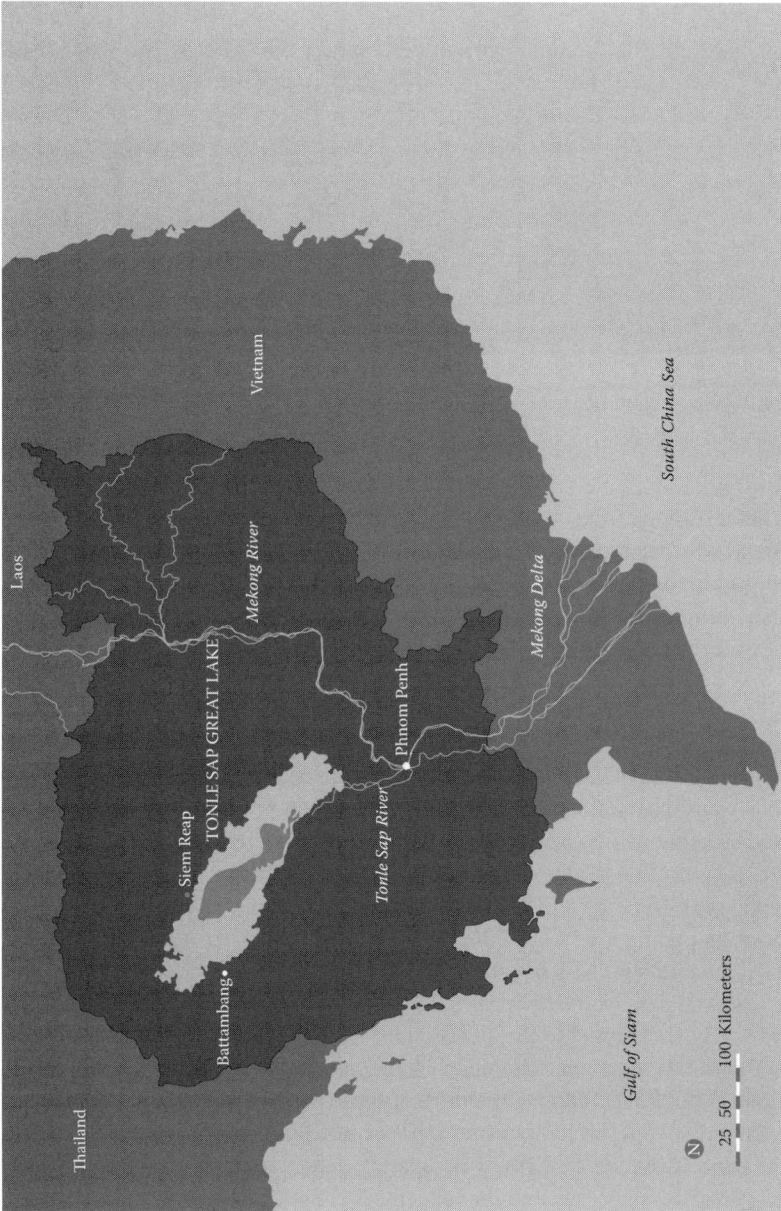

FIGURE 14.1 The Tonle Sap Great Lake increases its surface area fourfold when water backs up the Tonle Sap River from the Mekong River at Phnom Penh.

FIGURE 14.2   The submerged forest in Tonle Sap Lake.

largest concentration of religious buildings. Since 1992, the ruins have been managed within the 40 km² Angkor Archaeological Park, a World Heritage cultural site.

The Khmer Empire, like those in ancient Mesopotamia, was based on massive hydroengineering projects. Angkor has been referred to as a "hydraulic city" due to the dominating presence in the landscape of centralized storage reservoirs (Figure 14.6) which were also used for transport, aquaculture, and ritual. The multifunctional water management scheme also included many kilometers of canals (for transport and drainage) such as the one that leads from the Angkor ruins through Siem Reap to the lake and that today is referred to as the Siem Reap River. And, as in southern Mesopotamia, there is evidence that anthropogenically produced environmental problems may have contributed to a collapse in the Khmer Empire, in this case not due to overirrigation and salinization (see chapter 21) but to deforestation, erosion, and siltation, which progressively destroyed the water management system.

With an area of almost 77,000 km² (Anonymous 2004), the Tonle Sap drainage basin extends over 44 percent of Cambodia's land area (Figure 14.1) and is home to 32 percent of its population (ADB 2005c). No wonder, then, that the Great Lake is, like the Iraqi marshlands (Al-Khayoun 2007) and Mexico's Xochimilco (chapter 11), an inseparable and deep-rooted part of the very cultural identity of the nation.

The most striking parallel to the situation in the marshlands of southern Iraq are the tens of thousands of people living in what are referred to as "floating villages" on the Tonle Sap Great Lake. At one time most of these people were ethnic Vietnamese who, like the marshland dwellers in Iraq (Nicholson 2007), were regarded with hostility by the majority and referred to by derogatory names (Palmer and Martin 2005). There are over 150 villages around the lake, most consisting of a mixture of houses built on pontoons made of bamboo mats, metal drums, or boat hulls which are flush with the water level of the lake during the wet season but stand up on their stilts when levels recede (Figure 14.7). Several of these houses are hooked up to communication and television towers and have outside pens for keeping livestock and

FIGURE 14.3   Rice fields in the floodplains surrounding Tonle Sap Lake.

fish or crocodiles which they sell commercially (Figure 14.8). Otherwise, services are extremely limited, with refuse and sanitary waste discharged near where drinking water is obtained. Fishing is the major source of income for most of the villagers. These villages include many of the attributes common in normal Cambodian communities, such as brightly colored shops and markets, schools (in this case with fenced-in playing "fields"), pagodas or churches, casual restaurants, and dentists and barbers (Figure 14.9). And as was once the case in the Iraqi marshlands (Fulanain 1928), there is a floating economy of traveling merchants and craftsmen who make their way from village to village peddling their goods and services.

FIGURE 14.4   Bamboo and net fencing to contain fishes in the flooded forest. (*Continued*)

FIGURE 14.4 (*Continued*) Bamboo and net fencing to contain fishes in the flooded forest.

One difference that these overwater dwellers have from their marshland brethren, however, is their need for mobility in relation to the hydrological idiosyncrasy of the Tonle Sap. Most of the communities are situated near the mouths of inflowing streams, but during the dry season many of the fishermen and their families leave their stilted homes behind and use their houseboats (Figure 14.10) to move out and moor on the lake in order to be closer to their fishing areas, often deep within the flooded forests.

FIGURE 14.5 Khmer ruins and carvings surround the lake.

FIGURE 14.5 (*Continued*) Khmer ruins and carvings surround the lake.

FIGURE 14.6 The hydrologic city of ancient Angkor: small, medium, and large water reservoirs, the latter today given over to recreational boating. (*Continued*)

FIGURE 14.6 (*Continued*) The hydrologic city of ancient Angkor: small, medium, and large water reservoirs, the latter today given over to recreational boating.

FIGURE 14.7   Village homes are left dozens of meters in the air once the flood levels recede following the drainage of the lake.   (*Continued*)

FIGURE 14.7   (*Continued*) Village homes are left dozens of meters in the air once the flood levels recede following the drainage of the lake.

Tourism is growing at a rapid rate around Tonle Sap, today at levels approaching over half a million visitors a year, up from numbers of about three hundred thousand only half a decade ago. Most of this is related to the presence of the Angkor ruins, which are truly one of the most revered archaeological sites in the world. Increasingly, however, "overruined" tourists are seeking a break from staring at fellow tourists staring at bas-reliefs and are venturing out onto the lake to observe what some call "the people of the lake … a communion between man and nature" (Bailleux 2003). Unfortunately, for many this means nothing more than hiring a "tuk-tuk" or motorbike taxi for the short ride south from Siem Reap to the village of Chong Khneas situated at the base of Phnom Krom (Figure 14.11).

Here the squalor (Figure 14.12) must come as quite a shock to those tourists staying in thousand-dollar-a-night hotel rooms located only 15 km away. Although it is possible to see many of the characteristic features of the floating villages, proximity to the major tourist center of Siem Reap means that Chong Khneas is not really a typical fishing village. For the more adventurous and those with more time, it is possible to travel from the "Angkor tourist scene" via road and hired boat to visit more distant villages. Here tourists are few and it is still possible to catch a glimpse of lake life unadulterated by artificiality or tokenism (Figure 14.13). As is the situation in the floating villages of Lake Titicaca described below, such a visit on the Tonle Sap is probably the closest that one can come to experiencing what life might have once been like in the Iraqi marshlands. People paddling about on simple watercraft, smiling children swimming, and fishermen or -women going about their daily lives (Figure 14.14) all call to mind photos of the same from Iraq (e.g., Wheeler in Young 1977; France 2007).

FIGURE 14.8 The "floating villages" contain television and communication towers and holding pens for raising livestock as well as crocodiles and fish.   **(Continued)**

FIGURE 14.8 (*Continued*) The "floating villages" contain television and communication towers and holding pens for raising livestock as well as crocodiles and fish.

FIGURE 14.9 Villages also contain shrines, community meeting rooms, schools, recreation ball courts, small shops, markets, and local businesses such as dentists and barbers.

(Continued)

FIGURE 14.9 (*Continued*) Villages also contain shrines, community meeting rooms, schools, recreation ball courts, small shops, markets, and local businesses such as dentists and barbers.

FIGURE 14.9 (*Continued*) Villages also contain shrines, community meeting rooms, schools, recreation ball courts, small shops, markets, and local businesses such as dentists and barbers.

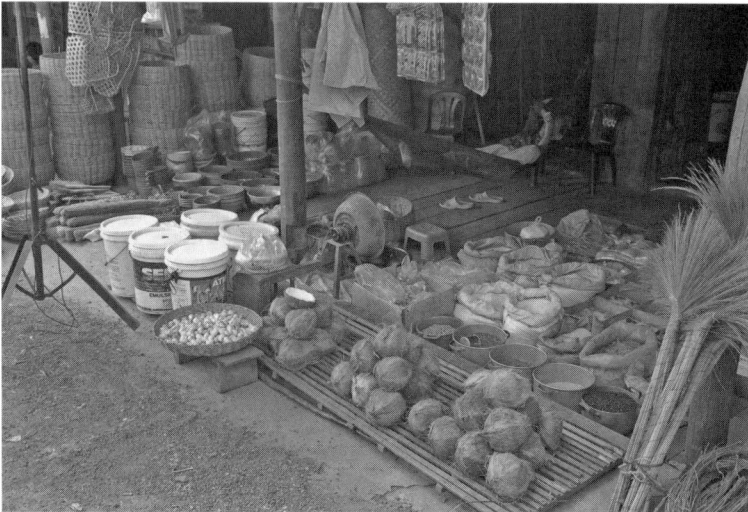

FIGURE 14.9 (*Continued*) Villages also contain shrines, community meeting rooms, schools, recreation ball courts, small shops, markets, and local businesses such as dentists and barbers.

FIGURE 14.9 (*Continued*) Villages also contain shrines, community meeting rooms, schools, recreation ball courts, small shops, markets, and local businesses such as dentists and barbers.

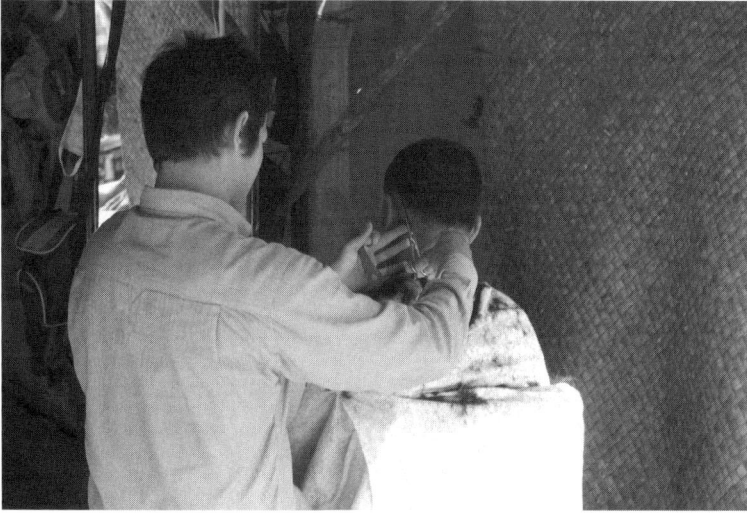

FIGURE 14.9 (*Continued*) Villages also contain shrines, community meeting rooms, schools, recreation ball courts, small shops, markets, and local businesses such as dentists and barbers.

The sad reality is that within Southeast Asia, Cambodia has the lowest GNP, the least infrastructure (particularly roads), the lowest life expectancy, and the highest rate of malnutrition (Liu and Syariffudin 2003). The situation is even bleaker when it is recognized that the floating villages on the Tonle Sap are among the poorest communities in the entire country (ADB 2005a). People here have little education, few livelihood options, no private land, and ironically, despite living over water, no access to safe drinking water. Diarrhea and other waterborne diseases due to lack of sanitation contribute to a situation in which the Tonle Sap region has an extremely high infant mortality rate (Poole 2005) of 124 per 1,000 live births (ADB 2005b).

The environmental integrity of the lake and its watershed is seriously compromised, leading some to state that "Cambodia faces environmental disaster if the Tonle Sap ecosystem is degraded further" (ADB 2004). Cambodia has lost one-fifth of its forests in the last two decades due to illegal logging (Palmer and Martin 2005). The inundated forests around the Tonle Sap are particularly vulnerable in this regard due to their being cut for agricultural expansion (the floodplain is one of the most productive areas in the country for growing rice) and for firewood (less than 5 percent of the rural homes in the area have electricity or gas). One-half of these floodplain forests have been lost (Evans, Marschke, and Paudyal 2004).

Loss of forest habitat has contributed to Cambodia losing thirty-eight species of birds since 1960 (Poole 2005). Biodiversity reduction is also related to the lingering presence of a lucrative trade in wildlife. Egg poachers continue to threaten waterbird populations, and the lake experiences the largest snake harvest in the world (Bailleux 2003). Today many of the species shown on the famous Churning of the Ocean of Milk carvings at the Bayon Temple and other Angkor ruins (Figure 14.15) have been locally extirpated (Poole 2005).

FIGURE 14.10   During the dry season, many fishing families leave the villages and go to the fishing grounds in their houseboats.   (*Continued*)

FIGURE 14.10 (*Continued*) During the dry season, many fishing families leave the villages and go to the fishing grounds in their houseboats.

Aquatic biodiversity is also threatened. Early accounts from the nineteenth century of the first Westerners traveling to Angkor describe fish being so abundant as to actually hinder the passage of boats. Fishing today is big business, with large concession areas given to wealthy businessmen at the expense of locals. Poole (2005) considers the need for the establishment of fish sanctuaries and research on sustainable fishing to be far more important a concern than issues related to deforestation since up to a half of the total animal protein intake for the entire country comes from the Tonle Sap.

Until recently, children could be seen fishing in moats in front of temples whose bas-reliefs (Figure 14.16) show similar species and harvest technologies being used eight hundred years previously. Today, however, the Asian Pacific Self-Development and Residential Association (APSARA), the organization responsible for management of the Angkor site, has been relocating people outside the park. Poole (2005) believes that the area could be better managed as a site of *living* history rather than as a restricted-access museum of the past. Such divergent opinions are a common issue of contention in the management of natural areas inhabited by indigenous people or longtime residents (see chapters 5, 11, 12, 15, and 16) and are of obvious importance to determining the future of the Iraqi marshlands.

After a month-long visit to the Tonle Sap and discussions with many development aid providers and conservation biologists working there, my take-away opinion is that rarely have I seen a location in more desperate need of immediate and comprehensively effective land-use planning on the scale and type discussed by Steinitz in chapter 4 and applied in chapters 5 and 15. The degree and speed of rampant hotel development in Siem Reap, much of it by foreign Asian investors, are shocking and seemingly unregulated in terms of location and ecological footprint. In a city in

FIGURE 14.11   Chong Khneas, the closest "floating village" to the tourist city of Siem Reap.   (*Continued*)

FIGURE 14.11 (*Continued*) Chong Khneas, the closest "floating village" to the tourist city of Siem Reap.

FIGURE 14.12   Ramshackle, tiny homes in Chong Khneas indicate the harsh living conditions there.

FIGURE 14.13   Two of the other floating villages located around Tonle Sap Lake.

FIGURE 14.14   Scenes of village life in the outlying communities around Tonle Sap Lake closely resemble those in the Iraqi marshlands before their drainage.   *(Continued)*

FIGURE 14.14 (*Continued*) Scenes of village life in the outlying communities around Tonle Sap Lake closely resemble those in the Iraqi marshlands before their drainage.

FIGURE 14.14 (*Continued*) Scenes of village life in the outlying communities around Tonle Sap Lake closely resemble those in the Iraqi marshlands before their drainage.

FIGURE 14.14 (*Continued*) Scenes of village life in the outlying communities around Tonle Sap Lake closely resemble those in the Iraqi marshlands before their drainage.

FIGURE 14.15 Local wildlife carved on the Khmer temples show threatened or extirpated species, such as the giant Mekong River catfish, the largest freshwater fish in the world.

FIGURE 14.16 Fishes shown on the temple carvings are still caught by locals using the same methods.

which most sewage is still not treated, it is inescapable to reach the conclusion that the absence of regulation of development is due to the well-known and widespread governmental corruption. Questions about carrying capacity, an important step in watershed development planning (France 2006), seem (if asked at all) to be conveniently ignored. In the meantime, rampant high-end tourism continues with the construction of luxury hotels, tacky theme parks (Figure 14.17), and ostentatious golf course resorts. One fears that the arrival of the casinos and the Las Vegas–ization of Siem Reap may only be just around the corner.

And there may be more serious threats on the horizon. The recent discovery of extensive offshore oil reserves will transform Cambodia (Mcleod 2006). Most of the revenue is expected to go to one of the world's most corrupt governments and help to create a select wealthy upper class (i.e., the "oil curse" paradox, where a massive influx of oil money causes economic stagnation and increased poverty, as in Nigeria). And it is these people who will become increasingly involved in developing Siem Reap as their history–nature–pleasure park. The situation looks bleak given that oil reserves have also been found around the Tonle Sap itself.

All this pales, however, in relation to what is in the works upstream. Just as the long-term future of the Iraqi marshlands is in the hands of the water managers in Turkey, so too does the ultimate survival of the Tonle Sap lie with China's hydroengineers. Plans exist for a series of hydroelectrical dams on Mekong River which would reduce the high-water floods and thus impact fish migration and production (Bailleux 2003; Varis, Kummu, and Keskinen 2006). It is feared that China, with its long history of meddling in Southeast Asia, can bribe the Cambodian government to turn a blind eye to the projected widespread environmental and sociological ramifications of altering the hydrological regime of the Tonle Sap. In contrast, Turkey, given its desire for future EU membership, might be able to be strong-armed into

FIGURE 14.17   Rich and timid tourists in Siem Reap can visit a replica floating village sanitized of unsavory elements characteristic of the real places (such as people).

loosening its tight grip on Iraq's water resources (European Parliament Member E. Nicholson, personal communication).

It is not all bad news by any means, however. In recognition that the Tonle Sap Great Lake is "one of the most significant environmental sites in Asia and the world" (ADB 2005a), the area was designated a UNESCO Biosphere Reserve based on its international importance for biodiversity and in containing the most productive floodplains found anywhere (ADB 2005c). Biosphere reserves are nominated by national governments and remain under their jurisdiction, and must meet a minimal set of criteria and adhere to a minimal set of conditions before being admitted to the UNESCO network (ADB 2002): "Each reserve is intended to fulfill three complementary functions: (i) a conservation function (preserve landscapes, ecosystems, species, and genetic variation); (ii) a development function (foster sustainable economic and human development); and (iii) a logistic function (support demonstration projects, environmental education and training, and research and monitoring related to local, national, and global issues of conservation and sustainable development)." The Tonle Sap Biosphere Reserve (TSBR) covers the entire lake and a significant portion of the floodplain and contains three management typologies: core zones, which are securely protected sites; a clearly identified buffer zone; and a flexible transition zone (ADB 2002; Anonymous 2004).

The core areas are for conserving biodiversity, monitoring minimally disturbed ecosystems, and undertaking nondestructive research and other low-impact uses such as education (ADB 2002). There are three such areas in the TSBR, two identified for bird-feeding areas and a unique canopy forest, but only the Prek Toal Core Area and its bird colonies are protected to any real degree (Anonymous 2004). The buffer zone usually surrounds or adjoins core areas, and is used for cooperative activities deemed compatible with sound ecological practices, such as environmental education, recreation, ecotourism, and research (ADB 2002). In the TSBR, this is an area of about 540,000 ha. Transition zones are areas in which existing stakeholders work together in a suite of economic and other activities designed to manage and sustainably develop the biosphere's natural resources (ADB 2002). For the TSBR, this is an area of about 900,000 ha.

The Prek Toal Bird Sanctuary, located in the northwest corner of the lake, is the last haven for large waterbirds in Southeast Asia (Bailleux 2003; Evans, Marschke, and Paudyal 2005; Goes 2005) and includes endangered species such as spot-billed pelicans and greater adjutant storks. Several companies operate tours to the Prek Toal Core Area. Cars pick up tourists from their hotels in the early morning and drive them to the harbor at Chong Kneas, where they embark on a boat that crosses the lake and enters the flooded forest of the reserve in time for the sunrise (Figure 14.18). There, the small boats make their way inside the labyrinthine waterways among the flooded forest to reach several observation towers from which the colonies of massive pelicans and storks can be seen atop their gray, guano-stained rookeries. It is also not uncommon to come across monkeys jumping from tree to tree as well as other wildlife such as large otters that resemble Midj, the famous one-time resident of the Iraqi marshes (Maxwell 1967).

FIGURE 14.18   A typical tree in the Prek Toal Bird Sanctuary that might house colonies of a variety of avian species.

Early planning (Nedeco and Midas 1998) suggested that once ecotourism is running at full scale in the Prek Tol Core Area (seventy-six tourists a day), an estimated gross income of more than a million U.S. dollars could be generated if an initial investment of over $100,000 was provided. Liu and Syariffudin (2003) later posited how the construction and maintenance of sustainable buildings in the community of Prek Tol (for health care, education, etc.) would be a sustainable source of income for local villagers and could be used to fund similar infrastructure projects in the surrounding villages. Today, the arrival of ecotourists to view the waterbird colonies has made the community of Prek Toal one of the most affluent floating villages on the lake. Proving that nature conservation can be linked to economic development through responsible ecotourism (see chapter 16), outside investment has built schools, a health center, and a pagoda. Tour groups often stop for breakfast or to pick up supplies in the village on their way to the bird colonies and can even spend a night at the visitor center for a true immersion experience (Figure 14.19).

In general, ecotourism may not be beneficial to either the environment or the local population, especially in developing nations, if it results in the exploitation of the local people, local culture, and natural resources, and/or the local peoples' exploitation of their own culture and resources (Liu and Syariffudin 2003; see also chapter 5). It is not unheard of, for example, that the impact of international organizations such as UNESCO and well-meaning nongovernmental organizations (NGOs) charged with enforcing regulations and implementing management practices has resulted in animosity and suspicion in the locals, whose cooperation may be merely superficial in nature, thereby driving a vicious cycle of continued exploitation of the land. Several observers offer cautionary notes to the seeming success of the Prek Tol story. Poole (2005) believes that the Prek Toal Core Area survives not due to park management but because it lies within some of the lake's most important fishing lots, which are closely managed

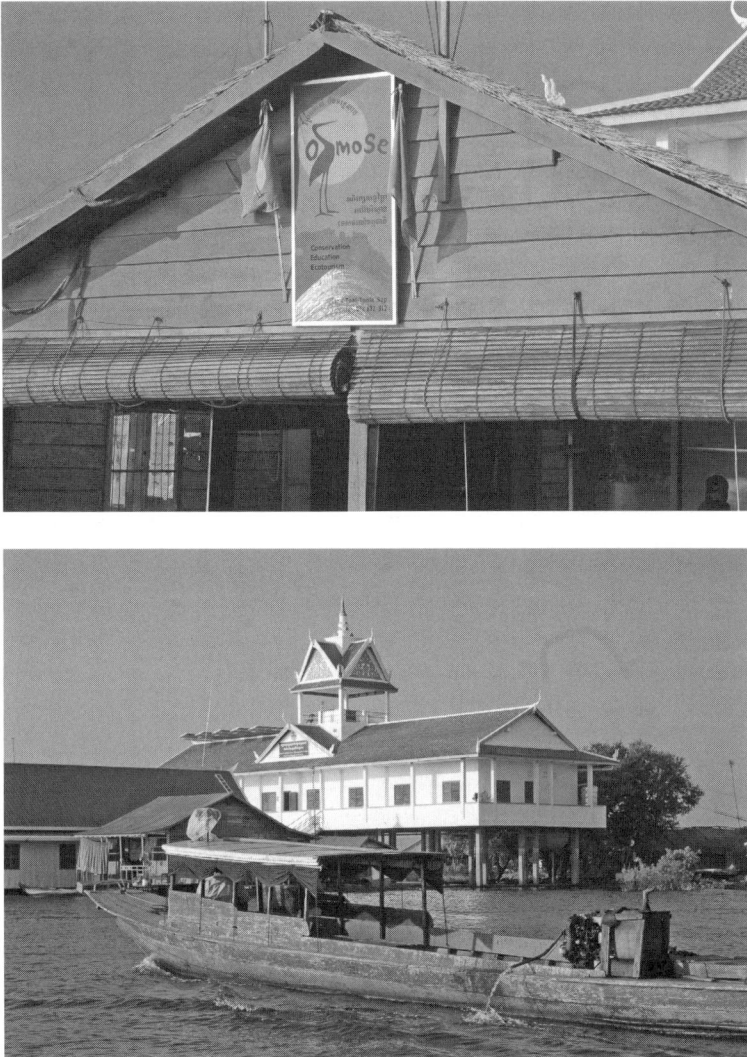

FIGURE 14.19   Ecotourism visitor center and park management office in Prek Toal village.

by their concession owners. In this respect, the most useful environmental management lesson emerging from Prek Toal is that fishing productivity can now be clearly shown to be dependent on the extent and quality of the flooded forest (Poole 2005). In the long term, commercial fishing as currently practiced may not be compatible with the protection of biodiversity, leading Poole (2005) to suggest that the area should eventually be managed as a fish sanctuary rather than as a consignment fishing lot.

To support his argument that industry needn't be seen as ipso facto disastrous to nature conservation, Poole (2005) offers the example of the Kulen protected area located on the outer boundaries of the Angkor archaeological site. Here, just like in many locations of Cambodia, although much foreign assistance was given to establish the reserve in association with the Ministry of the Environment, neither the government nor the donors provided the level of political or financial support needed to sustain management. As a result, boundaries were never legally demarcated and many of those living within actually have no idea of the area's "protected" status. Poole (2005) goes further when he states that better forest and healthier wildlife populations in Cambodia are more often found in logging concessions than in adjacent protected areas as a result of the timber barons having the political and financial muscle necessary to protect their investments, whereas the various environment ministries are marginalized and neutered in the corrupt centralized government.

Obviously, the situation is a complicated one. Poole (2005) offers the following balanced synopsis about the benefits and challenges in ecotourism that is worth quoting in its entirety:

> "Ecotourism" represents a great hope, but also a great threat, to the future of the Prek Toal Core Area. In a country like Cambodia, it is often offered as the solution to a myriad of conservation problems. The belief is that wealthy foreign tourists will provide long-term financial incentives to prevent short-term exploitation. In principle this sounds great, in practice in Cambodia, as in many developing countries, it is very difficult.
>
> In the context of Prek Toal, how to ensure that the poorest—the former bird collectors—benefit most? If bird collectors don't benefit, why should they stop hunting? However, those who gain first from the tourists are the shop keepers, restaurateurs and boat owners, already some of the wealthiest in the community. In situations where levels of governance are often poor, how to be certain that taxes on tourists do not end up in the wrong pockets? And in many areas of Cambodia, will foreign tourists ever generate sufficient income for ecotourism to be a viable alternative to logging?
>
> If ecotourism can work anywhere in Cambodia, it is in Prek Toal. Just a short boat trip across the Tonle Sap from Angkor, hundreds of thousands of tourists a year are only a day's excursion away from one of the world's most important waterbird colonies. During December and January, the peak tourist season, the colonies are easily accessible and represent an intense wildlife experience.
>
> Within the local community, the success of the conservation project to date provides a great foundation. Bird collectors turned conservation rangers are now being trained as guides. They may not be able to speak foreign languages, but who better to lead people through the maze of the swamp forest? And which villagers know more about the birds than those who used to live in the forest hunting them? …
>
> There is a real potential for a tourism boom in Prek Toal but conservation management systems are urgently required. Regulations need to define quotas for the number of boats and tourists, strict no-access zones that change with the season and the locations of specific colonies, designated access routes for boats and appropriate observation sites. Without these minimum requirements, tourists risk inadvertently disturbing and destroying the very thing they are coming to see.

Cambodia today, on the path of slow recovery from its volatile past (see epilogue, this volume), has another problem that one might imagine will also be part of the future cultural landscape of Iraq: groups of armed military forces concerned about "security issues." The Prek Toal Core Area came up with an ingenious way to defuse what might have become a severely limiting problem (Evans, Marschke, and Paudyal 2005). Local military commanders "concerned" about the security of tourists visiting the area proposed putting an armed guard on every boat for every trip, for which the soldier would, of course, be compensated. Because there have never been any security issues in the area, the statement was a thinly veiled threat disguising the fact that the military simply wanted their share of the $15–20 entrance fee that tourists have to pay. To the managers of the core area, the revenue loss was secondary to the unwanted presence of AK-47-toting uniformed guards on a boat with tourists who were going bird watching. The solution was to turn the problem into a conservation asset. A counterproposal was made that one officer from each military unit would join the conservation team and be paid the same rate as the other rangers. The core area management team also insisted on the following provisos: that the Environmental Station would select the officers from lake communities, and that the recruited officers must undertake the training and perform all tasks assigned to rangers. The benefits to conservation were that now the armed officers, who have a law enforcement authority, can be part of the management of the area and arrest any poachers they might come across (Evans, Marschke, and Paudyal 2005).

If the rampant development of Siem Reap is the first thing that strikes the environmentally minded tourist to the Tonle Sap, the rich assemblage of NGOs and other charity-related groups is a close second. Osmose, with a mandate of linking community-based conservation with ecotourism and community education, organizes tours to Prek Toal as well as educates villagers, turning one-time poachers into future rangers. They are housed in the same building as the Sam Veasna Center for Wildlife Conservation, an organization that promotes local wildlife conservation for both locals and foreigners. Another group, the Angkor Centre for Conservation of Biodiversity, runs a visitor center that rehabilitates animals rescued from the illegal wildlife trade and educates about biodiversity conservation (Figure 14.20). Krousar Thmy is a foundation that looks after deprived children and attempts to revive Cambodia's lost traditions; it runs the Tonle Sap Exhibition, an information center about people and the lake (Figure 14.21).

Development, in its widest definition, aims at improving standards of living. Many have criticized the superficiality of measuring developmental success with only economic yardsticks. Indeed, monetary wealth is only a small part of societal well-being (Abbott 2008, 2009), which is found in other benchmarks, including social and environmental health (Abbott 2005), education, security, and even the availability of recreational facilities (Adams 2005). Nevertheless, development cannot be divorced from financial considerations. It is not so much a matter of greed or misplaced priorities, but rather that economic wealth is often believed to be the surest guarantee of freedom and choice, as well as the ability to obtain other less tangible but more significant goals. Developing countries are at a disadvantage in that they are relegated to perpetually playing catch-up in terms of learning new technologies and knowledge constantly upgraded by other more developed countries. The benefit of not having experienced rampant industrialization, however, is that these countries have not been so quickly swept into the arena of globalization. As such, they have been able to maintain their unique geography and culture that are so attractive to tourists from

FIGURE 14.20   The Angkor Centre for Conservation of Biodiversity rehabilitates local birds, mammals, and reptiles rescued from illegal trade.   (*Continued*)

FIGURE 14.20 (*Continued*) The Angkor Centre for Conservation of Biodiversity rehabilitates local birds, mammals, and reptiles rescued from illegal trade.

cosmopolitan, yet homogenized urban centers around the unifying globe. Tourists are attracted to these idiosyncratic locations for various reasons: a natural and idyllic escape from their concrete jungles back at home, a feeling of a sense of connection with past civilizations, and a yearning to learn about unique cultures and ecologies. In the context of developing countries, ecotourism often takes place in communities where livelihood and survival are very real and dire concerns (Stevens 1995; Butler and Hinch 1996). Alternate sources of sustenance, namely agriculture and fishing/hunting, are a big part of the equation and cannot be separated from discussions of ecotourism (see chapters 11, 15, and 16 and the epilogue). Likewise, geographical idiosyncrasies, historical sensitivities, cultural diversities, and economic ambitions all need to be included in a thorough analysis of a site.

Chong Khneas is a floating community located 12 km south of Siem Reap, the tourist hub and fastest growing city in Cambodia, and is connected to the city by an all-weather, asphalt road. Chong Khneas is made up of seven small villages totaling about seven hundred households of five thousand people. Six of the seven villages are composed of assemblages of houseboats that move accordingly in relation to the seasonal flood cycle. During the wet season, the villages cluster around the base of Phnom Kraom, an isolated rocky outcrop rising 140 m above the otherwise flat terrain. During the dry season, the floating villages anchor in a small nearby inlet close to the fishing grounds. The remaining village lies alongside the road embankment. Port facilities do not currently exist at Chong Khneas because the shoreline moves as much as 5 km. An earthen road extends some 4 km south from Phnom Kraom alongside of which straddle many homes (Figure 14.11). A navigation channel, about 6 km long, runs alongside the road and connects to the small inlet mentioned above. Various parts of the road are submerged during seasonal flooding, and a temporary

FIGURE 14.21   The Tonle Sap Exhibition educates about the social ecology of the lake.

(Continued)

FIGURE 14.21 (*Continued*) The Tonle Sap Exhibition educates about the social ecology of the lake.

boat landing is moved along accordingly. When the lake is at its highest level, Phnom Kraom is accessible by fishing, cargo, and passenger boats even though the road is mostly submerged. During the dry season, the channel becomes nonnavigable for most of its length and is chaotically congested with fishing boats, floating houses, and tourist boats. Present also are medium-sized passenger boats (used by both locals and tourists) that travel between Chong Khneas and both Phnom Penh (the capital of the country) and Battambang (the industrial and second-largest city in Cambodia), each about a six-hour journey away. Chong Khneas is a tourist attraction given its proximity to Siem Reap and the Angkor ruins, and the UNESCO Biosphere Reserve and bird sanctuary, and also because of its novelty as a floating village.

Chong Khneas has been the busy shipping port for the Angkor area since Khmer times until the recent construction of new roads effectively put an end to the commodity shipping to and from Phnom Penh. Today the impromptu port has two major activities (Poole 2005). One is as the major landing area for fish caught in the northern part of the lake. Here fishermen arrive from small communities to trade or sell their fish, the best of which goes straight to Thailand or even North America (fifty thousand tons are exported from Cambodia each year). The other major role of Chong Khneas is as the passenger terminal for foreign tourists arriving after the leisurely and atmospheric six-hour trip from Phnom Penh. As Poole (2005) stated, "With the growth of this sector, and the development of more luxurious ferries and cruise vessels, once again the lure of Angkor is creating a port that is a crossroad for international travelers."

In 2002, the Asian Development Bank was interested in providing a multimillion-dollar loan for the development of a modern harbor at Chong Khneas, the Chong Khneas Environmental Improvement Project (CKEIP). The phenomenal flooding patterns of the Tonle Sap have always impeded water transportation to and from the city of Siem Reap. The movement of the shoreline by more than 4 km, corresponding to water level fluctuations of as much as 10 m, has made a permanent port location impossible. The new project proposed to dredge a canal from the lake to a permanent harbor basin of constant accessibility that would be constructed at the foot of Phnom Kroam hill. Here the dredged material would be used in a land reclamation project upon which villagers will be resettled. Many, however, were concerned about the impacts that such a massive intervention would create (Syariffudin 2006), including some NGOs who were critical that it would accentuate environmental and social problems by encouraging an influx of poor immigrants. As Poole (2005) again cautions, "Unregulated development will bring with it greater risk to the Tonle Sap ecosystem and therefore to the people living on it. The current floating village way of life may not be with us for much longer."

The ADB proposal was suspended in 2003 due to the concerns about its environmental impacts. Nevertheless, the dire need for a port on the site, related not only to the increasing level of tourism but also to fishery and trade needs, can be neither overlooked nor denied. As a result, Syariffudin (2006) developed a conceptual design proposal for her thesis that picked up where the ADB proposal left off. Inherent in the architectural thesis project is the proposition that multiple efficiencies could be achieved in multilayering hybridization. The meshing of architecture, landscape, and infrastructure; the untraditional juxtaposition of different programs (uses); as well as the mediation of vastly different scales promise conceptual, spatial, and material frugality, a basis for efficiency much needed in development work. She believed that one could use such a methodology to stimulate cooperation among separate development projects within a community. Very often, these projects are developed by different organizations (governmental, private, or nonprofit) that do not communicate in terms of sharing resources and information that would otherwise minimize costs—not only monetary costs, but also time, social, and environmental opportunity costs.

Syariffudin (2006) countered some of the concerns about the original ADB proposal by attempting to reinvent what a modern floating village might look like in her development of conceptual designs for the architectural transformation of the Chong Khneas port area. The stilted houses here would be situated along the dry-weather dirt road, which is itself closely parallel to the river to allow for boat navigation (Figure 14.22). As in other floating villages, the basic building typologies (Syariffudin 2006) would be houses on stilts or floating houses (Figure 14.23). Access to the latter reflects an adaptive use of space according to the different seasons (Figure 14.24). Houses would be divided into several parts for easy moving and later add-ons, though sometimes what constitutes a "house" is simply a roof over a large boat (Figure 14.25). Patios are important not only as social or functional space (e.g., for washing, bathing, and processing fish) but also for circulation among proximal houses (Figure 14.26).

Syariffudin's (2006) thesis attempted to address the dual issue of landscape and infrastructure through the mutual benefiting of one from the other. The earth dug from the canal construction would be used as landfill to create the mounds upon which an ecological park (as in chapter 13) would be sited. In addition to educational facilities, Syariffudin envisioned a restored landscape which included flooded forests for waterfowl and a treatment wetland into which wastewater would be discharged for polishing (as in chapter 18). Other infrastructural elements such as bridges, retaining walls, and steps became opportunities to create badly needed habitable spaces into which the passenger port and a fish market (Figures 14.27 and 14.28), as well as a wastewater treatment plant and operators' housing, could be inserted. Syariffudin's designs are based on the proposition that multiple efficiencies could be achieved in multilayering hybridization (Figure 14.29) that meshes traditional and untraditional uses in close juxtaposition.

If at first these conceptual designs seem incongruous to the present situation, that is the whole point. Siem Reap's high-end luxury hotels exist cheek-and-jowl with, but completely ignore, the Cambodian housing (resembling slums to western eyes) situated about them. Syariffudin, a student from Southeast Asia herself, was interested in opening a dialogue about how modern infrastructure could provide support to the growing tourism industry of Siem Reap at the same time as not only considering but also actually improving the living situations of the Chong Khneas residents. In this light, tourism should neither be viewed with such extreme skepticism and contempt as to pass up the opportunity for economic, cultural, and social developments, nor be allowed to dominate a society to the point of exploiting local resources or undermining genuine cultural, environmental, and/or social progress.

100 200   500   1000 Meters

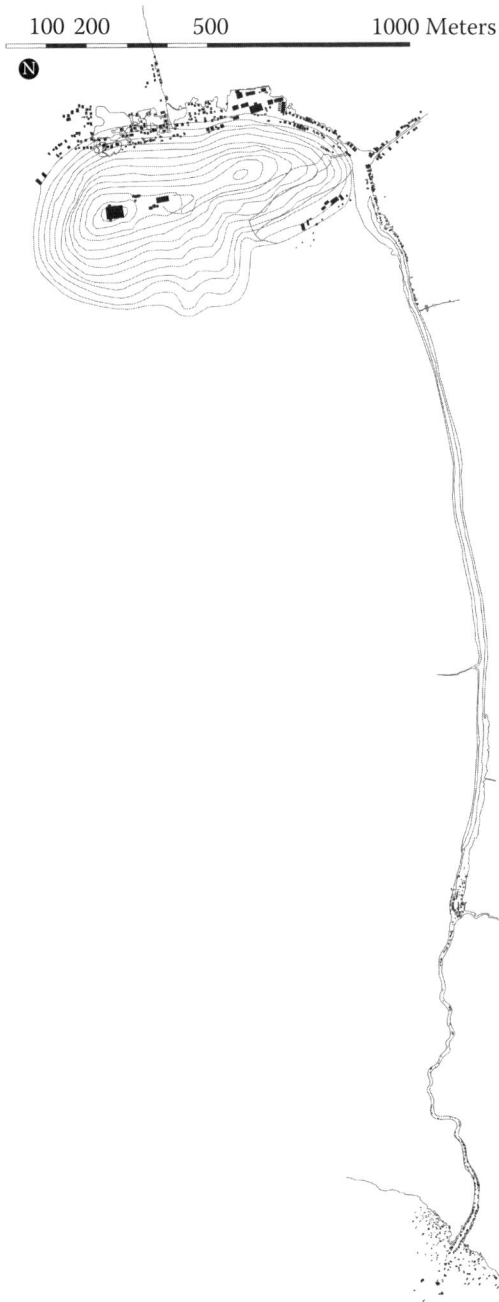

FIGURE 14.22   The road and navigation channel running south to the lake from the rocky outcrop of Phnom Kraom around which floating villages congregate.

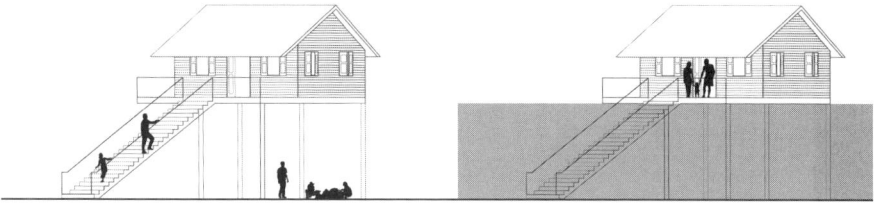

FIGURE 14.23 Schematic drawings of functionality of houses during low and high water levels.

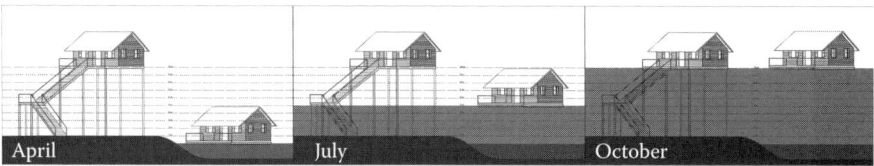

FIGURE 14.24 Water and land access from stilted and floating houses in relation to seasonal fluctuations in water level.

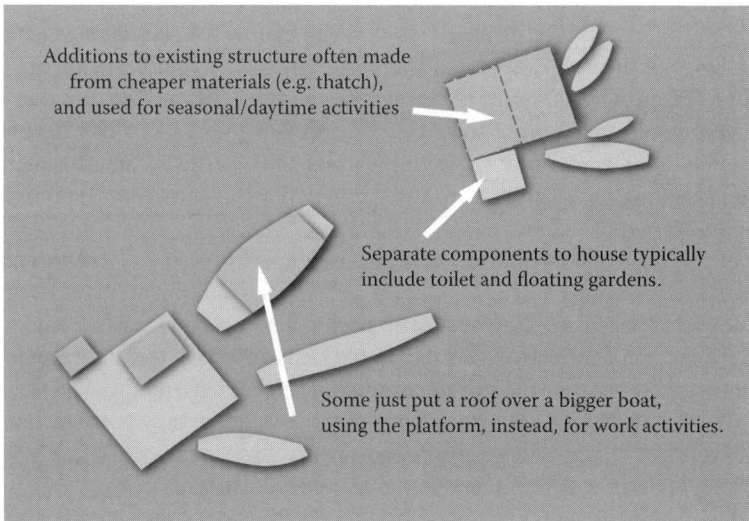

Additions to existing structure often made from cheaper materials (e.g. thatch), and used for seasonal/daytime activities

Separate components to house typically include toilet and floating gardens.

Some just put a roof over a bigger boat, using the platform, instead, for work activities.

FIGURE 14.25 Building typologies in Chong Khneas.

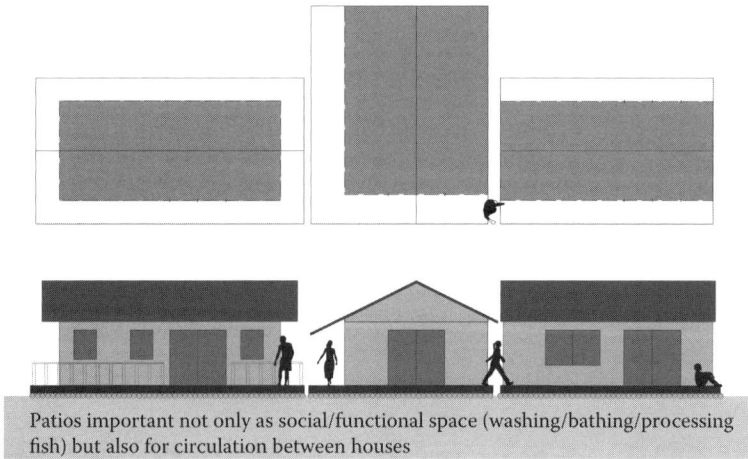

Patios important not only as social/functional space (washing/bathing/processing fish) but also for circulation between houses

FIGURE 14.26   Functional relationship of patios to buildings.

LAKE TITICACA

Lake Titicaca is located on the Andean Altiplano situated between Peru and Bolivia. The lake is about 200 km long and 65 km wide and covers a surface area of over 8,000 km². And at nearly 4,000 m in elevation, it is famously the world's highest navigable water body. Archaeological remains indicate several islands in the lake were inhabited an incredible ten thousand years ago, with over two thousand years of successive civilizations existing around the lake (Escandell-Tur 2003; Figure 14.30). Most significantly, the lake is integral to Andean culture and has been venerated as the birthplace of the first Incas (Stanish 2003). There is an important history of layered pilgrimage routes upon the landscape (Bauer 2001; Mayell 2001; Salles-Reese 1997). Mining has taken place since the Spanish Conquest, and today the lake and its shorelines are a center of intense commercial activities including fishing, alpaca and llama herding, production of crafts, and, increasingly, tourism (Jenkins 2003; Orlove 2002).

Today the major lakeside town of Puno is home to dozens of tourist agencies offering a plethora of packages to appeal to everyone from the affluent tour groups to independent backpackers, many of whom are on their way to or back from visiting Machu Picchu. Satiated with Peru's incredibly rich archaeological heritage, increasing numbers of foreigners are traveling out onto the lake in order to experience life on the islands. These outings offer contrasting views about how tourism can either support or impair indigenous culture, and in this respect provide lessons for how tourism might one day be regulated in the marshlands of Iraq.

Isla Taquile lies 30 or so km away from Puno at the edge of the central basin of the lake. Seven km long, the island rises like a ziggurat above the impossibly blue waters of the lake, with views of the mountains in Bolivia as a backdrop (Figure 14.31). The thermoregulatory effect of the large lake creates a microclimate favorable for agriculture. The twelve hundred residents work growing potatoes, oca, maize, and quinoa in terraces (Figure 14.32). These advantages have long been recognized as the island has been

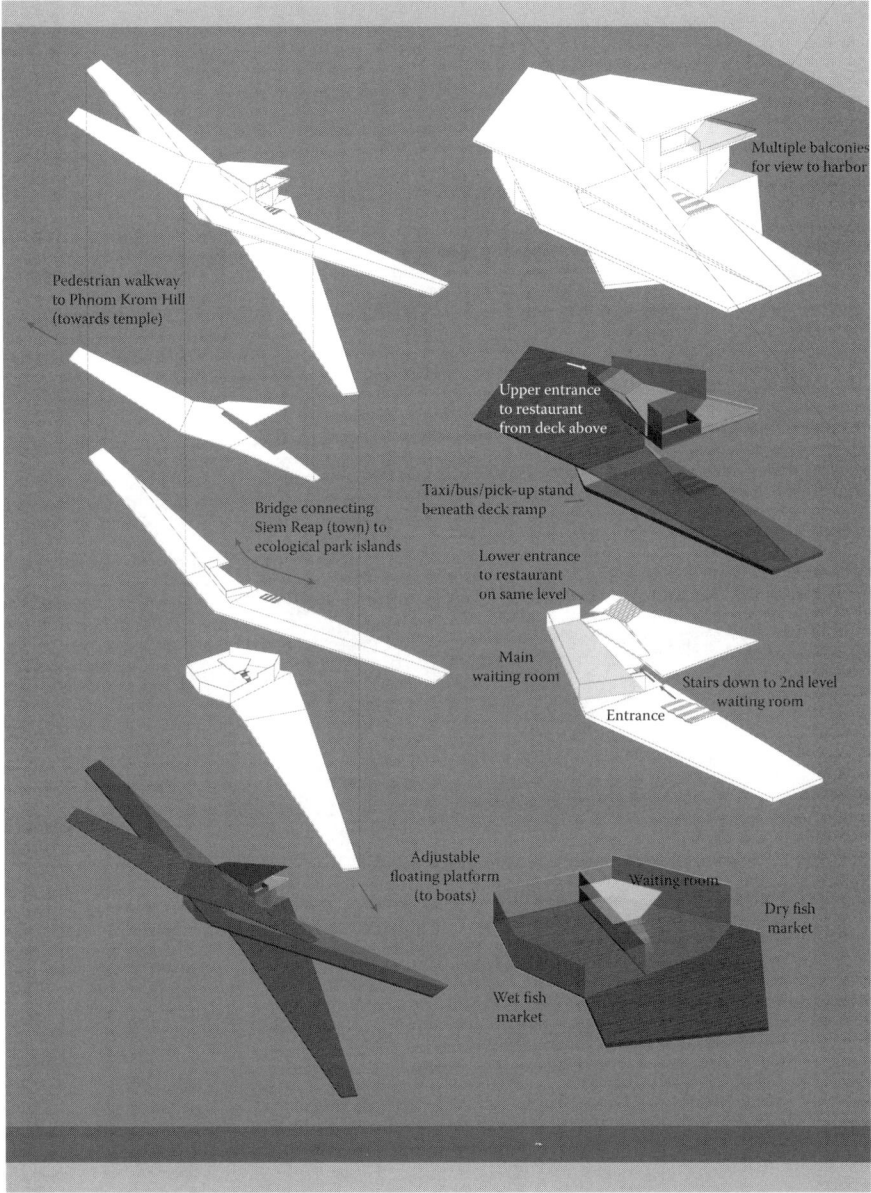

FIGURE 14.27 Conceptual designs of access for accommodating variable water levels for proposed passenger port and fish market.

FIGURE 14.28   Design concepts illustrating uses of proposed facilities.   (*Continued*)

FIGURE 14.28 (*Continued*) Design concepts illustrating uses of proposed facilities.

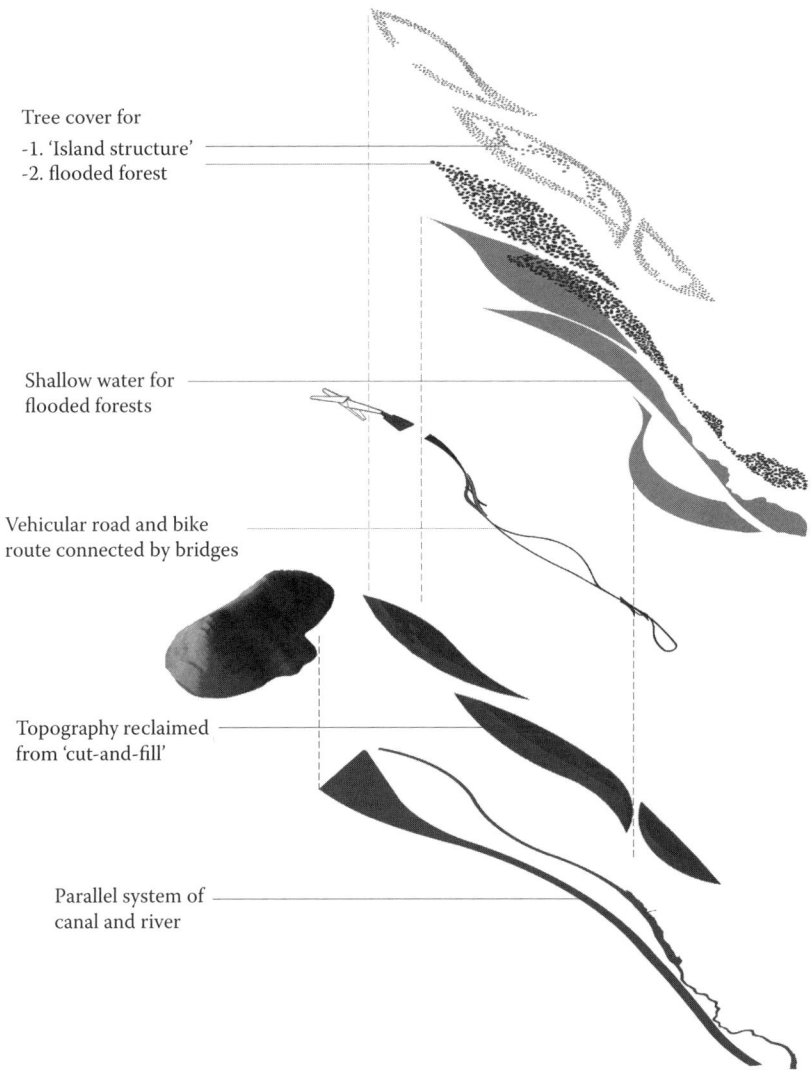

Tree cover for
-1. 'Island structure'
-2. flooded forest

Shallow water for
flooded forests

Vehicular road and bike
route connected by bridges

Topography reclaimed
from 'cut-and-fill'

Parallel system of
canal and river

FIGURE 14.29   Meshing of landscape and architecture in multilayering conceptual designs for an ecological park near Chong Khneas.   (*Continued*)

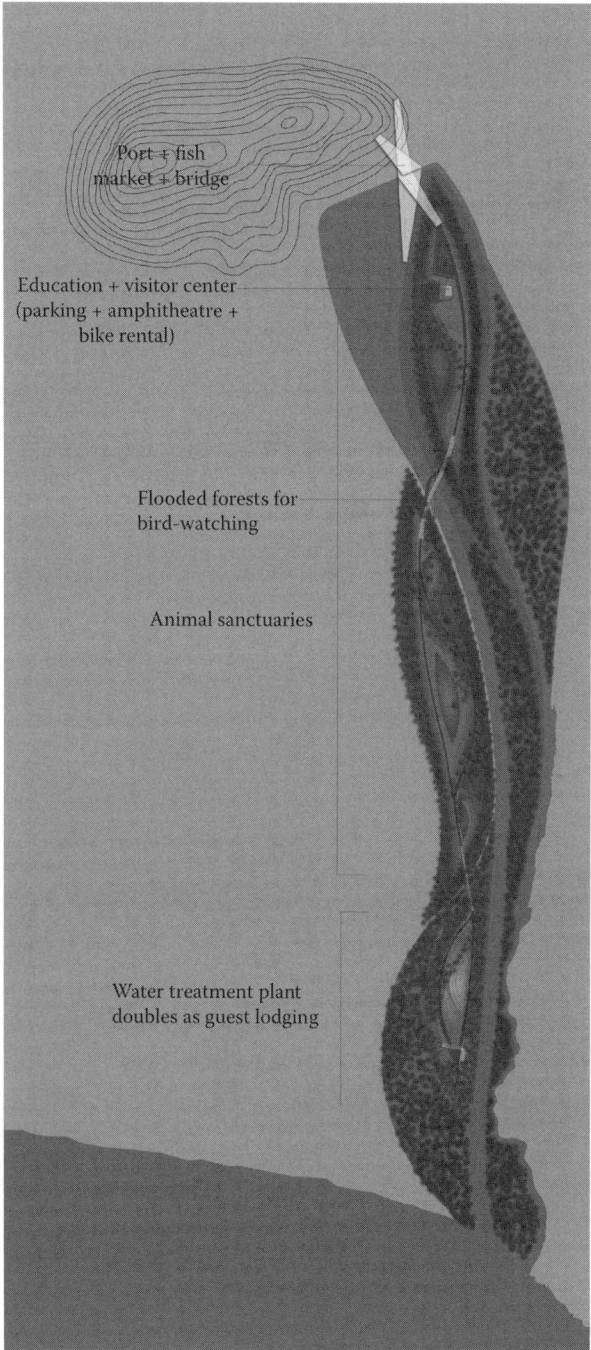

Port + fish market + bridge

Education + visitor center (parking + amphitheatre + bike rental)

Flooded forests for bird-watching

Animal sanctuaries

Water treatment plant doubles as guest lodging

FIGURE 14.29 (*Continued*) Meshing of landscape and architecture in multilayering conceptual designs for an ecological park near Chong Khneas.

FIGURE 14.30 Ancient fertility temple and another worship site around Lake Titicaca.

FIGURE 14.31   Isla Taquile in Lake Titicaca.

FIGURE 14.32   Terrace farming on Isla Taquile.   (*Continued*)

inhabited since about 8000 BCE, and agriculture established since 4000 BCE (Jenkins 2003; Escandell-Tur 2003). Most remarkably, despite being part of the ecotourism industry since the 1970s, the island still retains many of its local customs and traditions.

Visitors (termed "guests") are welcomed by islanders dressed in traditional clothing (this is apparently not done just for photo-ops but many locals really do still wear this garb even if no tourists are present) and greeted with a demonstration of local craft-making and agricultural methods (Figure 14.33). After this, of course, the

FIGURE 14.32 (*Continued*) Terrace farming on Isla Taquile.

opportunity exists to purchase craft souvenirs (but without pressure to do so). Due to
the high-quality craft products, Taquile residents are the most prosperous islanders
on the lake (Escandell-Tur 2003). With the exception of a few buildings with solar
power, there is no electricity on the island. And transportation is restricted to walking
where one can visit several villages, beaches, and ruins (Figure 14.34). The highlight
of the visit is the opportunity to spend the night with a family who has been selected
by rotation for this by the island's own governing council. All these arrangements are
facilitated by several of the more progressive travel agencies on the mainland in Puno

FIGURE 14.33 Demonstration of indigenous agricultural methods and craft making on Isla
Taquile.   (*Continued*)

FIGURE 14.33 (*Continued*) Demonstration of indigenous agricultural methods and craft making on Isla Taquile.

FIGURE 14.34   Foreign tourists to Isla Taquile wander on small paths and visit farming villages, pristine beaches, and ancient ruins.   **(Continued)**

working in close partnership with the local islanders. The islanders obtain revenue from the overnight guests, who stay in very basic accommodations (Figure 14.35) and are entertained under the stars by the family and neighbors. And the ecotourism company receives its revenue from the booking fee and through the hiring of one of its guides, the latter being a requirement to visiting Taquile and thereby excluding the onslaught of independent and unregulated foreigners who might have otherwise descended upon the island, thereby threatening the integrity of its unique culture.

FIGURE 14.34 (*Continued*) Foreign tourists to Isla Taquile wander on small paths and visit farming villages, pristine beaches, and ancient ruins.

If such a comprehensive immersion into spectacular nature and indigenous culture can be achieved in the future restored marshlands of Iraq, it could easily become one of the world's most treasured ecotourism experiences. Time spent on Isla Taquile remains one of the highlights of any trip to Peru, guaranteed to generate over-the-top, enthusiastic endorsements by all fortunate enough to have visited there.

Another, more up-market yet culturally limited form of ecotourism exists on the 48 ha Isla Suasi, which is located close to the border with Bolivia and is thus more remote than Taquile. Here a long-time private resident has developed a pioneer ecotourism industry where customers reside within comfortable, solar-powered, modern lodgings constructed of local material. The impression from photographs and conversations with travel guides and foreign visitors is of a situation similar in scale to those offered in Jordan by the Royal Society for the Conservation of Nature (RSCN) (see chapters 7 and 16). Local land management projects including reforestation are apparently underway with seemingly no attempts, however, to integrate this work with the paying foreign customers à la the Earth Watch model described in chapter 12.

It is Lake Titicaca's Isla de los Uros, often described as "one of the most exceptional attractions of Peru" (Escandell-Tur 2003), where the parallels to the Iraqi marshlands become most overt. The islands are, like those in Iraq, constructed from plants and wattle, in this case tortora reeds, which are also used to build the traditional boats called *caballitos*. Interestingly, when Thor Heyerdahl was looking to find reed boat builders to help with construction of the *Tigris* so that he could demonstrate that ancient Mesopotamians were capable of sailing such crafts from the marshes all the way to the Red Sea and from there begin the pyramid-building civilization along the Nile, he had to turn to Lake Titicaca dwellers for help because no one from the

FIGURE 14.35 Overnight guests stay in rustic accommodations in small family home sites.

marshes had any knowledge of such work. Images in his book (Heyerdahl 1984) of these alpaca-capped Bolivians building the marvelous boat in southern Iraq are wonderfully incongruous at the same time as showing the strong similarities between the two cultures and locations.

The Uros islands are an archipelago of ten to forty artificial reed structures, each island inhabited by up to fifty individuals (Figure 14.36). During the rainy season from November to February, islands are often pulled to different locations around the lake as occurs in the Tonle Sap in Cambodia. Traditional reed boats are kept for

FIGURE 14.36 The Isla de los Uros in Lake Titicaca closely resembles the tethered reed islands in the Iraqi marshlands.

tourists (Figure 14.37), while islanders increasingly use wooden rowboats or metallic motorboats. The island inhabitants are a mixture of Aymara and Quechua ethnic groups, the former being the descendants of those originally driven out into the lake to flee the violence of the latter's Inca forefathers. The primary activity of the islanders is fishing and making handicrafts.

Sustenance living only exists for those island archipelagos farthest out into the lake and thus least visited by day-trippers from Puno. Here, because the huts have only a single room and thus inhabitants are often outside, it is possible to see islanders

FIGURE 14.37 Traditional reed boat kept for tourists.

at work repairing fishing nets, drying fish, grinding maize, or working on crafts (Figure 14.38). These domestic Peruvian images look remarkably like those from earlier times in the Iraqi marshlands in Ochsenschlager (2004) and France (2007). From inside the few huts with solar panels, one can hear the din of the typically over-the-top Spanish soap operas, making for the most incongruous of visits. And just as in photos from the Iraqi marshlands (e.g., Wheeler in Young 1977), visitors observe

FIGURE 14.38 Just as was the situation in the pre-drainage Iraqi marshlands, reed island dwellers in Lake Titicaca can be observed fixing their fishing nets, drying their catch of fishes, and bundling reeds together for construction purposes.   **(Continued)**

FIGURE 14.38 (*Continued*) Just as was the situation in the pre-drainage Iraqi marshlands, reed island dwellers in Lake Titicaca can be observed fixing their fishing nets, drying their catch of fishes, and bundling reeds together for construction purposes.

endearing children reveling in their semi-aquatic existence (Figure 14.39) and can hire a shallow boat and be poled about deep within the atmospheric Titicaca marshes (Figure 14.40). Today, with travel to Iraq being too dangerous for foreigners, a visit to these outer floating islands in Lake Titicaca is the closest one can experience to the culture of the former and hopefully future fabled Mesopotamian marshlands.

However, the situation in the main group of islands most proximal to Puno (Figure 14.41) is instructive in being a warning about what hopefully might be

FIGURE 14.39   In a scene closely reminiscent of many of those captured in photographs by travelers visiting the Iraqi marshlands, amphibious children abound.

FIGURE 14.40   Being poled through the dense reed marshes in Lake Titicaca.

FIGURE 14.41   Traditional reed and modern metal housing on the reed islands closest to the mainland city of Puno.

avoided in the future restored marshlands of Iraq. Here many of the Uros Indians actually live on the mainland and travel out to the islands early in the morning before the tourists begin to arrive, and there get dressed up to sell their handicrafts. There is a Hollywood-like unreality to the whole "show" (and that word is chosen deliberately) that is off-putting if one has experienced the untarnished island villages found in the remoter sections of the lake (this is similar to the situation in the Tonle Sap as described above). When an arriving tour boat from Puno is spotted by a villager situated atop one of the observation towers (Figure 14.42), bells are rung and the actors

FIGURE 14.42   The islands closest to Puno are characterized by signage, observation towers, and large groups of tourists.   *(Continued)*

assume their quaint positions as if in a living museum diorama. Jenkins (2003) in his travel guide to Peru is very critical of the charade, lamenting the rapid devastation of the traditional values of these islanders from being the one-time proud guardians of the lake as recently as only twenty years ago. How much of this cultural erosion can be related to the prominent presence of the evangelical Christian school and active missionaries is open to debate. More likely, it is simply the proximity to the major mainland city and a consequent succumbing to the lucrative seduction of group tourism that is the biggest culprit. Today, a cusp may be approaching as many foreigners

FIGURE 14.42 (*Continued*) The islands closest to Puno are characterized by signage, observation towers, and large groups of tourists.

are beginning to be put off by the artificiality of the manufactured experience as well as the in-the-face begging of children for candies and money.

These near-shore floating villages also suffer in other ways due to their proximity to Puno. The embayment in which they are located has experienced increasingly severe eutrophication (Figure 14.43) to such an extent that islanders now have to travel out into the lake to obtain freshwater. No smiling amphibious children frolicking about here.

Some good news is that the islanders and group tourism companies on the mainland are beginning to realize that their livelihoods might be threatened as the perception of the fabled floating villages of Titicaca becomes one of a polluted Disneyland. Efforts are now underway to make better inroads in treating the water discharged into the lake from the city of Puno. And a consortium of progressive ecotourism companies has made alignments with island leaders to better plan tourism in a more sustainable and less culturally damaging manner.

Similar concerns exist for the archaeological sites around the lake, particularly the historic Islands of the Sun and Moon where the Inca culture was born (Stanich 2003). Stanich fears what tourism development might bring: "The pressures of large tourism companies to remake [the islands] into a Disneyland kind of place will be immense.... It's really important that the local people have a stake in the tourism economy. If they do, the sites will be preserved" (quoted in Mayell 2001).

Such a local people = good, tourism companies = bad analogy is, however, too dichotomous and possibly simplistic. The lesson for the Iraq marshlands from the floating islands of Lake Titicaca is twofold: if left unregulated or at least unsupervised, tourist companies can certainly contribute to ruining tourism (there are many examples of this around the world); but if on the other hand, all tourism decisions

FIGURE 14.43   Eutrophication in the bay near the city of Puno has produced blooms of floating and benthic (bottom-attached) algae (shown respectively as white and black in the photos).

reside with the empowered indigenous locals alone in the absence of an external master plan created by experts (as, for example, for the island of Coiba in Panama, as described in chapter 15), the seduction of short-term economic gains may override long-term sustainability goals. In the end, success will come about through finding a location-specific power-sharing balance between local and outside, bottom-up and top-down, knowledge (see chapter 5). In this respect, the fledgling attempts by the progressive ecotourism companies of Puno, like those discussed above for Siem

Riep in Cambodia, offer promise in finding ways to implement the sort of balance that Lu discusses and champions in chapter 5 for her Taiwanese example.

CONCLUSION

As the case studies of Tonle Sap and Lake Titicaca introduced in this chapter suggest, and in support of a growing body of detailed research (e.g., Stevens 1995; Butler and Hinch 1996; chapter 5), ecotourism can be an important vehicle for preserving both protected areas and indigenous culture.

REFERENCES

Abbott, R. M. 2005. Into the great wide open: Rethinking design in an era of economic, social, and environmental change. In *Facilitating watershed management: Fostering awareness and stewardship*, ed. R. L. France. Lanham, MD: Rowman & Littlefield, 29–42.

———. 2008. *Uncommon cents: Thoreau and the nature of business*. Winnipeg, MB: Green Frigate Books.

———. 2009. *Conscious endeavors: Essays on business, society and the journey to sustainability*. Research Triangle Park, NC: Scriptorium/Palimpsest Press.

Adams, C. A. 2005. Watershed management: The never-ending story. In *Facilitating watershed management: Fostering awareness and stewardship*, ed. R. L. France. Lanham, MD: Rowman & Littlefield, 71–89.

Al-Khayoun, R. B. 2007. Experiences and hopes of the people of the Al-Ahwar marshes. In *Wetlands of mass destruction: Ancient presage for contemporary ecocide in southern Iraq*, ed. R. L. France. Winnipeg, MB: Green Frigate Books.

Anonymous. 2004. *Experience and lessons learned brief for Tonle Sap*. Manila: Asian Development Bank.

Asian Development Bank. 2002. *Report and recommendation of the President to the Board of Directors on a proposed loan and technical assistance grant to the Kingdom of Cambodia for the Tonle Sap environmental management project*. Manila: Asian Development Bank.

———. 2004. *Technical assistance to the Kingdom of Cambodia for establishment of the Tonle Sap basin management organization II*. Manila: Asian Development Bank.

———. 2005a. *Grant assistance to the Kingdom of Cambodia for improving the access of poor floating communities on the Tonle Sap to social infrastructure and livelihood activities (financed by the Japan Fund for Poverty Reduction)*. Manila: Asian Development Bank.

———. 2005b. *Proposed Asian Development Fund grant Kingdom of Cambodia: Tonle Sap sustainable livelihoods project*. Manila: Asian Development Bank.

———. 2005c. *Technical assistance to the Kingdom of Cambodia for the study of the influence of built structures on the fisheries of the Tonle Sap (financed by the government of Finland)*. Manila: Asian Development Bank.

———. 2006. *Future solutions now: The Tonle Sap Initiative*. Manila: Asian Development Bank.

Bailleux, R. 2003. *The Tonle Sap Great Lake: A pulse of life*. Rome: Food and Agriculture Organization.

Bauer, B. S. 2001. *Ritual and pilgrimage in the Ancient Andes: The Islands of the Sun and the Moon*. Austin: University of Texas Press.

Butler, R., and T. Hinch, eds. 1996. *Tourism and indigenous peoples*. London: International Thomson Business Press.

Escandell-Tur, N. 2003. *Todo Titicaca*. Lima, Peru: Tierra Firme.

Evans, P. T., M. Marschke, and K. Paudyal. 2004. *Flood forests, fish and fishing villages: Tonle Sap, Cambodia*. Bohol, Philippines: Asia Forest Network.

France, R. L. 2006. *Introduction to watershed development: Understanding and managing the impacts of sprawl*. Lanham, MD: Rowman & Littlefield.

———, ed. 2007. *Wetlands of mass destruction: Ancient presage for contemporary ecocide in southern Iraq*. Winnipeg, MB: Green Frigate Books.

———, ed. 2008. *Handbook of regenerative landscape design*. Boca Raton, FL: CRC Press.

———. 2010. Veniceland Atlantis: The bleak future of the world's favorite city. Libri Books, London.

Fulanain. 1928. *The marsh Arab*. Philadelphia: Lippincott.

Goes, F. 2005. *Four years of waterbird conservation activities in Prek Toal Core Area, Tonle Sap Biosphere Reserve (2001–2004)*. Phnom Penh, Cambodia: Wildlife Conservation Society.

Heyerdahl, T. 1984. *The Tigris expedition: In search of our beginnings*. New York: Doubleday.

Jenkins, D. 2003. *The Rough Guide to Peru*. London: Rough Guides.

Liu, H.-L., and E. Syariffudin. 2003. Eco-tourism in Cambodia: The Tonle Sap Great Lake. Penny White proposal, Graduate School of Design, Harvard University.

Maxwell, G. 1967. *Ring of bright water*. New York: Penguin.

Mayell, H. 2001. Pilgrimage route uncovered at South America's Lake Titicaca. *National Geographic News*, June 4. http://news.nationalgeographic.com/news/2001/06/0604_boliviashrines.html.

Mcleod, G. 2006. Cambodia finds oil and changing fortunes. *Sunday Bangkok Post*, December 17.

Nedeco, T., and T. Midas. 1998. *Potential for ecotourism in the Tonle Sap area, Cambodia*. Phnom Penh: Mekong River Commission.

Nicholson, E. 2007. Human rights issues in the Iraqi marshlands: A case for genocide. In *Wetlands of mass destruction: Ancient presage for contemporary ecocide in southern Iraq*, ed. R. L. France. Winnipeg, MB: Green Frigate Books.

Orlove, B. 2002. *Lines in the water: Nature and culture at Lake Titicaca*. Berkeley: University of California Press.

Ochsenschlager, E. L. 2004. *Iraq's marsh Arabs in the Garden of Eden*. Philadelphia: University of Pennsylvania Museum of Archaeology and Anthropology.

Palmer, B., and S. Martin. 2005. *The rough guide to Cambodia*. London: Rough Guides.

Poole, C. 2005. *Tonle Sap: The heart of Cambodia's natural heritage*. Bangkok: River Books.

Salles-Reese, V. 1997. *From Viracorha to the Virgin of Copacabana: Representation of the sacred at Lake Titicaca*. Austin: University of Texas.

Stanish, C. 2003. *Ancient Titicaca: The evolution of complex society in southern Peru and northern Bolivia*. Berkeley: University of California Press.

Stevens, S., ed. 1995. *Conservation through cultural survival: Indigenous peoples and protected area*. Washington, DC: Island.

Syariffudin, E. 2006. Canal, mounds, bridges ++: The economics of infrastructure in a shifting landscape. March. Master's thesis, Graduate School of Design, Harvard University.

Varis, O., M. Kummu, and M. Keskinen, eds. 2006. Integrated water resources management on the Tonle Sap Lake, Cambodia. *Water Research Development* 22(3, special issue).

Young, G. 1977. *Return to the marshes: Life with the marsh Arabs of Iraq* (photo. N. Wheeler). London: Collins.

Ecology and Economics

15 Ecotourism, Local Economy, and Ecological Strategies in Planning Nature Reserves in Panama[*]

CONTENTS

Introduction 297
Alternative Futures Models in Land Use Planning 298
Case Study: Coiba, Panama 298
Public Policy in Environmental Planning 303
Acknowledgment 303
References 304

INTRODUCTION

As early as the 1970s, the pre-Saddam government in Iraq was exploring ways in which to develop a tourism industry associated with the southern marshlands. For example, photographs by Nik Wheeler in Gavin Young's *Return to the Marshes* (1977) show Iraqi tourists comfortably sitting at tables in front of guest houses designed to resemble the reed-constructed *mudhifs* of marsh dwellers. And Assam Alwash, field director of the Eden Again nongovernmental organization (NGO) involved in restoring the marshes and a former Iraqi who was relocated to southern California as a child, has frequently stated his dream of being able to bring his American friends to the restored marshes for vacations that would involve kayaking from village to village. It is certainly not an exaggeration, if one has read the engaging and seductive prose of the British adventurers who penetrated this remote area in the middle of the last century (Maxwell 1966; Thesiger 1964; Young 1977), to believe that the southern Iraqi marshes could very well become one of the world's premier ecotourism and adventure tourism destinations. However, in other areas where these more sustainable forms of tourism exist, there are still problems associated with the more

[*] Adapted by Robert L. France from Karish, K., and R. Faris. 2004. Ecotourism development models in the planning of nature reserves. Paper presented at the Mesopotamian Marshes and Modern Development: Practical Approaches for Sustaining Restored Ecological and Cultural Landscapes conference, Cambridge, MA, October.

traditional tourism development (hotels, commercialism, restaurants, social inequities, crime, etc.) that are also attracted to these special areas for the very same reasons (see chapter 14, for example, for a discussion about the floating marsh villages in Lake Titicaca). It will become essential, therefore, to find ways to achieve some sort of balance between tourism development and environmental conservation.

ALTERNATIVE FUTURES MODELS IN LAND USE PLANNING

Gaps often exist in the production of scientific information and the delivery of those insights in a useful form for people to act upon (France 2006). Furthermore, for tourism planning there is a need to develop place-based relationships between environmental policy and implementation among ecological, social, and economic aspirations, and in a form that not only supports but also actually encourages objective discussion about what the future might look like on the ground.

One approach that is receiving increased interest among land use planners, particularly those dealing with large-scale, highly contentious water-driven issues, is alternative futures scenario modeling (France 2006; see also chapter 4). This technique enables predicting impacts of land use alterations (such as tourism development) on ecological processes, integrating human dimensions into effective planning, and developing an understanding of the uncertainty of impacts and associated risks of various development scenarios (e.g., Hulse et al. 2000; Van Sickle et al. 2004). And, most importantly, the alternative futures approach provides a framework to effectively incorporate science into the community-based decision-making process (France 2006). Major components (Hulse, Branscomb, and Payne 2004) include the following:

- Characterization of the trajectory of landscape changes and formulating these as a series of defined alternative future scenario assumptions
- Development of spatially explicit alternative future landscapes through models that reflect varying assumptions about land and water use and the range of stakeholder viewpoints
- Modeling and evaluating the likely effects of the landscape trajectory and alternative futures on key and valued ecological and socioeconomic end points
- Characterizing and synthesizing the differences among the alternatives

The alternative futures approach has its roots in build-out analyses that have been used for decades by land use planners as a tool to predict future development patterns and their effects. The build-out methodology is directed toward promoting sound land management decisions by providing growth projections, assessing the impacts of existing growth, and encouraging actions to reduce the impacts of growth (France 2006).

CASE STUDY: COIBA, PANAMA

One of the limitations facing any land use planning process is recognizing the difficulty or perhaps impossibility of coming up with perfect plans or perfect public policy choices. Nevertheless, due to the rarity of these types of studies, Kimberley

Karish and Robert Faris (2004) believe that ideas and experiences from one particular location can often be generalized to public policy and land management decisions in other areas such as, for example, the marshlands of southern Iraq. One such location is the 500 km^2 island of Coiba located 22 km off the coast of Panama. Fourteen scenarios and alternative futures were developed for the region (Steinitz et al. 2005; Karish 2007). In this case, the study area included not only the island of Coiba itself but also the surrounding mainland in order to include the primary and secondary impacts from economic and ecological changes.

Prior to the twentieth century, the Island of Coiba had been largely untouched by human influence. A penal colony was established in 1915 which reached its height of occupation in the 1970s and 1980s with several thousand inmates. It was an unusual prison colony in that the only locks were for the guards' facilities. The prisoners themselves lived in a half dozen very informal settlements on the southeastern side of the island, where they engaged in subsistence agriculture. Domestic animals such dogs, cats, cows, and water buffalo were introduced, which led to some significant impacts in immediate areas surrounding the encampments. Despite this, in retrospect, it was the existence of the penal colony that prevented more widespread ecological changes on the island due to land clearing, such that today the vast majority of the island still remains covered in primary forest, constituting a valuable ecological asset for the entire region (Karish and Faris 2004) and representing one of the most valuable tropical forest ecosystems in Central America (Steinitz et al. 2005). Presently, the penal colony is being phased out (only a handful of prisoners are left) and the island is to be developed as a national park. The basic policy questions (Steinitz et al. 2005) are as follows: what and how much development should be allowed, and where should development be allowed?

Coiba, like many Western Hemisphere subtropical islands, is extremely beautiful with very nice beaches and dense forests. The communities on the nearby and sparsely populated mainland coastline are very small, with many being accessible only by boat. The economic base for the mainland coast, which is one of the poorest areas in all of Panama, is low-productivity agriculture and small-scale fishing. With the sole exception of tourism, there are really very few sources of potential growth. There is currently substantial interest within the private sector tourism industry in developing areas on both the mainland and the island. As a result, major changes in the landscape are forecast in association with the projected increase in tourism pressure. To understand the dynamics of these potential landscape changes, Karish and Faris (2004) undertook a study based on creating scenarios based upon different levels of tourism development. The principal questions addressed in the study were as follows:

1. How to manage the national park, which is both the island as well as its surrounding waters
2. How to shape and promote a tourism industry that is conducive to the responsible use of the resources at the same time as stimulating economic development in the area

Alternative models were developed to predict the ecological impacts of different projections of future tourism development. The impacts of alteration to the landscape were examined with change scenarios in relation to modeled policy changes (these are described in Steinitz et al. 2005). The selection of scenarios was intended to represent the widest reasonable range of possibilities concerning development and conservation (Karish and Faris 2004). Each model assumed a different but escalating level of development as measured by the number of beds of new tourism facilities that were applied in the scenarios as being a representative marker of tourism growth. Beds were allocated to different locations on the island using GIS analysis along consequent developmental attributes such as new roads, recreational facilities, and so on.

In order to evaluate these development changes on the land, the environmental impact of the alternative futures was assessed in relation to their influence on estimated species loss on Coiba (Karish and Faris 2004; Steinitz et al. 2005). Different species were selected based on their rarity or endemism in the area (Karish 2007). Some of the species, such as the scarlet macaw, were only found on Coiba Island, whereas others were also found or predicted to occur on the mainland. The amount of habitat and its quality (the latter broken down into the three levels of primary, secondary, and tertiary) were mapped. Habitat was thus tied to the predicted success of a suite of species under current land use conditions as well as the varying future development impacts. Coral species were also modeled in the same fashion.

An ecological impacts model was used to evaluate the influence of each scenario's land use changes on the potential habitats of each of the selected species as well as on the summary index of species richness (Karish and Faris 2004). The proposed changes in land use for the scenarios were aggregated into groups based upon their levels of disturbance associated with construction, maintenance, or landscape modification. These groups were then assessed for the potential disturbance in the area surrounding the site of the allocation. So, for example, when a road was built, the greatest impact may be right beside the road, but there are also additional impacts due to runoff beyond the site itself (Forman et al. 2003; see also chapter 17). These surrounding areas and their associated disturbance values were combined with the species habitat models to calculate the areas of habitat that would be affected by the changes in each scenario. The calculated map grid contained a range of values based on the cross between the sensitivity of the habitat for that species and the severity of the projected impacts. The resulting outputs could then be grouped into five general impact levels (Karish and Faris 2004). The impact could be beneficial, meaning there's a positive improvement in habitat quality. This occurred for some species that actually exist better in edge habitats or make use of agricultural lands. There could be a compatible impact, in which case no perceivable changes are likely to occur in the landscape quality. There could be a moderate impact, meaning that maybe a natural mitigation procedure might be necessary to alleviate the deleterious effects. Severe impacts are possible, such that there would have to be some sort of engineering type of solution in order to alleviate the impacts. And then, of course, terminal impacts might occur in which case there is no possible mitigation. An example of this would be the immediate footprint of a building which would destroy the underlying habitat completely.

The modeling process resulted in an impacts model specific to each species for each scenario as well as in a summed species richness impact (Steinitz et al. 2005). For one example, the highest development impact scenario, both the mainland and Coiba Island are severely affected by the spread of development. The projected impacts were found to mostly range from low to very high, except in several locations on the island that have resident endemic species and would consequently be predicted to have extreme impacts (Karish and Faris 2004). Overall, there were more severe levels of impacts predicted from tourism development on Coiba compared to the mainland. Corals were found to be specifically sensitive to development due to the impacts of sedimentation or recreational overuse.

Ecological impact results can be displayed in graphic format to enable comparisons across the alternative scenarios (see Steinitz et al. 2005). Impacts on terrestrial species result from deforestation, and effects on the marine species are due to either sedimentation from runoff associated with the different land use changes or direct damage from recreational activities such as snorkeling. The impacts in these different categories were weighted so they could be combined into a single index to enable predictions of ecological repercussions to be tracked to the various levels of development (Karish and Faris 2004). One scenario examined what would happen if the national park was determined by decree but not passed into law, in which case unregulated development on the island would bring about very severe alterations due to the presence of large-scale logging operations with the corresponding massive influxes of people coming over from the mainland for work in the industry. In this case, it's a good example of a scenario that has a minimum number of beds, in this case only fifty, but a maximum extent of negative effects.

The economic analysis was very straightforward. Investment requirements from private and public sources were examined for each one of these possible scenarios (Karish and Faris 2004). Other variables included the distribution of income and the government revenue stemming from each one of the scenarios in order to estimate how much investment would be needed from private and from government sources and how many jobs would be created under each scenario of tourism development. Direct and indirect labor incomes and returns on capital were also determined, as well as estimates of how much of the income accrues to locals within the study area compared to Panamanians from outside of the study area.

Modeling revealed that economic subtleties existed within similar levels of tourism development depending on where that development occurred (Karish and Faris 2004). For example, the economic analysis showed that when comparing development solely on the mainland versus development solely on Coiba, it is far better off to develop the mainland, thus preserving the island's greenspace. And interestingly, the alternatives with the best returns for a given level of development would often entail a little development on the island and a little bit on the mainland.

Such analyses have direct relevance to the sustainable development marshlands of Iraq in terms of providing an easily adaptable methodology for future tourism planning. For example, what will the correct balance be in terms of developing tourism facilities within the marshes themselves compared to along the dryland edge of the marshes in terms of revenue generation for the local populations and the like?

The ecological results and the economic results can be combined in graphical format to create a matrix of easily comprehendible possible alternatives for eco-tourism planning (Karish and Faris 2004). An extreme example is the "Leave it to the wolves scenario," which leads to an ecological disaster. This is the highest build-out scenario with high economic returns, and it performs very poorly in ecological terms. One would, of course, have to question whether such an alternative is really sustainable in the long run. On the other extreme is the scenario based on preserving the island completely free of tourists for the benefits of conservation and science. Of course, such a strategy is going to perform the best in ecological terms, but one has to seriously question whether there will be adequate political and financial support in order to sustain such an exclusionary approach. In this light, just how much of the future restored Iraqi marshes should be preserved as an ecological "Eden" versus how much should be kept as a working landscape that is welcoming to tourists?

The most realistic development pathways for Coiba will almost certainly lie somewhere between these two extremes, recognizing that there will always be trade-offs between economic and ecological outcomes. Although this might just appear to be common sense (and of course it is), the methodology described here is most useful in being able to quantify that balance and to provide warnings as the tipping point is approached (Karish and Faris 2004). It is important to recognize that what one can gain from examining such alternative futures graphs is not what the right solution is. The right solution depends, of course, on values in terms of how society balances ecological versus economic outcomes, and the only way that this can really be settled is through the political process. What can be ascertained from analyzing the graphical outputs from alternative futures modeling exercises is what the really bad ideas are that should be avoided at all costs (Karish and Faris 2004). It is, therefore, an approach that is very useful in early planning strategizing through helping to immediately screen out the most unlikely development pathways from future consideration (see chapter 4). And as an outside research group, this is where the consultants should back off and leave the nuanced decision making to the locals.

What Karish and Faris (2004) wanted to do in the Coiba project was to create a useful public policy tool and to foster a better informed dialogue about the future of the valuable area. At the time of their study, there were a few different versions of legislation being considered by the government for the national park. The first version of the law would have severely restricted any development on the island, and this was subsequently vetoed by the president. Two further versions were introduced. Another version was a terrible piece of legislation which thankfully was turned down in favor of what is presently a very sensible piece of legislation. And it was this particular alternative futures study, marrying predictions of development with economic and ecologic benefits, that was influential in helping governmental officials to reach that sane decision of moving toward a balanced and thus *sustainable* tourism plan for the nature reserve. And it is through such a process that those concerned with the sustainable future of the Iraqi marshlands should direct their energies (Karish and Faris 2004).

PUBLIC POLICY IN ENVIRONMENTAL PLANNING

It is worth concluding with a few reflections on the public policy choices that are available for places like Coiba or the Iraqi marshlands, and the role of researchers in trying to contribute to these public policy choices (Karish and Faris 2004). It is a common misconception that the role of the public policy analyst should be to come up with perfect public policy solutions based on good ideas. And many such individuals truly believe that they do come up with a lot of really good ideas, even brilliant and sophisticated ideas, that are very well elaborated. However, it is often the case, of course, that many of the people who hear these ideas don't share the same enthusiasm for them. Certainly a much more effective approach seems to be to forget this idea of coming up with good solutions, but instead look at what people are actually considering on the ground and give them a better idea of how to evaluate the ideas that exist already (Karish and Faris 2004). And the best way to do this, of course, is to put *all* the ideas out on the table. In evaluating these ideas, the government of course can't do it alone, and what is needed is an active, engaged, and motivated populace in order to assist with these ideas. Governments will typically have very short-term planning horizons barely beyond three to five years into the future. And this is where NGOs and private organizations can contribute to the public policy exercise. The role of researchers is really to make research accessible and also to foster the development of these constituencies and institutions so that they can pressure the public policy decision makers, including even making the development choices part of elections.

The alternative futures framework is so strong and compelling that it is certainly very worthy of consideration of being applied to planning the restorative redevelopment of the Iraq marshlands (see chapter 4). Because the analysis can be accomplished very quickly with a minimum amount of information (Karish 2007), there is a lot to recommend its adoption. In making public policy choices, it is often recognized that the science may not always be there in order to answer questions that society would like to know. However, society often doesn't put that fact into context, such that the science is also generally far ahead of the public policy. As a result, the institutions, constituencies, processes, and mechanisms by which good public policy is made can often be very, very poorly developed. Indeed, one of the principle measures of economic development is how well these institutions have developed in order to help make such public policy choices (Karish and Faris 2004). And the role of research (such as that described in chapters 10 and 12) can be to contribute to fostering a dialogue that can create the constituencies for change.

ACKNOWLEDGMENT

Adapted from Karish and Faris (2004).

REFERENCES

Forman, R., Daniel Sperling, John Bissonette, Anthony Clevenger, Carol Cutshall, Virginia Dale, Lenore Fahrig, Robert France, Charles Goldman, Kevin Heanue, Julia Jones, Frederick Swanson, Thomas Turrentine, and Thomas Winter. 2003. *Road ecology: Science and solutions*. Washington, DC: Island.

France, R. L. 2006. *Introduction to watershed development: Understanding and managing the impacts of sprawl*. Lanham, MD: Rowman & Littlefield.

Hulse, D., J. Eilers, K. Freemark, C. Hummon, and D. White. 2000. Planning alternative future landscapes in Oregon: Evaluating effects on water quality and biodiversity. *Landscape Journal* 19:1–19.

Hulse, D., A. Branscomb, and S. Payne. 2004. Envisioning alternatives: Using citizen guidance to map future land and water use. *Ecological Applications* 14:325–41.

Karish, K. 2007. *Theory rich, data poor: Ecological principles for biodiversity conservation planning with limited data: The case study of Coiba Island, Panama*. Cambridge, MA: Harvard Design School.

Karish, K., and R. Faris. 2004. Ecotourism development models in the planning of nature reserves. Paper presented at the Mesopotamian Marshes and Modern Development: Practical Approaches for Sustaining Restored Ecological and Cultural Landscapes conference, Cambridge, MA, October.

Maxwell, G. 1966. *People of the reeds*. New York: Pyramid.

Steinitz, C., R. Faris, M. Flaxman, K. Karish, A. D. Mellinger, T. Canfield, and L. Sucre. 2005. A delicate balance: Conservation and development scenarios for Panama's Coiba National Park. *Environment* 47:24–39.

Thesiger, W. 1964. *The marsh Arabs*. New York: Penguin.

Van Sickle, J., Joan Baker, Alan Herlihy, and Peter Bayley. 2004. Projecting the biological condition of streams under alternative scenarios of human land use. *Ecological Applications* 14:368–80.

Young, G. 1977. *Return to the marshes: Life with the marsh Arabs of Iraq*. London: Collins.

16 Sustainable Development and Nature Conservation in Jordan*

Jordan, although a country generally poor in natural resources, is actively working toward conserving its natural biodiversity, a task made difficult by its small economy.This chapter reviews some environmental issues in Jordan, describes the role of the Royal Society for the Conservation of Nature (RSCN) organization, and focuses on the business approach to nature conservation and protected areas management particularly in relation to the Dana Reserve, demonstrating that protected areas can be important mechanisms for the sustainable development of indigenous human populations (Anonymous 1997). As Johnson (1997) summarized, "The Dana project revolutionalised the approach to conservation in Jordan and, in so doing, has become one of the most significant and acclaimed conservation initiatives ever undertaken in the Middle East." Specifically, it is this successful conjoining of ecology and economy that offers a realistic model that could be adapted to the development of future eco-tourism programs in the eventually restored marshes of southern Iraq.

Like Iraq, Jordan has an ancient culture, an extensive archaeological heritage, and the presence of a rich and diverse array of native species. In terms of biodiversity, Jordan, given its location at the juncture of Asia, Africa, and Europe, is vitally important as part of the major flyway for migrating birds between Europe and Africa along the northern Rift Valley. Interestingly, Jordan has only 1 percent of the world's plant species but contains almost 10 percent of all the medicinal plants that have been identified.

* Adapted by Robert L. France from Irani, K. 2004. The nature conservation programme in Jordan: Helping nature help people. Paper presented at the Mesopotamian Marshes and Modern Development: Practical Approaches for Sustaining Ecological and Cultural Landscapes conference, Cambridge, MA, October.

According to Khalid Irani (2004), former director of the RSCN (now minister of the environment for the government of Jordan), one of the most important environmental issues facing Jordan is its limited water resources. Indeed, Jordan is actually the tenth poorest country in the world in terms of water, and by 2015 is expected to be tapping into all its available water supplies and beginning to look elsewhere for additional sources. In the meantime, Jordan has started to mine the last of its fossil waters in a nonsustainable way. There is thus an intense water demand pressure on all natural habitats (see chapter 6, for example), and sometimes the degree of water withdrawal means that the landscape becomes unsustainable for agriculture. To counter this tendency, the RSCN promotes the message that "yellow is nice" in terms of maintaining natural deserts rather than attempting to turn them into unsustainable green pastures (Irani 2004).

There are also many threats to the wildlife such as overhunting, which occurred to such a severe degree from the 1960s to 1980s that it led to the formation of the RSCN. Today, the mission of the RSCN is to conserve nature, wildlife, and wildlife habitat, and to integrate these with socioeconomic programs at the same time as generating public support (RSCN 2009). This new mission, therefore, focuses on public action directed toward achieving environmental sustainability in terms of linking economy with ecology (Boyce, Narain, and Stanton 2007).

Jordan is in the process of building a comprehensive, computerized database about national biodiversity based on RSCN research which will be linked to a GIS system (Irani 2004) with reference to transborder concerns (Johnson 2007). The public is reached through the RSCN website and different publications (such as *Al Rem* magazine), some of which are specifically targeted to students and children to encourage them to venture outside and use and enjoy the protected areas through engaging in various activities. The RSCN also uses different interactive techniques to promote biodiversity conservation such as the "Birds Know No Boundaries" project which links the migration of birds through the Jordan Rift Valley to satellite data and the Internet, thereby allowing viewers to follow the birds' movement in real time. The whole bird migration story is fascinating for students because they also learn about geography, different habitats, avian behavior, and so on. RSCN also participates in developing nationwide school curricula about both birds and water conservation issues (see chapter 6). Part of this process involves the use of innovative pedagogical approaches such as designing board games and memory cards.

In addition, the RSCN has a training program in Arabic which is relevant to concerns of the greater region, including running ten workshops per year which have been attended by Syrians, Palestinians, Lebanese, Saudi Arabians, Egyptians, and, most recently, Iraqis. The top training priority for this outreach program for various environmental institutions was identified as being capacity building rather than specific technical training in biodiversity monitoring. The RSCN believes that it is very important to work with institutions to build their capacity by allowing them to be able to adapt the process to conserve nature and biodiversity in their own particular way once they return home (Irani 2004). Additionally, the RSCN has also established a partnership with local police to strengthen their environmental enforcement role. A network of enforcement was created by teaching an ongoing course at the police academy which eventually led to the establishment of an environmental police unit. It is obvious that all these programs could find easy adoption to the evolving situation in southern Iraq in terms of helping to protect the restored marshlands and their returning fauna.

The RSCN has developed a national plan for protected areas—the "jewels of the Kingdom," in Johnson's (1997) parlance—based on the representation of the key eco-system types of Jordan (Irani 2004). The reason why an NGO in Jordan is in the business of managing Jordan's nature reserves is due to the fact that when the RSCN was formed in the 1960s, there was no governmental institution at that time that dealt with nature conservation and protected-areas management. It is therefore a unique situation in that the RSCN is not controlled by the government yet does in fact manage public land. And the government is very happy with this situation since it comes at no cost to their own strapped budgets. And, as mentioned above, the former director of the RSCN is now head of the new governmental Ministry of the Environment.

One of the protected areas managed by the RSCN is the Dana Nature Reserve located near the Dead Sea and billed as the lowest elevation protected land on Earth. Recently, several plants new to science have been discovered in the area, and the RSCN is working hard to protect the unique wadi landscape. The Dana story is important not only for its obvious success in its own right but also through its role as serving as the school from which the RSCN learned how to adopt an integrated approach to protected lands. This approach has subsequently been applied to most of the other protected areas in the country, and by extrapolation could be imagined to be integrated into future management plans for the Iraqi marshlands.

With its scenery, wildlife, culture, and history, Dana, like the Iraqi marshlands, is a paradise for tourism. The success of the Dana Nature Reserve has transcended the local environmental community in Jordan and is recognized internationally through winning four international awards for sustainable development (Johnson 2007). I (R. France) first learned about the Dana project not from an environmental publication but rather from two general guidebooks for tourists, both of which were effusive in their praise. The *Jordan Footprint Handbook* (Mannheim 2000) describes the site as being "one of the most breathtaking experiences in Jordan" in terms of natural beauty, as well as being home to seven hundred species of plants, of which one hundred are rare and eight are endemic. Further, Dana is "one of the most impor-tant non-wetland areas for birdlife in the Middle East" with more than two hundred species being present. What is most interesting is the attention paid to sustainable development such that the reserve "has become something of a model for integrated conservation and development projects. A strong emphasis has been placed on mak-ing the nature reserve of tangible economic benefit to local people, on the basis that this is the only way of giving conservation efforts any long-term prospects of success" (Mannheim 2000). And in an easily understood parallel to the present-day situation of empty villages along the edge of the Iraqi marshlands, the Dana ecotourism efforts have been instrumental for local regeneration by renovating the abandoned Dana Village such that one-time inhabitants have now returned and are again harvesting and selling fruit from their orchards.

The second tourist guidebook, the *Rough Guide to Jordan* (Teller 1998), is equally filled with praise about Dana, calling it "unique in the Middle East, a posi-tive, visionary programme combining scientific research, social reconstruction and sustainable tourism. The nurturing of traditional crafts and the rejuvenation of a village economy by creating an array of long-term jobs for local people have managed to proceed successfully hand-in-hand with active protection of the environment." Given the pre-Saddam importance of sustainable agriculture in communities located on the edges of the Iraqi marshes as well as the age-old tradition of marketing crafts by dwellers from deep within the marshes (Ochsenschlager 1995, 2004), the Dana Nature Reserve is a model that could be transferred to Iraq. It is to here, in this small location in neighboring Jordan with the RSCN's landmark creation and development of a sustainable local economy in association with nature protection and ecotourism, where I think those involved in the future management of the Iraqi marshlands should turn for inspiration and instruction. In order to make this point, it is worth quoting Teller's (1998) independent summary of the Dana project before we return to Khalid Irani, the former director of the RSCN, for his more detailed description:

> Dana lay semi-abandoned for a decade or more, its handful of impoverished farmers forced to compete in the local markets with bigger farms using more advanced methods of production. This was what a group of twelve women from Amman discovered in the early 1990s as they traveled across the country to catalogue the remnants of traditional Jordanian culture. Realizing the deprivation faced by some of the poorest people in the country, these "Friends of Dana" embarked on a project to renovate and revitalize the fabric of the village under the auspices of the RSCN. Electricity, telephones and a water supply were extended to the village and 65 cottages renovated. People started to drift back to Dana. The RSCN quickly realized the potential of the secluded Wadi Dana for scientific research, and, in a new $3.3 million project funded partly by the World Bank and the UN, turned the area into a protected reserve, built a small research station next to the village and, in 1994, launched a detailed ecological survey. Dana's small-scale agriculture was clearly no longer economically viable and thousands of domesticated goats, sheep and camels had overgrazed the wadi for decades; continued grazing couldn't be reconciled with the need for environmental protection and was banned, and studies were made into the feasibility of creating sustainable opportunities for villagers to gain a livelihood from the reserve without destroying it. The ingenious solution came in redirecting the village's traditional crops to a new market. Dana's farmers produced their olives, figs, grapes, other fruits and nuts as before, but, instead of going to market, they sold everything to the RSCN, who employed Dana villagers to process these crops into novelty products such as organically produced jams and olive-oil soap for direct sale to relatively wealthy, environmentally aware consumers. Medicinal herbs were introduced as a cash crop to aid the economic recovery, and the last Dana resident familiar with traditional pottery-making has been encouraged to teach her craft to a younger generation. Dana soon hit the headlines, and in 1996 the RSCN launched low-impact tourism to the reserve, with the traditional-style Guesthouse going up next to the research buildings. Local villagers—some of whom were already employed as research scientists—now also work as managers and guides.

The Dana Nature Reserve is an area of 320 km^2 surrounded by a handful of villages having a population of about six thousand, including some Bedouin communities located inside the protected area. The first survey conducted dealt with socioeconomics even before flora and fauna (Irani 2004). Work initiated in the mid-

1990s focused on the surrounding people's needs, working with them to determine how they were affected by the protected area and how they in turn were influencing the protected area. The RSCN developed Jordan's first protected area management plan for the Wadi Dana which outlined its objectives, strategies, and priorities for the area, all directed toward attempting to find a balance between protecting Dana's natural attributes and meeting the needs of the local population. Over a hundred specific management options were prescribed, many based on concepts of zoning in terms of defining certain areas where particular activities either could or could not occur (e.g., a core conservation zone, a grazing zone, and a recreation or tourist zone; see chapter 11 for another example of this strategy in Mexico).

Local people were at the forefront of the Dana project from the very beginning (Johnson 1997, 2007). The previous state of the village was dreadful, with years of deprivation and poverty having forced most of the residents to abandon the village. The Friends of Dana saw the great potential of the site in terms of its beauty, history, culture, and unique location. They worked with the RSCN to raise money to restore the crumbling fabric of the village and to bring new life and hope to the residents by renovating houses, building a new mosque, paving the main road, and, as in other examples of regenerating rural Arab villages (e.g., Wessels and Hoogeveen 2007), renovating the water management system including constructing a new spring head (Anonymous 1997; and see Figure 16.1). And in the true spirit of the present book on "restorative redevelopment" and as a lesson to the situation in southern Iraq, the project became an immediate success: "As a result of these improvements, village life is being restored and people are returning" (Anonymous 1997).

Nature reserves around the world often flourish at the expense of local people. Because of this, alternative means of livelihood had to be created for Wadi Dana once land use restrictions were implemented (Salti 1997). In particular, the Dana area is heavily grazed by over ten thousand goats, which has led to vegetation removal and, in consequence, severe erosion (Johnson 1997; Irani 2004). Other threats included limestone quarrying and the potential reopening of an ancient copper mine. The challenge, then, became how to protect the area without damaging the livelihood of the people. The RSCN approached this challenge by developing a socioeconomic model with respect to nature conservation wherein they tried to identify and establish new businesses that could be linked with nature conservation and the livelihood of people and their culture. For example, the RSCN worked in the production of fruits grown on the formerly neglected trees and sold the nicely packed health foods to people in Amman and other cities (Figure 16.2). These products have higher value and are marketed as being linked with conservation. One such product is frames made from goat leather (Figure 16.3; a motto in the RSCN office is "If you can't beat them, eat them" in terms of dealing with removing as many goats as possible; Irani 2004). Other products include metal jewelry (brooches, earrings, necklaces, and so on) which are based on archaeological themes or inspired by nature, and also decorated ostrich eggs (Figure 16.4). A further product is packaging of medicinal herbs, T-shirts, as well as reused pop-drink cans that, in a move similar to production of the book *Wetlands of Mass Destruction: Ancient Presage for Contemporary Ecocide in Southern Iraq* (France 2007), highlight ancient inscriptions that are over three thousand years old.

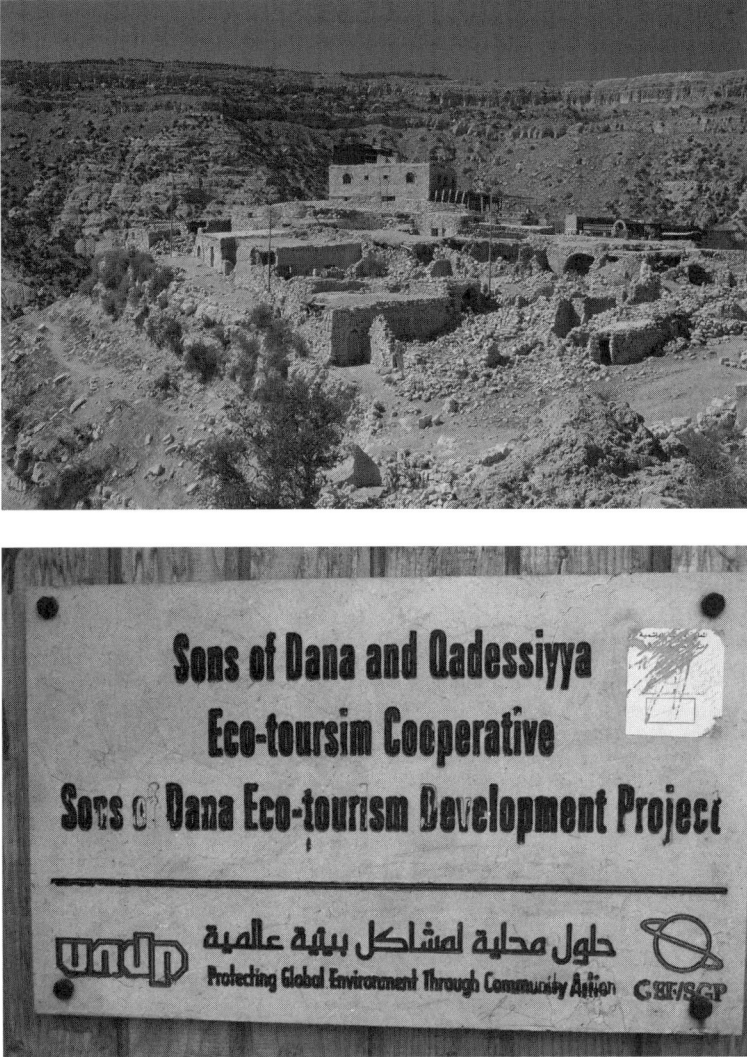

FIGURE 16.1   Surrounded by spectacular scenery, Dana Village is one of the world's most interesting ecotourism and restorative redevelopment projects. Work included constructing a new village well head and renovating many formerly abandoned buildings.   (*Continued*)

All of RSCN's crafts and products are linked to conservation (Salti 1997). In other words, the philosophy sells the products and the products sell the philosophy. This novel socioeconomic and ecological project, Jordan's first experiment in ecotourism and sustainable development for protected-area management, has completely revitalized the RSCN, giving it a clear vision and a cadre of committed and trained staff (Johnson 1997, 2007).

FIGURE 16.1 (*Continued*) Surrounded by spectacular scenery, Dana Village is one of the world's most interesting ecotourism and restorative redevelopment projects. Work included constructing a new village well head and renovating many formerly abandoned buildings.

Many jobs have been generated in craft production as well as in the creation and promotion of the niche market. And the program has been very successful such that today more than 250 such products are being sold from the protected areas. This is important in demonstrating to local people, Jordanians from across the country, and especially decision makers that nature *does* have a quantifiable value and that is why it needs to be protected (Irani 2004). In this regard, it was only after nature

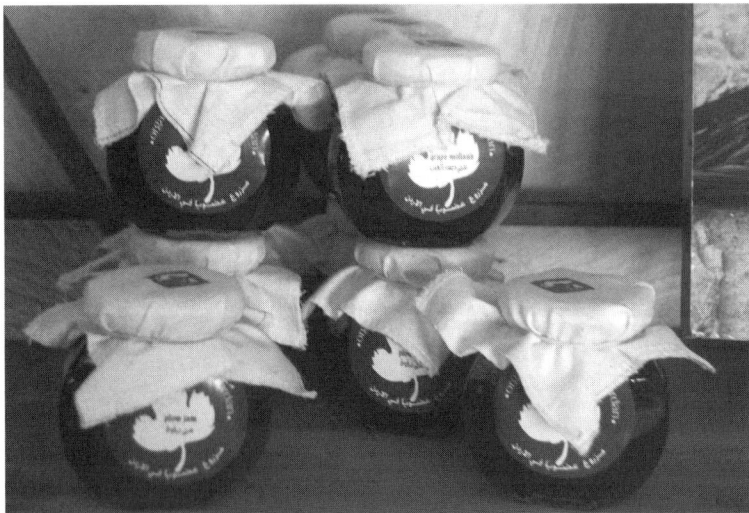

FIGURE 16.2   Locally produced fruit spreads sold in the RSCN Guest House shop.

conservation had been shown to be directly linked to local economic development that the political decision makers became interested in the RSCN's efforts. And it also represented an important shift in how the RSCN came to be regarded in the country. At one time, the organization was marginalized as nothing more than "a group of bird watchers and nature nuts" (Irani 2004). Now, however, they are considered a serious group involved with creating jobs and regional economic regeneration. The RSCN was so successful in convincing decision makers that protected areas were needed that today these former critics not only support the RSCN's efforts but also even occasionally go as far as to solicit their help in suggesting which other protected areas should be established about the country. And this all comes about from the important observation that in impoverished communities, protected areas can indeed support the local economy.

Dana began a decade ago with only a few dozen visitors a year and now receives more than fifty thousand. Today, tourists brush shoulders with local farmers working the newly renovated irrigation channels that sustain the fruit orchards (Salti 1997; Figure 16.5). It is important to note that these visitors are carefully managed so as not to exceed determined carrying capacities. The area is zoned and conservation has not been sacrificed and is still the overriding bottom line in all management decisions. In fact, wildlife has actually increased in abundance as the managed influx of tourists has pushed out the hunters (Johnson 2007). In the end, it was the once-perceived limitations that became the best promotion. Every protected area in the country now has a business plan based on the Dana success (Irani 2004). All decision making is firmly based on a model that integrates economic, social, and cultural with environmental considerations (Khouri 1997). Recently, in the Dana Reserve for example, the villagers are returning to their formerly neglected houses and renovating them into bed-and-breakfast establishments or small hotels with help

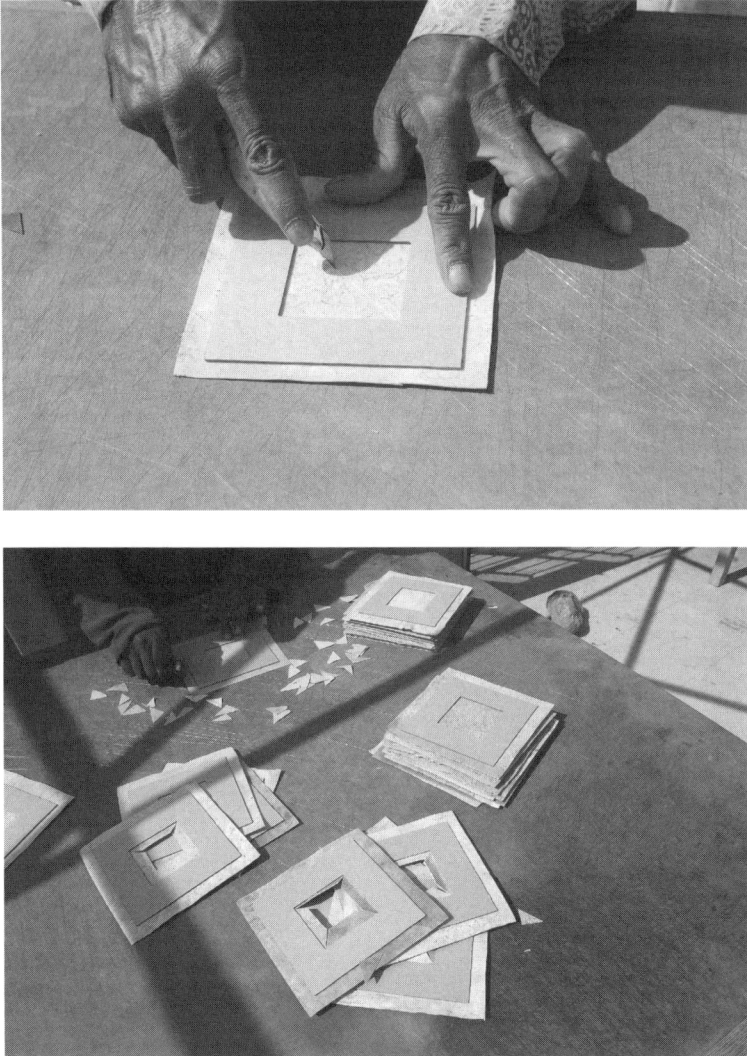

FIGURE 16.3   Locally produced crafts: photo frames fashioned from goat leather.

from some governmental grants (Figure 16.6). And the RSCN has also moved in this direction by establishing ecologic-economic infrastructure in terms of creating campsites (Figure 16.7) and an eco-lodge in several of its protected areas. The focal point for tourism is the impressive Dana Guest House, which sits in the village perched on the lip of the wadi and whose breathtaking scenery Queen Noor once referred to as a "10-star view" (Anonymous 1997; Figure 16.8). The building also houses craft workshops, a research center, meeting rooms, guest accommodation, a multimedia center, the reserve management offices, and a gift shop selling locally produced crafts.

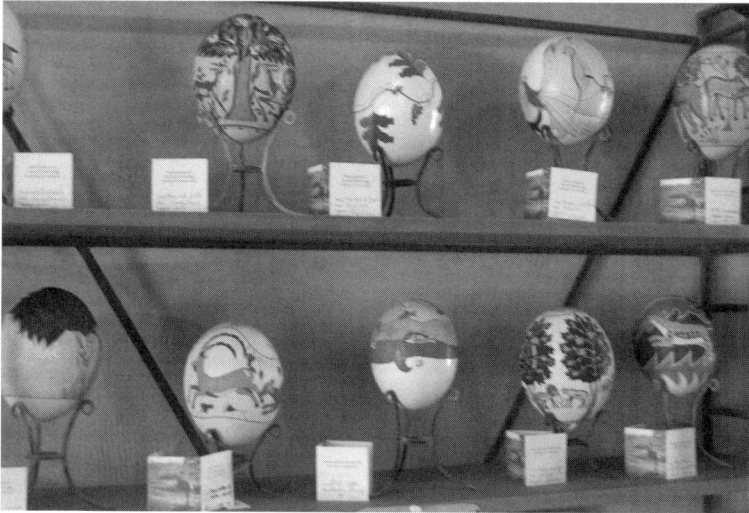

FIGURE 16.4　Locally produced crafts: metal jewelry and decorated ostrich eggs.

The lower end of Wadi Dana contains one of the largest and most important archaeological sites in southern Jordan (Findlater 1997), including remains of Bronze Age villages and Byzantine churches, cemeteries, and water works (Figure 16.9). The site is most famous, however, for its rich mining history (Figure 16.10), and its role as a major source of copper throughout the ancient Near East. In particular, the four-thousand-year-old remains of a copper mine are located here which at one time the government had wanted to reopen. Because this might have destroyed the reserve, the RSCN, aided by Queen Noor, protested against the plan and provided an alternative which convinced the government cabinet to abandon the idea. In its place the RSCN built a small and very

FIGURE 16.5   Renovated irrigation channel and restored fruit orchards on the edge of Dana Village.   (*Continued*)

rustic eco-lodge at the remote site (accessible by hiking) which uses solar power and is lit by candles, which are of course another income-generating activity for the local women (Irani 2004; Figure 16.11). So the site was saved for ecotourism at the same time as providing alternative income for those living in the area. Importantly, women have been empowered in the process and are now making decisions for their families because they are bringing money into the household. Also, such a strategy of increasing eco-literacy by targeting women through training is based on Gandhi's famous maxim that "if you educate a man, you educate one person, but if you educate a woman, you educate the whole family."

FIGURE 16.5 (*Continued*) Renovated irrigation channel and restored fruit orchards on the edge of Dana Village.

FIGURE 16.6 Tourist B & B's in Dana Village provide rustic yet comfortable accommodation. (*Continued*)

FIGURE 16.6 (*Continued*) Tourist B & B's in Dana Village provide rustic yet comfortable accommodation.

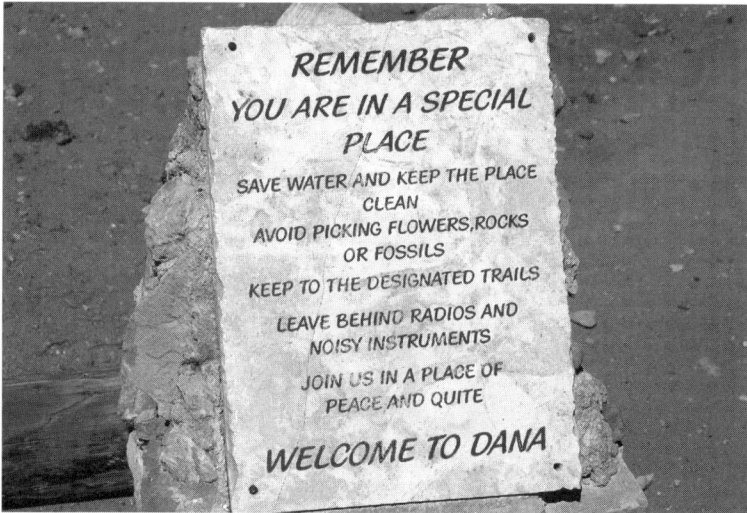

FIGURE 16.7 A remote RSCN campsite perched on the rim of the wadi offers sleeping in modern canvas tents and dining in traditional camelhair Bedouin tents while enjoying some of the most amazing desert scenery in the Middle East.   **(Continued)**

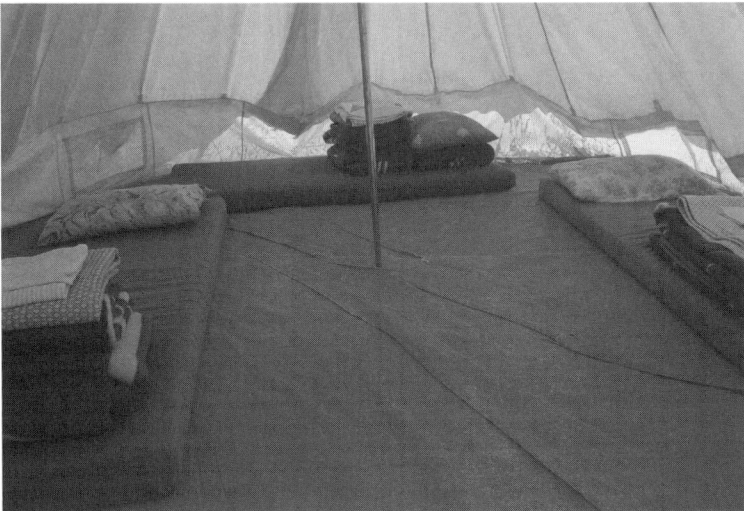

FIGURE 16.7 (*Continued*) A remote RSCN campsite perched on the rim of the wadi offers sleeping in modern canvas tents and dining in traditional camelhair Bedouin tents while enjoying some of the most amazing desert scenery in the Middle East.

FIGURE 16.7 (*Continued*) A remote RSCN campsite perched on the rim of the wadi offers sleeping in modern canvas tents and dining in traditional camelhair Bedouin tents while enjoying some of the most amazing desert scenery in the Middle East.

FIGURE 16.8   The RSCN Guest House in Dana Village is almost always completely booked due to the attraction of what Queen Noor once described as its "10-star view."   (**Continued**)

FIGURE 16.8 (*Continued*) The RSCN Guest House in Dana Village is almost always completely booked due to the attraction of what Queen Noor once described as its "10-star view."

FIGURE 16.9 At the bottom of the impressive Wadi Dana can be found numerous archaeological ruins including those of a Bronze Age village and a Roman-era church and water storage reservoir.   (*Continued*)

FIGURE 16.9 (*Continued*) At the bottom of the impressive Wadi Dana can be found numerous archaeological ruins including those of a Bronze Age village and a Roman-era church and water storage reservoir.

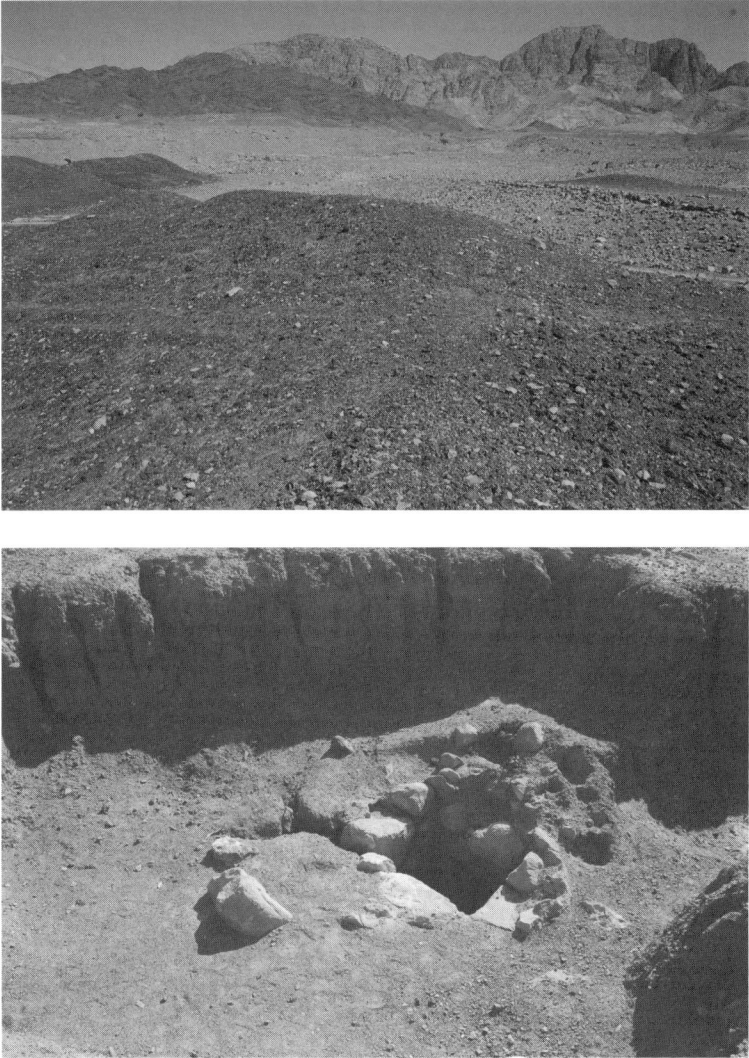

FIGURE 16.10   Ancient slag heaps and shaft openings of copper mines located in Wadi Feynan into which Wadi Dana runs.   *(Continued)*

    For all the protected areas in Jordan, virtually all the new tourism-related jobs (tour guides, rangers, campsite managers, cooks, cleaners, receptionists, shop managers, etc.) are for the local communities (Johnson 1997). Experts are not brought in from Amman. If there are no local individuals with the technical capability, the RSCN works on the counterpart system in which they send off someone for a few months or a year to another protected area location from which, fully trained, they then return back to their homes and work within their own community. In this approach, it is significant to understand that ecotourism is therefore being used as a tool for conservation rather

FIGURE 16.10   (Continued) Ancient slag heaps and shaft openings of copper mines located in Wadi Feynan into which Wadi Dana runs.

than being the first objective in itself (Irani 2004). And people are certainly coming to recognize the success of the program. In fact, the government has recently defined its own tourism strategy, and for the first time in an official document, ecotourism and adventure tourism are listed as recommended approaches to be used in Jordan.

Working from the business perspective, the Wild Jordan Café was opened in the heart of Amman to sell all the projects (Figure 16.12). Marketing was quickly realized to be the key strategy in order to compete with other attractive products being created by other companies in Jordan. The strategy here was that, because in any large national capital, citizens rarely go out into the countryside for nature, the RSCN would instead bring nature to them. Nature conservation is promoted through the sale of products. By coming to the marketing center, people learn about nature, they buy the nice products, and they know that by buying the product they are supporting both nature conservation *and* the local economy (Irani 2004). The marketing center also has a restaurant (Figure 16.13) in which Jordanians have another opportunity to learn about their country's natural resources at nightly events about nature. In this respect, the RSCN, through the many varied activities that occur at the marketing center, is reaching urban Jordanians who they had previously never hoped would become interested in nature.

The chairman of the World Commission on Protected Areas believes that nothing short of a revolution is needed in the way protected areas are regarded. In particular, it is only through developing a program of capacity building that this can be brought about. Such areas should be treated

- not as places shut away from the rest of society but instead as places valued for their contributions to society,

FIGURE 16.11   The spectacularly located Feynan Eco Lodge is popular with not only international adventure tourists but also urban Jordanians seeking out romantic weekend getaways in the atmospheric desert.   (*Continued*)

FIGURE 16.11 (*Continued*) The spectacularly located Feynan Eco Lodge is popular with not only international adventure tourists but also urban Jordanians seeking out romantic weekend getaways in the atmospheric desert.

FIGURE 16.11 (*Continued*) The spectacularly located Feynan Eco Lodge is popular with not only international adventure tourists but also urban Jordanians seeking out romantic weekend getaways in the atmospheric desert.

FIGURE 16.12 The Wild Jordan Café, situated on a large *jebel* overlooking downtown Amman, sells handicrafts manufactured in the various RSCN nature reserves. (***Continued***)

FIGURE 16.12   (Continued) The Wild Jordan Café, situated on a large *jebel* overlooking downtown Amman, sells handicrafts manufactured in the various RSCN nature reserves.

- not as places planned against people but rather as ones created with and for them,
- not as places that exist solely for their value for wildlife and scenery but in recognition of their larger contribution to society, and
- not as places designed only for tourists but also as ones taking into consideration the needs of the local community.

In this regard, the RSCN considers that its greatest success is in developing a strong component of capacity building with other institutions throughout the overall process. The RSCN started this strategy when they first began the Dana project and it has enabled them to move the project forward and to continue the work by putting these mechanisms in place such that the work becomes sustainable in the long term. Issues addressed in this procedure included strategic planning, task delegation, organizational decentralization, and business management, all of which were very important in enabling the creative part of the team to take the lead in developing the conservation process (Irani 2004).

The lessons from the Dana experience for desert tourism (Johnson 2007) include the following:

- Tourism can sustain protected areas and offers the best potential for bringing financial benefits to local people.
- Attempts have to be made to shift the economy away from environmentally destructive activities such as goat grazing.
- Planners of nature reserves should not succumb to compromise in terms of their initial conservation objectives, and all zoning land uses must be strictly enforced.

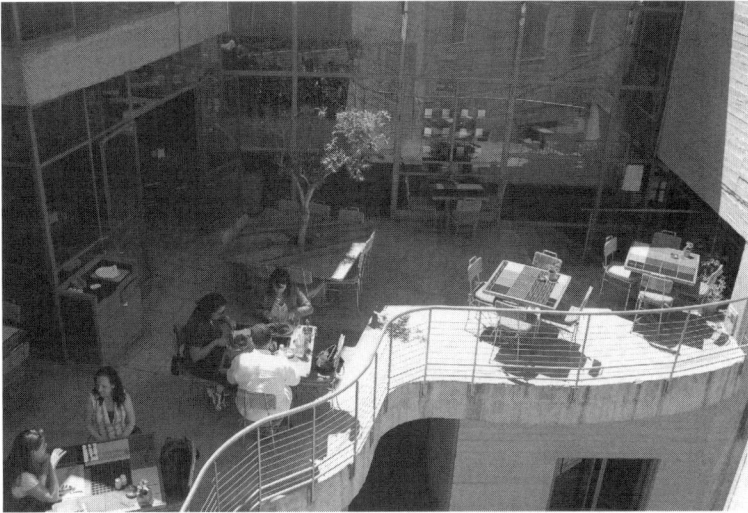

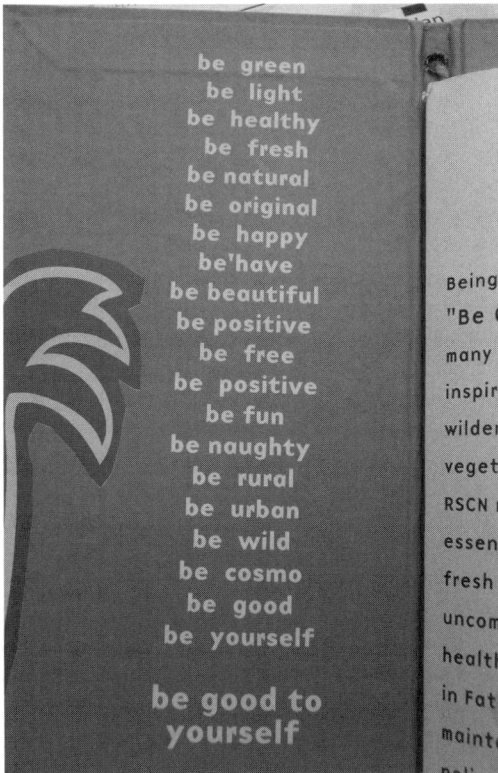

FIGURE 16.13   Many affluent Jordanians from Amman visit the café for its interesting cuisine and begin to learn about the RSCN's environmental and business mandates either while flipping through the menu or through attending various music or social events. (*Continued*)

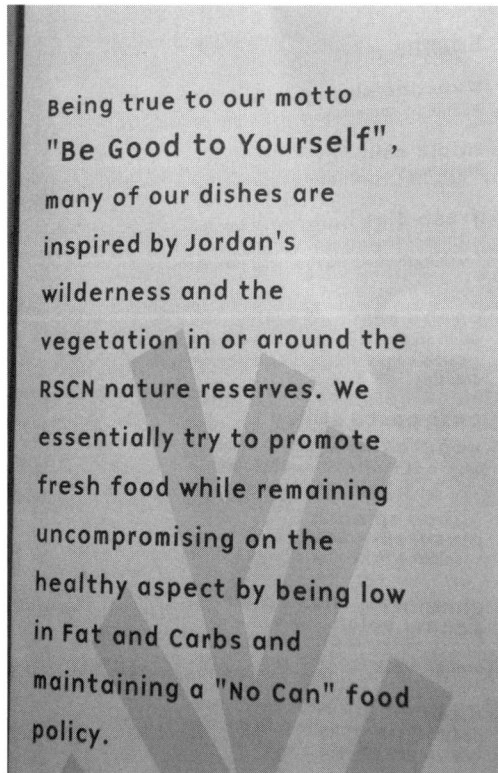

Being true to our motto
"Be Good to Yourself",
many of our dishes are
inspired by Jordan's
wilderness and the
vegetation in or around the
RSCN nature reserves. We
essentially try to promote
fresh food while remaining
uncompromising on the
healthy aspect by being low
in Fat and Carbs and
maintaining a "No Can" food
policy.

FIGURE 16.13 (*Continued*) Many affluent Jordanians from Amman visit the café for its interesting cuisine and begin to learn about the RSCN's environmental and business mandates either while flipping through the menu or through attending various music or social events.

- Tour operators need to be educated about what ecotourism is and how it can be developed.
- Greatest success comes about from mixing ecotourism with archaeological tourism.

The take-home message is that under such a regime, tourism can very easily compensate for changing livelihoods among locals.

The RSCN plans are to develop and finalize a protective network of eighteen protected areas, of which seven currently exist. Importantly, and again useful as a model for the large spatial scale of the Iraqi marshlands, the RSCN wants to move from managing these protected areas as isolated islands toward a concept based on larger regional areas such as the entire Jordan Rift Valley as a single ecological corridor in which conservation is integrated into agriculture toward an end of creating the county's first regional land use plan.

Today, Dana Village is almost at the point of becoming a victim of its own success, threatening to be turned into a sort of "mini-Arabic Las Vegas" (Johnson 2007; Figure 16.14) with several entrepreneurs installing satellite televisions and such.

FIGURE 16.14 Part of modern Dana Village decked out as an atmospheric Levantine village for an Italian television film about the life of Christ.

This has had the expected effect that tourists (both foreign travelers and Amman residents) are beginning to bemoan the loss of the rustic charm that at one time had characterized the ramshackle village perched at the edge of the wadi. Such concerns about balancing modernization with traditional values and tourism are important for the management of areas such as the Iraqi marshlands (France *in* Reed 2005) as well as elsewhere (see chapters 4 and 5).

Tourism can be a positive force for economic progress and cross-cultural understanding, but only if it is tightly regulated and monitored so that it benefits the indigenous culture as much as it brings enjoyment to the foreign tourist (Khouri 1997; see chapter 5). Unfortunately, commercial tourism in such popular locations as Petra and the Dead Sea, most overtly discerned through the rapid expansion of luxury hotels (Figure 16.15), is thought to be damaging to Jordan's natural heritage. But alternative models do exist based on the success of the Dana project. One such is the Taybet Zaman tourist village near Petra, which has won an international award for achievement in cultural heritage and environmental preservation and is becoming recognized as a signature case study in green development in the Middle East (Anonymous 2006) with possible transference to other locations in the region (Anonymous 2007).

FIGURE 16.15   Jordan is suffering from the sprawl of massive luxury hotels along the Dead Sea coast and in Petra.

FIGURE 16.16 The dilapidated ruins of an Ottoman village near the famous tourist destination of Petra (this is what Dana Village looked like a decade ago).

Taybet Zaman is a luxury, five-star tourist "hotel" that was actually a dilapidated Ottoman-period village (Figure 16.16) that had been sold to a Jordanian tourist company and whose one-time residents were paid to relocate into new dwellings located nearby (Teller 1998). Former small cottage-homes have now been renovated, and the old village transformed into an assemblage of self-contained accommodation suites, restaurants, Turkish baths, and shops that are decorated with traditional furniture and overlook the Petra Mountains. A large, corridored hotel has been replaced by an open-air tourist village with cobbled streets and the atmosphere of a nineteenth-century Jordanian town (Figure 16.17). Concepts of such innovative tourist accommodation based on regeneration that began with the RSCN's demonstrable success with the Dana Nature Reserve have therefore infiltrated and influenced the private tourism industry in Jordan and could be adapted to southern Iraq in relation to the remarkable ecological and archaeological heritage of the marshlands located there.

FIGURE 16.17 The alternative to the hotels shown in Figure 16.15: transformation of ruins from Figure 16.16 into the five-star tourist hotel Taybet Zaman, which includes shops arranged into a faux *souq* or Arab market, individual suites accessed by the old network of streets serving as hallways, comfortable accommodation, outdoor private courtyards, and dining in an atmospheric restored cellar.   **(Continued)**

FIGURE 16.17 (*Continued*) The alternative to the hotels shown in Figure 16.15: transformation of ruins from Figure 16.16 into the five-star tourist hotel Taybet Zaman, which includes shops arranged into a faux *souq* or Arab market, individual suites accessed by the old network of streets serving as hallways, comfortable accommodation, outdoor private courtyards, and dining in an atmospheric restored cellar.

FIGURE 16.17 (*Continued*) The alternative to the hotels shown in Figure 16.15: transformation of ruins from Figure 16.16 into the five-star tourist hotel Taybet Zaman, which includes shops arranged into a faux *souq* or Arab market, individual suites accessed by the old network of streets serving as hallways, comfortable accommodation, outdoor private courtyards, and dining in an atmospheric restored cellar.

ACKNOWLEDGMENT

Adapted from Irani (2004).

REFERENCES

Anonymous. 1997. The road to Dana. *Al Rem* 60:8.
———. 2006. Green buildings of the future: It's not just a concrete jungle out there! *Abu Dhabi Environment Agency Al Dhabi Magazine* 3:30–35.
———. 2007. *Plan Abu Dhabi 2030 urban structure framework plan*. Abu Dhabi: Abu Dhabi Urban Planning Council.
Boyce, J. K., S. Narain, and E. Stanton. 2007. *Reclaiming nature: Environmental justice and ecological restoration*. London: Anthem.
Findlater, G. M. 1997. Discoveries in Wadi Faynan. *Al Rem* 60:20–23.
France, R. L., ed. 2007. *Wetlands of mass destruction: Ancient presage for contemporary ecocide in southern Iraq*. Winnipeg, MB: Green Frigate Books.
Irani, K. 2004. The nature conservation programme in Jordan: Helping nature help people. Paper presented at the Mesopotamian Marshes and Modern Development: Practical Approaches for Sustaining Ecological and Cultural Landscapes conference, Cambridge, MA, October.
Johnson, C. 1997. Dana: Helping nature, helping people. *Al Rem* 60:11–14.
———. 2007. Desert tourism as a vehicle for nature conservation: The Jordan experience. Presentation at the Harvard Design School conference Desert Tourism: Delineating the Fragile Edges of Development, Cambridge, MA, April.
Khouri, R. 1997. Promoting tourism and protecting our natural and cultural environment. *Al Rem* 60:26–28.
Mannheim, I. 2000. *Jordan Footprint handbook*. Bath, UK: Footprint Handbooks.
Ochsenschlager, E. L. 1995. Carpets of the Beni-Hassan village weavers in southern Iraq. *Oriental Rug Review* 15:12–20.
———. 2004. *Iraq's marsh Arabs in the Garden of Eden*. Philadelphia: *University of Pennsylvania Museum* of *Archaeology* and *Anthropology*.
Reed, C. Paradise lost? *Harvard Magazine* 107:30–37.
Royal Society for the Conservation of Nature. 2009. [Home page]. www.rscn.org.jo.
Salti, R. 1997. Dana: The human perspective. *Al Rem* 60:17–19.
Teller, M. 1998. *Jordan: The Rough Guide*. London: Rough Guides.
Wessels, J., and R. Hoogeveen. 2007. Renovation of Byzantine qanats in Syria as a water source for contemporary settlements. In *Handbook of regenerative landscape design*, ed. R. L. France. Boca Raton, FL: CRC Press, pp. 237–62.

Section III

Innovative Approaches and Technologies

Overview: Habitation and Livelihood Context for the Restorative Redevelopment of the Iraqi Marshlands

Section 3 comprises six chapters that deal with innovative approaches and technologies to promote sustainable living in association with the restorative redevelopment of the Iraqi marshlands and other landscapes devastated by conflict or natural disasters. Together, these chapters describe means to ensure that residents can move about, treat their sanitary effluent, and husband their farms in ways that will not be deleterious to the restored environment. The most important lesson from these chapters is the understanding that the marshlands are very much living landscapes that have been and hopefully will be inhabited by thousands of people going about their daily lives amidst the wonderful nature.

In terms of transportation planning and design, chapter 17 by Ilze Jones, Ints Luters, Rene Senos, and Robert France describes an award-winning case study wherein a highway was sensitively built with the landscape rather than against it, and through the cooperation of the indigenous peoples as opposed to against their values and wishes.

For alternative wastewater treatment, reuse, and infrastructure, chapter 18 provides a comprehensive review of technological and ancillary aspects involved with treating water in a way to integrate wetland restoration into communities. David Austin's chapter (chapter 19) reviews the issues of cultural attitudes to wastewater and its decentralized treatment and harvesting for reuse. In chapter 20, Scott Wallace summarizes the planning and management challenges of operating varying types of small-scale wastewater treatment infrastructure.

Considering agriculture and water management, chapter 21 reviews the considerable problems associated with irrigation in arid landscapes and introduces a number of relevant case studies showing how salt can be effectively managed. Finally, in chapter 22, Theib Oweis and Ahmed Hachum review the particular challenges of dryland agriculture in terms of harvesting, distributing, and effectively using scarce water resources in such environments.

Transportation Planning and Design

17 Transportation Corridors as Vehicles for Cultural and Ecological Regeneration

Ilze Jones, René Senos, Ints Luters, and Robert L. France

CONTENTS

Introduction 347
Case Study—Montana's Highway 93 348
  The Landscape 349
  Relationship between People and Place 350
  Landscape Changes Impacting Cultural Values 352
  Design Approach 358
    Client as Design Partner 359
    Design and Alignment Concepts 360
    Design Guidelines and Recommendations 363
    Wildlife Crossings 364
    Revegetation Guidelines 365
    Experience of Place 367
  Cultural Landscapes 369
Conclusions 370
References 370

INTRODUCTION

Inhabitants of the remote reaches of the once extensive marshes in southern Mesopotamia/Iraq used more or less the same type of bitumen-covered boats for over seven millennia (Pournelle 2003; Ochsenschlager 2004; France 2007). Transport, however, between cities and along the margins of the large wetland has always been by roads (Young 1977), the oldest such structures in existence anywhere in the world.

347

Predictions of the maximum possible extent that the marshes might be reflooded in relation to the dramatically diminished supply of water available due to upstream diversions, are less than half the surface area that existed a century ago (France 2007, 2011). This observation, as well as the desires by contemporary marshland dwellers for increased transportation ease (for health care, selling of crafts, etc.), necessitates the presence of an extensive network of roads linking the future restored marsh patches to urban centers.

Roads alter vegetation patterns, impede wildlife movement, produce erosion, and cause chemical pollution, in addition to a myriad of wetland effects, including severed connectivity and destruction of wildlife habitat, altered hydrology and nutrient flows, and widespread contamination (e.g., Salvesen 1994; Zug 1997; Findlay and Houlahan 1997; Voss and Chardon 1998; France 2003). Therefore, while it might be possible to run straight roads across the "empty" desert bordered by a few roadside shrubs for purposes of aesthetics or reduction of blowing sand (as is common practice in Abu Dhabi) with only minor deleterious ecological effects, wetlands on the other hand are particularly sensitive to the presence of roads. It makes little sense then to invest so much in the restoration of the marshes of southern Iraq if their ecological integrity is to be immediately compromised by the widespread construction of environmentally insensitive roads.

Fortunately, new ways of regarding roads as being ecological entities in their own right (Forman et al. 2003) and as being designed to be part of, rather than in opposition to, the landscape, are beginning to take hold. And nowhere has this paradigm shift, combining ecological sensitivity and landscape architecture, been more successfully accomplished than in the award-winning redesign and reconstruction of a section of U.S. Highway 93 in Montana by Jones & Jones Architects and Landscape Architects.

Roads are of course cultural constructs that shape and are shaped by human minds just as much as they are by human hands (e.g., Route 66—the "Mother Road," the Appian Way, the Camino de Santiago, etc.). The U.S. Highway 93 story is particularly relevant to the restorative redevelopment of southern Iraq in that it is a road built not only for, but also with, the resident dwellers. Engaging indigenous or Native populations in ecocultural landscape regeneration is an important attribute to the long-term, sustainable success of such projects (e.g., Kadlecik and Wilson 2008; Senos 2008, Egan and Howell 2001) and one that should be adopted in southern Iraq.

CASE STUDY—MONTANA'S HIGHWAY 93

Design and planning projects on Native American tribal lands offer opportunities to integrate cultural values with ecological restoration and thus offer a model for the ecocultural restorative redevelopment in southern Iraq. U.S. Highway 93 is a highway project Jones & Jones worked on in western Montana, on the Flathead Indian Reservation, the homeland of the Confederated Salish and Kootenai Tribes (CSKT). The project focused on four major efforts: (1) advocating for Tribal cultural values as the foundation for design decisions; (2) finding a middle ground between the stakeholders and formalizing this in a Memorandum of Agreement; (3) developing design guidelines for stormwater treatment, wildlife crossings, revegetation, corridor

FIGURE 17.1   Regional context of the Pacific Northwest.

signage, and detailed community designs to facilitate consistency between the eight segments of the project; and (4) designing and constructing the new highway in an innovative way that fits the landscape and the culture. This project has been of key importance in opening the door to restoring vital landscape systems that support the cultural, spiritual, and subsistence life of the Confederated Salish and Kootenai Tribes.

THE LANDSCAPE

The Flathead Indian Reservation is located within the Rocky Mountain Plateau in northwestern Montana, an area where the Maritime Pacific Northwest, Intermontane West, and Great Plains ecoregions converge, creating a rich landscape mosaic (Figure 17.1). The watersheds of the Flathead and Jocko Rivers define the formal boundary of the Flathead Indian Reservation (Figure 17.2). The landscape of the Reservation is comprised of diverse ecosystems ranging from kettle pond complexes, wetlands, and aquatic zones, to forested mountains and alpine tundra, to semi-arid grasslands, to wet meadows, prairies, and rivers. The numerous habitat niches within these ecosystems support a great variety of wildlife species. At least 67 mammal species including grizzly bear, elk, moose, lynx, bobcat, fox, antelope, and bison are found within the boundaries of the Reservation. There are nine known amphibian and reptile species, and 308 resident and migrating birds inhabit the Reservation.

The section of highway that runs through the Reservation begins in Evaro at the south end of the Jocko Valley and ends in Polson where the Mission Valley meets Flathead Lake. In the Jocko Valley, U.S. Highway 93 runs roughly parallel to the Jocko River. In the Mission Valley, it runs due north to Polson, between the Flathead

FIGURE 17.2   Reservation boundaries of the Flathead Indian Reservation,

River on the west side of the valley and the Mission Mountains which define the eastern edge of the valley. It is important to note, therefore, that this road alignment runs perpendicular to the river tributaries and wildlife corridors that are used for seasonal movement of wildlife from the riverine lowlands in the winter to the mountains in the summer. Hence the north–south road cuts against the grain of major ecological forces that flow in an east–west direction across the valley floors between major continental mountain ranges (Figure 17.3).

Relationship between People and Place

The historic Tribal Homelands and trail network of the Salish, Kootenai, Pend d'Oreille and other indigenous peoples covered much of what is now known as western Montana, eastern Washington, the Idaho Panhandle, and the southern Canada provinces of British Columbia and Alberta. These indigenous groups have inhabited this region for 12,000 years or so (Figure 17.4). In this diverse landscape both hunting animals and foraging for edible plants and fruits required an intimate knowledge of the natural systems and seasons to survive. The semi-nomadic bands followed

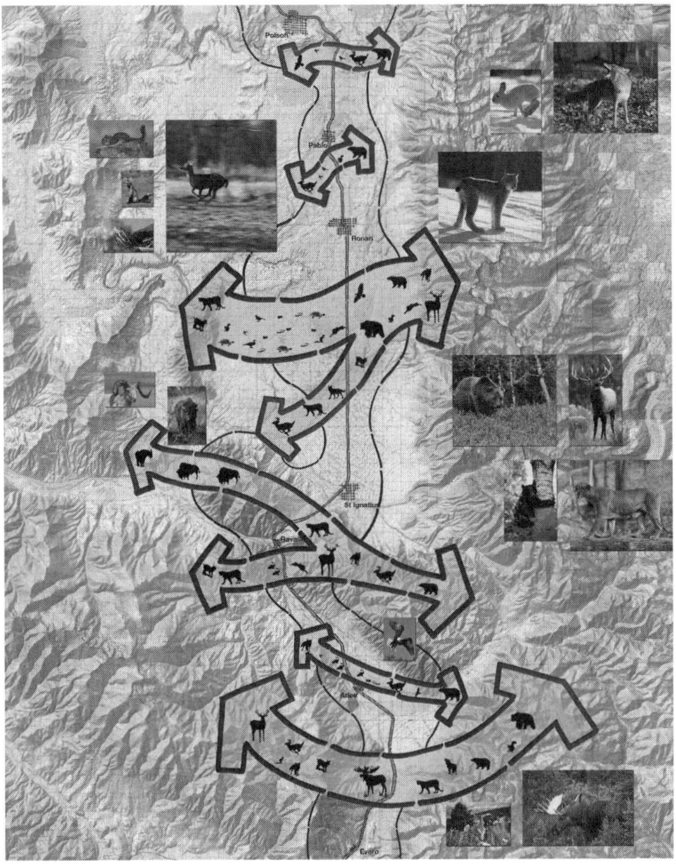

FIGURE 17.3   Wildlife connectivity across the U.S. 93 corridor.

the seasonal growth of plants and wildlife movement from the valley bottom in the winter to the mountains in the summer. Dependent on the landscape, the Native Amerindians cultivated a direct, intimate relationship with the natural environment, which met their physical, cultural, and spiritual needs (Figure 17.5).

The connection between tribal culture and the landscape continues today, although the tools have changed in response to evolving technologies. In this regard, the key concept to keep in mind is that tribal life and ties to the land are not romantic remembrances to the past. As a Washone leader observed: "We are here to stay because we have never left." Therefore, highway-related wildlife deaths represent not just a biological loss, but also a cultural loss. Traditionally animals, birds, fish, and plants provided food, medicine, clothing, and other forms of sustenance, and were the basis of daily lives in stories, songs, and spiritual beliefs. These cultural practices and spiritual connections still prevail today. Oral tradition and artwork promote the role of animals as kinfolk. As one Salish elder stated: "To us, wildlife are not just animals. The game is our medicine, our clothes, our food. Wildlife are sacred" (Figures 17.6, 17.7, and 17.8).

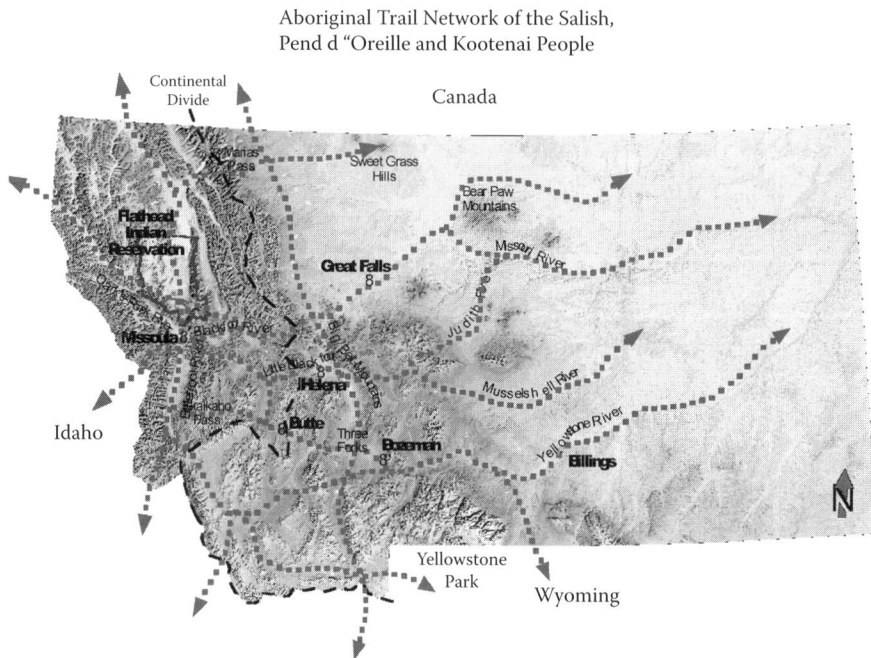

FIGURE 17.4   Historic geographic range of the Rocky Mountain Plateau peoples.

Landscape Changes Impacting Cultural Values

The ecocultural landscape of western Montana has been greatly modified over the last 150 years. The 1855 Hellgate Treaty established the Reservation, and expedited the conversion of landscape and culture from a traditional subsistence lifestyle to agrarian, homestead settlement. The 1910 Allotment Act formalized this transition by parceling the Flathead Indian Reservation into 110- or 220-acre allotments for individual tribal members to farm. Today this legacy in terms of issues of government and fragmented land ownership makes large-scale ecological planning very challenging (Figure 17.9).

During the last century agriculture and grazing have replaced prairies, forests, and riparian areas. The extensive irrigation canals and drainage systems developed to establish agriculture in this landscape significantly modified the hydrology, which in turn affected vegetation, water quality, and wildlife habitat (Figure 17.10).

One of the biggest intrusions on the landscape was the 1930s construction of U.S. Highway 93, a roadway that slices its way through the middle of the Reservation (Figure 17.11).

With each subsequent improvement, the road became wider, flatter, and straighter. These improvements allowed more traffic to move faster along the roadway, but they also required the road alignment to ignore local landform variations, cutting into hillsides and filling low-lying areas. The predecessor road was therefore an affront to the land as it sliced through ponderosa pine savannas and cut off rivers and streams from their floodplains (Figure 17.12). It was a straight line bisecting

Changing Seasons

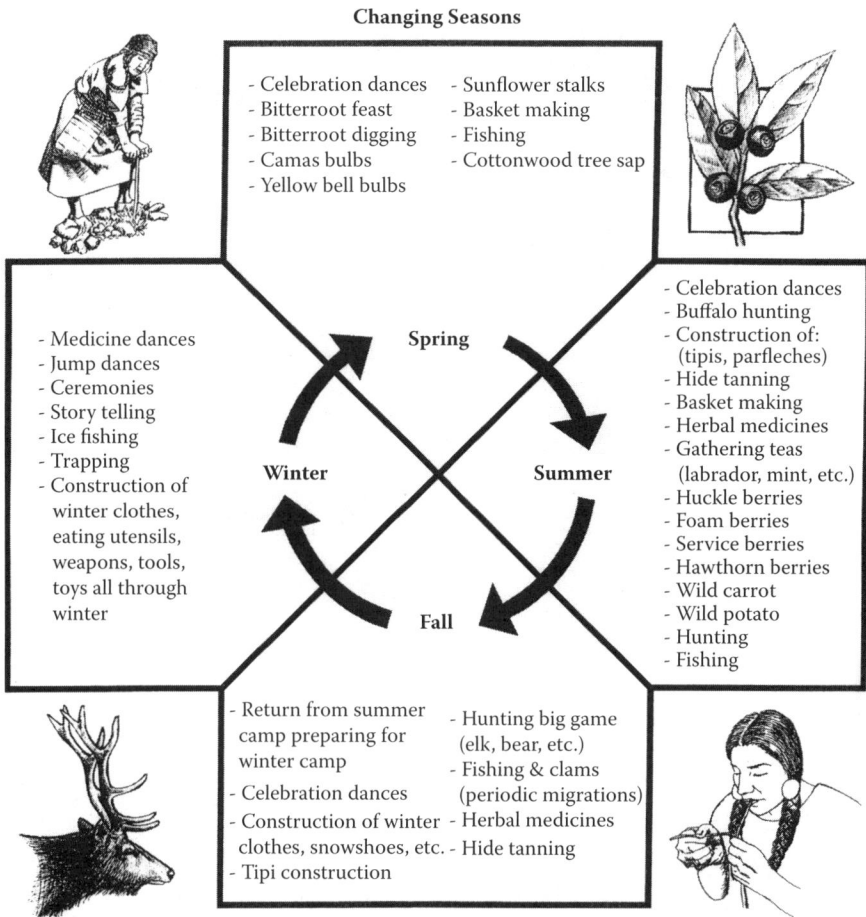

- Celebration dances   - Sunflower stalks
- Bitterroot feast   - Basket making
- Bitterroot digging   - Fishing
- Camas bulbs   - Cottonwood tree sap
- Yellow bell bulbs

- Medicine dances   **Spring**
- Jump dances
- Ceremonies
- Story telling
- Ice fishing
- Trapping   **Winter**   **Summer**
- Construction of
  winter clothes,
  eating utensils,
  weapons, tools,
  toys all through
  winter   **Fall**

- Celebration dances
- Buffalo hunting
- Construction of:
  (tipis, parfleches)
- Hide tanning
- Basket making
- Herbal medicines
- Gathering teas
  (labrador, mint, etc.)
- Huckle berries
- Foam berries
- Service berries
- Hawthorn berries
- Wild carrot
- Wild potato
- Hunting
- Fishing

- Return from summer   - Hunting big game
  camp preparing for   (elk, bear, etc.)
  winter camp   - Fishing & clams
- Celebration dances   (periodic migrations)
- Construction of winter   - Herbal medicines
  clothes, snowshoes, etc. - Hide tanning
- Tipi construction

FIGURE 17.5   Seasonal calendar.

critical ecosystems such as the extensive kettle pond complex in the Ninepipe region and severed wildlife migration corridors between the Mission Mountains and the Flathead Valley (Figure 17.13). Road impacts extended far beyond the paved road, affecting wildlife, habitat quality, connectivity, and communities along the corridor (Figure 17.14).

High accident rates, safety issues, and increased traffic volumes prompted the Montana Department of Transportation (MDT) to begin a redesign for this stretch of highway. Their preferred solution was a divided four- or five-lane highway, separated by a wide median that remained on the same straight path through the heartland of the Flathead Indian Reservation. The adverse impacts of this proposed highway design to the landscape and tribal cultural values inspired the Confederated Salish and Kootenai Tribes to develop their own preferred alternative, a two-lane highway with the necessary safety improvements. The vast difference between opposing pre-

FIGURE 17.6   Western turtles, deer, and grizzly bears are among the broad range of wildlife species whose encounters with vehicles turn deadly.

FIGURE 17.7   Western turtles, deer, and grizzly bears are among the broad range of wildlife species whose encounters with vehicles turn deadly.

FIGURE 17.8   Western turtles, deer, and grizzly bears are among the broad range of wildlife species whose encounters with vehicles turn deadly.

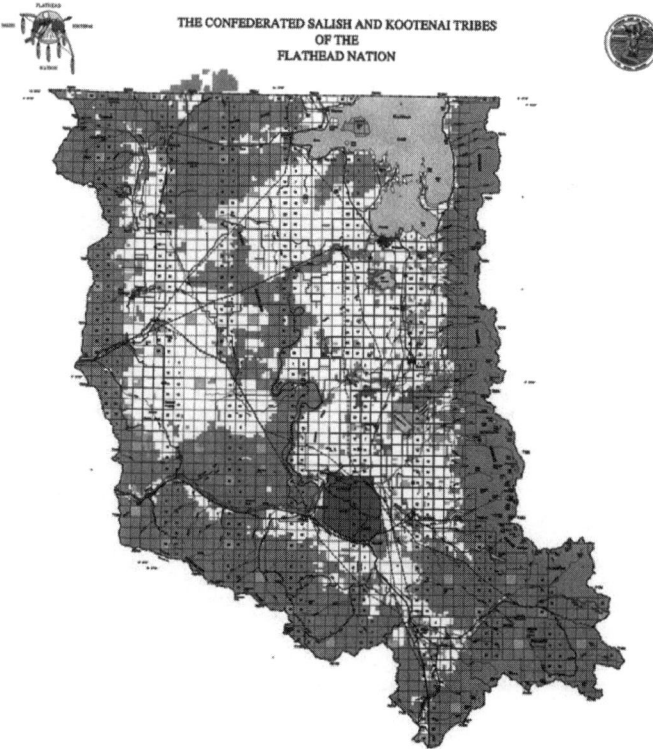

FIGURE 17.9   Land ownership and land use patterns on the Flathead Indian Reservation.

FIGURE 17.10   Farming and irrigation projects significantly transformed the landscape.

FIGURE 17.11   Aerial view of existing U.S. 93 corridor prior to highway redesign.

FIGURE 17.12   Road cut along U.S. 93 in the Jocko Valley is both an aesthetic and cultural affront to traditional tribal values.

FIGURE 17.13   Existing highway bisects kettleponds and the sensitive hydrology of the Ninepipe wetland complex.

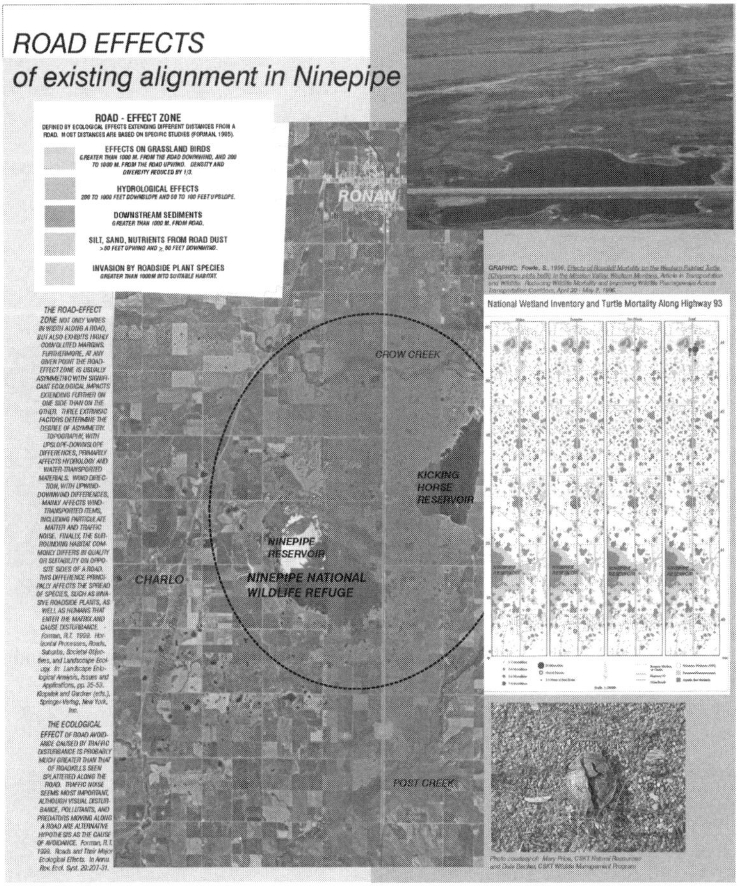

FIGURE 17.14   Road impacts extend far beyond the immediate right-of-way.

ferred alternatives sparked a ten-year stalemate between the Tribal government and Montana Department of Transportation.

DESIGN APPROACH

Because of these long-standing and unsettled problems, the Tribes asked Jones & Jones Architects and Landscape Architects to be their advocates in designing a new highway that respected tribal values. As Grant Jones explained: "This is what rarely gets spoken. There is a sense of place here, because there are spirits here in the landscape, in the ground. When does the spirit stop coming to a Place? Ask yourself: Will the road, when it's built through the valley, drive the spirit away? What can I do so the spirit is not driven away?"

FIGURE 17.15   Design collaboration with tribal government.

Client as Design Partner

Working very closely with the tribal government and state and federal transportation departments, Jones & Jones created a framework for the design process, a set of goals that captured how the road could better fit with the Tribes' homeland from respecting the character of the place, people, and wildlife to revitalizing small communities along the corridor to restoring fragmented habitat areas (Figure 17.15). These design principles ranged from respecting the character of the place, people, and wildlife, to revitalizing small town communities along the corridor, to restoring fragmented habitat areas. Specifically, these principles included:

- Develop understanding of the land and relationship of CSKT to the land
- Develop concepts that respect integrity and character of place, people, and wildlife
- Create a better visitor understanding of the CSKT homeland
- Respect and restore the way of life in small communities along the road
- Explore ways the land can shape the road rather than the other way around
- Restore habitat areas fragmented by the road

As a result of this collaboration, the design team of landscape architects, civil engineers, and traffic engineers produced a Memorandum of Agreement (MOA) for the redesign of the highway. The MOA established a three-government decisions making process with the Confederated Salish and Kootenai Tribes (CSKT), Federal Highway Administration (FHWA), and the Montana Department of Transportation (MDT) as equal partners. The MOA ended the ten-year stalemate, and established the Design and Alignment concepts, Design Guidelines and Recommendations, Wildlife Crossing Workbook, and the necessary safety and performance criteria for the project. The MOA therefore became the visionary and operational guidebook for the redesign and construction of U.S. Highway 93 (Figure 17.16).

FIGURE 17.16   U.S. 93 Memorandum of Agreement between tribal, state, and federal governments outlined the fundamental design principles for U.S. 93.

Design and Alignment Concepts

The basic premise resulting from Jones & Jones' work with the Tribes was that of the "road as a visitor." As such, the road needed to accommodate the land, people, plants, and wildlife, as well as reflecting the "Spirit of Place." The process included an evaluation of cultural and historic resources including sensitive tribal or ethnographic resources that cannot be put on any map, as well as vernacular or historic rural resources such as old homesteads. These resources were identified so impacts could be avoided as best as possible. Sensitive discussions regarding cultural and historic resources were possible because the tribes trusted Jones & Jones as design partners from the beginning of the project (this was one reason why it was especially important that the CSKT staff and tribal members were design partners from the beginning of the project).

"Spirit of place" became the mantra to promote a context-sensitive design approach that considered the full environmental and social context within which a transportation improvement project would exist. It was important to look at the whole landscape continuum and its distinct bioregions (Figures 17.17 and 17.18). And in this regard, it is important to note that because the Salish, Kootenai, and Pend

SPIRIT OF PLACE—The Landscape Continuum

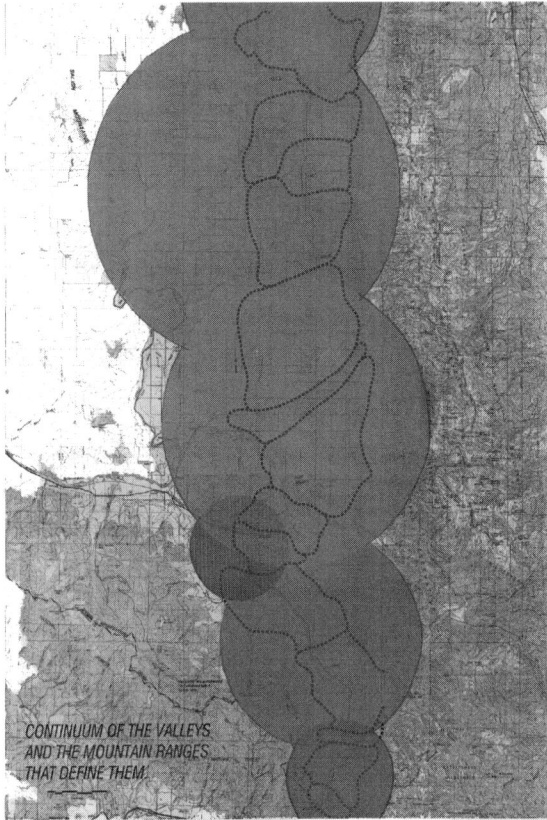

CONTINUUM OF THE VALLEYS
AND THE MOUNTAIN RANGES
THAT DEFINE THEM.

US 93 DESIGN DISCUSSIONS

Project Committee: Evaro to Polson, Montana
Montana Department of Transportation
Federal Highway Administration
The Confederated Salish & Kootenai Tribes of the Flathead Nation

Prime Consultant: Skilings-Connolly, Inc. - Consulting Engineers

December 20, 2000

JONES

JONES

Architects & Landscape Architects

FIGURE 17.17   Intrinsic qualities such as distinct landforms aggregate into discrete landscapes; together these landscapes constitute a complete landscape continuum.

d'Oreilles peoples look at impacts to land in terms of seven generations, holding a long-term view became essential.

By applying curvilinear design and flexible lane configurations, Jones & Jones were able to design a roadway that met safety and traffic concerns without drastically increasing pavement cross-section or grading. Rather than a straight, unforgiving four-lane road, a scenic corridor was created that largely avoided sensitive natural and cultural resources. Less than 50 percent of the total 55 mile corridor is being constructed as a multi-lane highway, a far cry from the 100 percent four- or five-lane superhighway originally proposed by transportation officials (Figure 17.19).

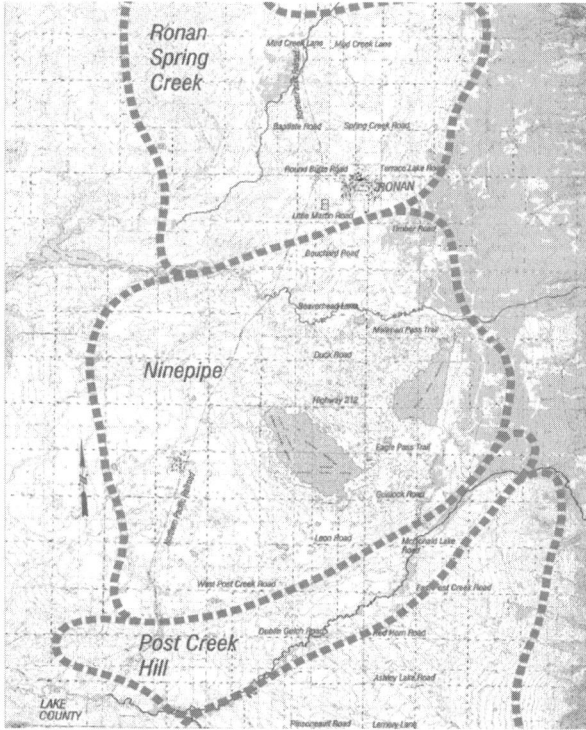

FIGURE 17.18 Intrinsic qualities such as distinct landforms aggregate into discrete landscapes; together these landscapes constitute a complete landscape continuum.

FIGURE 17.19 New curvilinear design of U.S. 93.

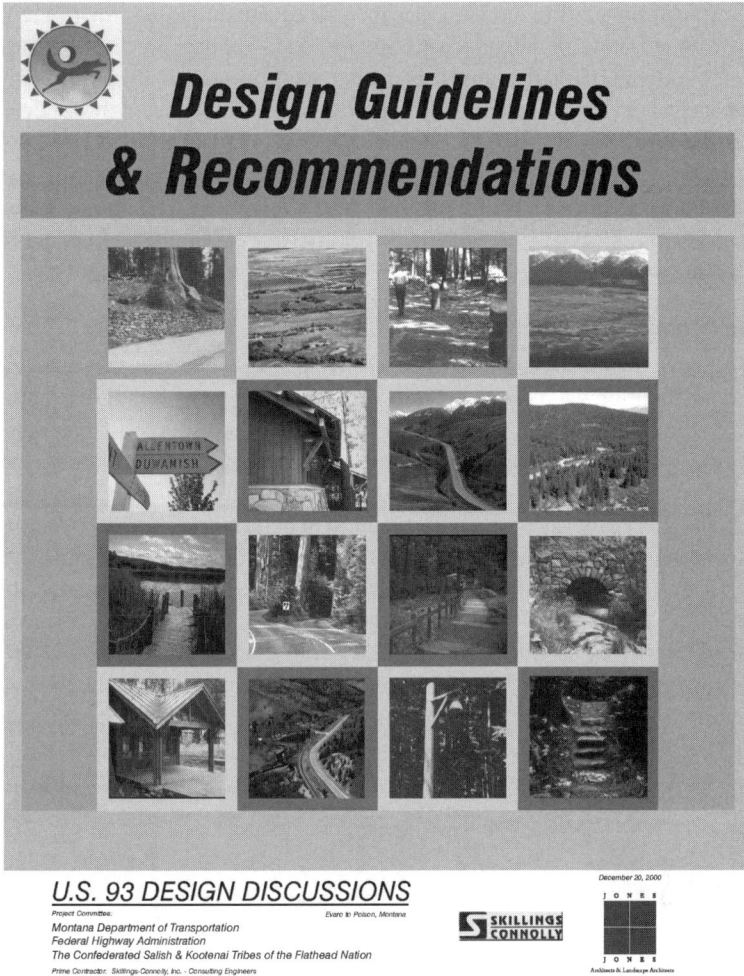

FIGURE 17.20   Wildlife Crossing Guidelines helped set the parameters for ecocultural restoration of U.S. 93.

Design Guidelines and Recommendations

The U.S. 93 design guidelines addressed road alignment, aesthetics, cultural interpretation, small town design, restoration, vegetation, wildlife crossings and habitat, hydrology, pedestrian and bike access, interpretive overlooks, signage, and visitor centers. These guidelines were aligned with tribal values as well as transportation goals, and strongly directed the design and construction of the highway facility and associated developments (Figure 17.20).

Since the existing highway bisected multiple riparian corridors and wetland systems transecting the Jocko and Mission Valleys, our design concepts incorporated hydrological and floodplain restoration. For example, small bridges or metal culverts restricting water and wildlife flows were replaced with significantly larger spans or box culverts. Wetland mitigation was also integrated into the plans. Stormwater guidelines were based on regional standards and included appropriate Best Management Practices. And water quality and fish habitat were improved through daylighting buried waterways in both rural and urban areas. Detailed concepts were developed for each wildlife crossing which included restoration plans to restore native vegetation, improve hydrologic functioning, facilitate wildlife migration, and enhance habitat structure. Also, the cultural uses of these sites were taken into consideration; i.e., the major creeks and rivers were and still are networks for traveling, harvesting, hunting, and camping.

Wildlife Crossings

Simply improving hydrology and connectivity of waterways is not enough. Throughout the highway corridor, there was a need to provide wildlife habitat for cover, forage, and movement. Forty-two wildlife crossings were incorporated into the 56-mile highway stretch, strategically located in response to known patterns of wildlife migration such as along drainage paths. Tribal wildlife data on animal behavior and mortality rates were instrumental in determining proper structure type and placement. Wildlife crossing structures ranged from a large overpass near the town of Evaro to bridges to large box culverts to smaller 1.2 × 2 m undercrossings (Figures 17.21 and 17.22). Throughout the agriculture-dominated corridor, there was a need to provide wildlife habitat for cover, forage, and movement. Native vegetation typifying one of the sixteen distinct plant communities along the highway was specified by the design team to be planted along the roadside, and clustered

FIGURE 17.21   New Jocko River bridge.

FIGURE 17.22   Large culvert at Mission Creek.

around each wildlife crossing. All these crossings and the associated site restoration activities will remediate damage caused by the existing highway by restoring wildlife passage, stream and river floodplains, and native vegetation. Given the rarity of such an extensive system of wildlife crossing structures in the United States, the Western Transportation Institute (Hardy 2007) has conducted pre-construction monitoring and will continue with several years of post-monitoring to assess wildlife use and survival (Figure 17.23).

Revegetation Guidelines

Grazing and clearing have eradicated most of the native vegetation, especially sensitive plant communities such as palouse prairies. Working with CSKT staff and botanists, Jones & Jones developed restoration plans for 16 different plant communities identified within the highway corridor. The revegetation of the corridor was based on the full complement of species found in each of the native plant communities bisected by the road. To support cultural values consideration was given not just to the plant species that will survive corridor conditions but also to where opportunities might exist to use a broader plant species palette that will restore a particular cultural landscape type. By identifying the target plant communities and projected plant quantities well ahead of construction, time-sensitive activities such as seed collection, plant salvage, and propagation were started early enough to meet the need for revegetation materials once the road construction was completed.

The U.S. Highway 93 Revegetation Guidelines were therefore developed to identify key revegetation strategies to support cultural values, and to ensure successful

FIGURE 17.23   Cougars captured by wildlife monitoring camera at Ravalli Hill underpass near National Bison Wildlife Refuge.

native revegetation of a harsh roadside environment. These guidelines included salvage, preservation, restoration, and maintenance recommendations. Preservation and salvage were the primary means used to minimize revegetation costs, first by minimizing the area disturbed during construction, and second by salvaging vegetation from areas that required disturbance. The salvaged material provided an excellent source of locally adapted plants that could be installed along the re-constructed roadside.

Revegetation provided other functions in addition to the standard roadside erosion control. At the wildlife crossings plantings reduced the visual and noise impacts of the highway, and provided improved habitat to facilitate structure use. This concept is illustrated in the planting plans for wildlife crossing undercrossing at Ravalli Hill, which clearly show the connection between vegetation, wildlife fencing, jump-outs, and wildlife structure (Figures 17.24). These elements work together as a system to promote safe wildlife passage through the structure, rather than across the road.

Revegetation also restored the visual character along the corridor, a landscape characterized by open agricultural areas punctuated by rural towns. Native landscape plantings helped to integrate the highway within the six communities along the U.S. 93 corridor, and also enhance the identity of these distinct towns or cities by reinforcing the particular regional plant palette. For example, the town of Pablo is the tribal government center and is located in a unique dune and Ponderosa Pine landscape complex. This dune and pine system will now flow

FIGURE 17.24   Wildlife habitat connectivity and transportation corridor converging south of Ravalli.

across the road rather than be bisected by the corridor as it was previously. Thus the tribes' political center will be seated within a landscape context that better fits tribal values.

Experience of Place

The Salish, Kootenai, and Pend d'Oreille are very interested in instilling visitor respect and appreciation of their homelands. The design team developed a series of interpretive features along U.S. 93 to enhance both residents' and visitors' experience of place. Visitor education and appreciation of the landscape and culture is essential to cultivate respect. Interpretive overlooks, place name signage, and community signs now introduce travelers to the tribal homeland and cultural values. Key to enhancing the visitor's experience of place were interpretive overlooks designed to introduce visitors to the Reservation. There are two overlooks, one at Ravalli Hill overlooking the Mission Valley and the National Bison Range, the other at Polson Hill overlooking Flathead Lake. Visitor education and appreciation of the tribal homeland was essential to cultivate respect of Amerindian culture. The overlooks are constructed of native materials and nestled into the landscape and will incorporate native artwork and oral traditions to tell the story of the landscape and its people (Figure 17.25).

Jones & Jones addressed community design and how to preserve rural character while integrating transportation corridor issues with small town planning. In one community, for example, the design team incorporated aesthetics, community identity, and open space design with road improvements. One project included capitalizing on an opportunity to daylight a buried creek running through the town center.

FIGURE 17.25 Interpretive features introduce visitors to the majestic homeland and values of Confederated Salish & Kootenai Tribes.

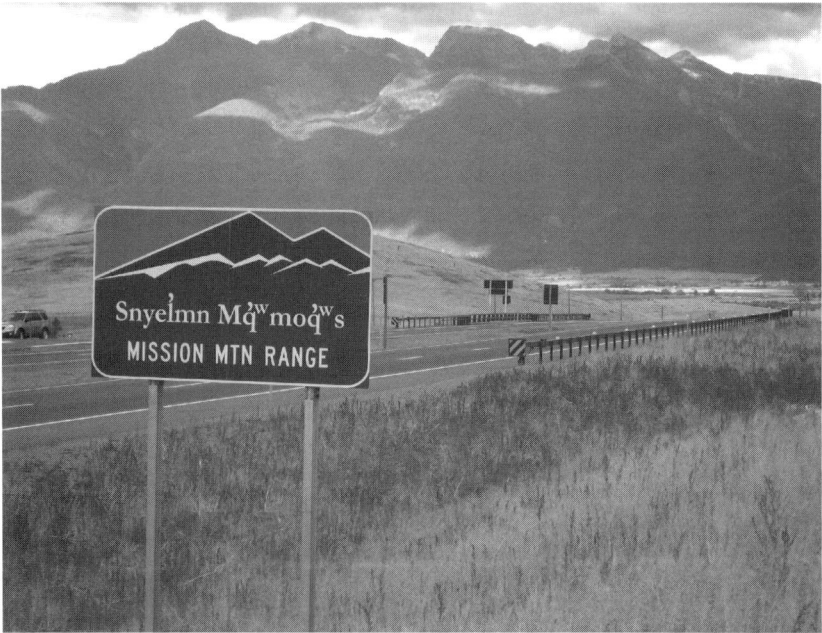

FIGURE 17.26 Interpretive features introduce visitors to the majestic homeland and values of Confederated Salish & Kootenai Tribes.

FIGURE 17.27   Ethnobotanist Joanne Bigcrane propagating native plants in the tribal nursery.

Signage reinforces a visitor's perception of entering and moving through a landscape defined by tribal cultural values. Along the road corridor, a consistent signage program was established to identify culturally significant place names, Reservation boundaries, and community entry signs (Figure 17.26). In some communities, enhancing the experience of place included designing formal streetscapes for urban commercial areas.

CULTURAL LANDSCAPES

Prairies are key cultural landscapes on the Reservation. As a result, culturally utilized plants were purposely included in the restoration design. Close collaboration with CSKT Tribal Preservation staff and tribal elders helped to understand the plants of traditional value (Figure 17.27). As the diversity of plants and animals decreases, the cultural base also becomes depleted. As one tribal member put it: "Culture is dependent upon the environment…Without the wilderness and natural environment, human beings cannot get advice and information to pass to youth…Without the environment there is no culture."

Furthermore, holistic landscape restoration needs to include cultural practices, e.g., foraging and burning, as part of the repertoire of restoration. There are five main ways that First Nations Peoples have influenced the nature of North American vegetation: burning, herbivore predation, food gathering, agricultural practices, and plant introductions (accidental and intentional). In restoration it is important to look to Nature as teacher, but also to move ahead to seeing nature as a partner by including these cultural practices to maintain health. Holistic landscape restoration therefore needs to include cultural practices such as for example foraging and burning as part of the repertoire of restoration.

CONCLUSIONS

The redesign of U.S. Highway 93 is an award-winning example of how required infrastructure improvements developed in partnership with a place and its inhabitants can restore vital landscape systems that support cultural, spiritual, and subsistence values. The adaptability of this approach, sensitive to the ecocultural landscape of indigenous peoples, is an obvious model to use for transportation planning in the marshlands region of Iraq. Key lessons to be learned from this project include:

- Design and planning projects on tribal lands provide an opportunity to restore damaged land and cultural resources. Hundreds of generations of a people living in one place has by necessity aligned CSKT cultural values with ecological values. Their long-term perspective is embodied in the CSKT concept of considering impacts for seven generations. Advocacy of CSKT cultural values promoted ecological values.
- Following the example set by the CSKT, look to Nature as Teacher and pay attention to the land, listen to the plants and animals, and hear what they are telling you. Then embrace Nature as Partner and incorporate cultural practices into the ongoing care of the land.
- Cultural values can drive ecological restoration. "Wildlife as sacred" became the impetus for whole landscape restoration. This strategy avoided the limitations of the usual approach that typically focuses only on threatened or endangered species. Instead, the full range of species were considered.
- Honoring the "Spirit of Place" structured the design process, requiring a holistic approach for the redesign of U.S. Highway 93.
- Emphasize partnership and collaboration by finding solutions that benefit multiple stakeholders. Only solutions that were acceptable to all were included in the Memorandum of Agreement.

REFERENCES

Egan, D., and E.A. Howell, eds. 2001. *The historical ecology handbook: A restorationist's guide to reference ecosystems.* Island Press.

Findlay, C.S., and J. Houlahan. 1997. Anthropogenic correlates of species richness in southern Ontario wetlands. *Conserv. Biol.* 11:1000–1009.

Forman, R.T.T., and 13 coauthors (including R. France). 2003. *Road ecology: Science and solutions.* Island Press.

France, R.L. 2003. *Wetland design: Principles and practices for landscape architects and land-use planners.* New York: W.W. Norton.

———, ed. 2007. *Wetlands of mass destruction: Ancient presage for contemporary ecocide in southern Iraq.* Winnipeg, MB: Green Frigate Books.

———. 2011. *Restoring the Iraqi marshlands: Potentials, perspectives, practices.* Sussex, UK: Sussex, Academic Press.

Hardy, Amanda. 2007. Developing the integrated transportation and ecological enhancements for Montana process: Applying the "eco-logical" approach. In *Transportation Research Journal.* Transportation Research Board of the National Academies. Vol. 2011, pp. 148–156.

Kadlecik, L., and M. Wilson. 2008. Cultural and environmental restoration design in northern California Indian country. *Handbook of regenerative landscape design,* ed. R. L. France. Boca Raton, FL: CRC Press. pp. 315–341.

Montana Department of Transportation, Confederated Salish & Kootenai Tribes, and Federal Highway Administration. 2001. Memorandum of Agreement: US 93 Evaro to Polson. Pablo, MT.

Ochsenschalger, E. L. 2004. *Iraq's marsh Arabs in the Garden of Eden.* Philadelphia: University of Pennsylvania Museum Anthropology and Archeology.

Pournelle, J.R. 2003. Marshland of cities: Deltaic landscapes and the evolution of early Mesopotamian civilization. PhD diss., University of California, San Diego.

Salvesen, D. 1994. Wetlands: Mitigating and regulating development impacts. Urban Land Institute.

Senos, R. 2008. Rebuilding salmon relations: Participatory ecological restoration as community healing. *Handbook of regenerative landscape design.* ed. R.L. France. Boca Raton, FL: CRC Press. pp. 205–235.

Senos, R., F. Lake, N. Turner, and D. Martinez. 2006. Traditional ecological knowledge and restoration practice. *Restoring the Pacific Northwest: the art and science of ecological restoration in cascadia,* ed. D. Apostol, and M. Sinclair. Island Press. pp. 393–426.

Voss, C.C. and J.P. Chardon. 1998. Effects of habitat fragmentation and road density on the distribution pattern of the moor frog. *J. Appl. Ecol.* 35:44–56.

Young, G. 1977. *Return to the marshes: Life with the marsh Arabs of Iraq.* London: Collins.

Zug, L.S. 1997. Habitat, water quality, and wetland preservation: U.S. Route 220 (I-99) replacement wetlands, Blair County, Pennsylvania. *Wetland J.* 9:3–7.

Alternative Wasterwater Treatment, Reuse, and Infrastructure

18 Linking Water Treatment with Wetland Restoration

Engineering Challenges and Associated Benefits*

CONTENTS

Introduction... 375
Natural Treatment Processes.. 378
Hydrology and Hydraulics... 381
Planting Design.. 382
Public Use and Operations and Maintenance 383
Case Studies in Water Treatment and Restoration Benefits............................ 385
Conclusion 387
Acknowledgment 389
References... 389

INTRODUCTION

There have been many discussions about the human needs and some of the benefits associated with restoring the marshes in southern Iraq (e.g., France 2007), but the issue that James Bays from CH2MHILL is most concerned with regards the specific tools that can be employed to undertake and sustain those restoration efforts through the long term. And one of these tools is a natural system-designed philosophy that involves exploitation of natural processes for contaminant removal and habitat improvement.

Treatment wetlands (Bays 2002; France 2003) as a technology and restoration tool have been in application for over fifty years. Of the many technologies available (see chapters 19 and 20; Steinfeld and Del Porto 2007), one relies on the use of surface-flow wetlands designed to receive pretreated effluents for water "polishing" and habitat restoration. The natural processes that occur in wetlands are both robust and malleable (Bays 2004), with dozens of species that can be used in any typical treatment technology and many hundreds more potentially available from the rich natural heritage and diversity.

---

* Adapted by Robert L. France from Bays, J.S. 2004. Wastewater and salinity treatment in wetlands. Paper presented at the Mesopotamian Marshes and Modern Development: Practical Approaches for Sustaining Restored Ecological and Cultural Landscapes conference, Cambridge, MA, October.

The assessment and analysis techniques that are now available, including hydro-dynamic modeling, hydrologic water balance models, and landscape appraisals, far exceed in complexity, precision, and power those available even just a decade ago. Importantly, with regard to the Iraqi marshlands, treatment wetland technologies are scalable (i.e., they can be built for small sites and then scaled up to a large scale). At the same time, this becomes a good way to build knowledge and move forward, and in this regard stakeholder input is an important tool in terms of deciding what is desired and creating a plan to achieve this (see chapter 4). The well-known educational, recreational, and wildlife benefits of wetland construction (France 2003) are tools in themselves, useful in that they help persuade others that these technologies can be applied.

The conceptual model is based on building a "treatment train" (*sensu* Apfelbaum 2005) where initially there are some pretreatment basins where the heaviest contaminant loads are removed through aeration and settling. Next the water flows into a treatment marsh of alternating shallow and deep zones, perhaps to a larger system for habitat enhancement, followed by either discharge to a receiving water body or recycling back to the beginning of the system. The goal is to design these wetlands within their tolerance range in terms of constituent and hydraulic loadings (Bays 2002).

There are really several types of wetland technologies that can be engineered (Bays 2004). The simplest is a natural wetland where effluent is applied to a natural system and the engineering is simply constructing a distribution system and letting the system process that wastewater naturally. In the real world, there is an open marsh and the distribution system might be a gated pipe perhaps located on a boardwalk for access and maintenance. Another type of technology is surface-flow constructed wetlands which more or less resemble what people think that a traditional marsh should look like. Typically the system is enclosed with berms and contains plants in shallow water. One such example is the Dupont Nylon Treatment Wetlands in Victoria, Texas (Bays 2002), which receives industrial wastewater from a treatment plant and processes it through a series of treatment wetland "cells." Significantly, this particular treatment wetland also contains habitat islands and open-space features, and thus represents a good example of compatibility between treatment as a remediation tool and habitat creation as a restoration tool. The point is that such surface-flow wetlands can offer significant recreational educational opportunities (Bays et al. 2002; France 2003). In the United States there are many examples of such systems that have been built as outdoor laboratories or educational and bird-watching facilities. Industrially contaminated waters are also treated by surface-flow wetlands such as the system in Richmond, California, which receives and removes petroleum refinery runoff effluent. One example of a surface-flow wetland with particular reference to the environmental conditions in southern Iraq is the Sweetwater Wetlands in Tucson, Arizona (Bays 2002), another arid region where there are just a few centimeters of rainfall a year. This wetland is designed to be a flow-through treatment wetland to improve the quality of the reclaimed water prior to its use to recharge the groundwater as well as provide a wildlife habitat marsh with educational and recreational benefits (Figure 18.1).

FIGURE 18.1 CH2MHILL's Sweetwater Wetlands in Tuscon, Arizona. An example of a multipurpose wetland park in an arid region for water purification, wildlife enhancement, and human recreation.

The final type of constructed treatment wetlands is based on subsurface flow (see chapters 19 and 20). These are either gravel- or soil-based systems where the applied water flows below ground and through the roots of all the plants. There are a number of types of these systems but all share in common reliance upon the complex microbial communities that exist between the soil particles and within the vicinity of the roots. Subsurface-flow wetlands can look like gardens or like traditional wetlands depending on the art of the designer and the requirements of the project (Bays 2004).

There are as many as six thousand constructed wetlands currently in operation in North American and Europe. In addition to municipal wastewaters, constructed wetlands are used to treat stormwater, a wide range of industrial effluents including pulp and paper, food processing, landfill leachate, and acid mine drainage, as well as agricultural runoff from both feeding-lot operations and crop fields (France 2003).

Wetland design and performance data are readily available. Three books that CH2MHILL engineers regularly consult include *Constructed Wetlands for Waste Water in Europe* (Vymazal et al. 1998); *Treatment Wetlands*, 2nd ed. (Kadlec and Wallace 2004); and *Natural Systems for Waste Water Treatment: WEF Manual of Practice* (Water Environmental Federation 2005). All these documents provide ready guidance for anyone charged with having to apply this information, and Bays (2004) believes that if any wetland engineer went to Iraq today, he or she would want to bring books such as these in order to begin progress.

There is a set of questions that people often ask about treatment wetlands. The first: do the wetland systems need to be harvested? In point of fact, it is actually desirable and important for plants to accumulate in the wetlands in order for them to build up a new layer of soil. On the other hand, it is possible, and sometimes necessary for vector and hydraulic purposes, to periodically harvest plants. For the marshes in southern Iraq, the harvesting and commercial use of the reeds is actually an important part of the heritage (Ochsenschlager 2004) which could represent another somewhat novel end product of constructed treatment wetlands (Bays 2004). The point is that it is certainly possible to harvest plants successfully if there are reasons for doing so but it is not necessary to undertake this task in relation to the performance of the treatment system.

Do accumulating sediments need to be removed? Because accumulation rates are very slow (usually in order of mm or perhaps a cm a year at the most in the most extreme case) the need for removal is extremely rare for municipal systems. In the case of stormwater treatment wetlands, the high inorganic load of sand and silt particles is usually pretreated and removed in a special forebay, and as such, there is no need to remove them from the marsh itself.

Is there an ecological risk to these wetlands? Most treatment wetlands receive water in which the toxic constituents are relatively dilute and not at levels that would bioaccumulate or be harmful. Industrial effluents, however, may deserve extra attention. And selenium, which has been identified in some of the waters in the Iraq marshes (France 2011), may be a concern that requires special assessment and consideration (Bays 2004).

Mosquitoes and wildlife can be problems that have resulted in decades of wetland destruction in attempts to try to control these vectors. But we know from experience that vector control is a tractable problem which can be solved by management and design approaches (Bays 2004). Odors are not a problem in treatment wetlands; they simply smell like marshes and ponds and not like sewage.

NATURAL TREATMENT PROCESSES

Consider a shallow wetland where there's about a foot or so of water overlying a layer of soil (Bays 2004). As the plants grow and die through their annual cycle of renewal, they are building really a layer of soil by laying out a new layer of detritus, which becomes the substrate for the microbes that provide the basis for much of the treatment. Physical settling processes also take place. The important point is that though we see the plants and the plants provide a layer of this fuel for the system, most of the processes are happening at the microscopic level. The big issue of concern for designers is exactly what elements are adjustable in a system like this, and how can one really manage and maintain and operate these systems? It turns out that there are really many factors that are readily amenable to direct manipulation such as water depth, the concentration of the pollutant input relative to the size of the system, the hydraulic loading rate in terms of how much water is being applied, the configuration and the type of wetlands, and the abundance and type of plants. All these variables can have a big impact on the type of performance that can be expected from any constructed wetland. As Bays (2004) stated, treatment wetlands do have a few "knobs" that can be easily turned one direction or another in order to obtain the desired outflow quality goal.

Past performance provides preliminary design loading criteria. The first question with a treatment wetland design is how large does the system have to be to achieve the goals, either in terms of receiving the flow that is available or for processing the wastewater to the desired goal (Bays 2004)? If contaminant loading versus outflow concentration from hundreds of treatment wetlands is plotted, spanning a range of input loads on the order of magnitude, a proportional change in response is observed. In other words, the more the system is loaded, the higher the outflow concentration. The important point is that it is a predictable relationship that produces design information (i.e., simply fitting a regression equation into this provides coefficients and constants and a loading rate).

For biological oxygen demand (BOD), a number of processes occur within wetlands that are important (Bays 2004). Dissolved oxygen is supplied to the wetlands through passive aeration and through the re-aeration of soils by the plants themselves. But dissolved oxygen is taken up in proximity to the sediments through the microbial activity that is consuming it at a great rate. Longitudinally within a wetland, the consumption of the oxygen produces a marked gradient in concentration. Construction of weirs between adjacent cells is a good strategy for re-aeration. Oxygen is important not only in its own right but also through its influence on the nitrogen cycle. The oxygen content of water is also dependent upon temperature and salinity, and this is a factor worthy of consideration in Iraq (Bays 2004) (i.e., the higher the temperature, the lower the oxygen saturation; the higher the salinity, the lower the oxygen content). And so, if water has been present on the landscape for a long time it is already going to have a low oxygen content which will be a factor of potential significance for fish populations.

A well-established wetland has a good cover of plants that shade the water column and as a result limit algal growth, which in effect controls the pH of the system. Influent pH is often very variable, whereas outflow pH is much more stable. Wetlands are, therefore, actually buffering systems.

Wetlands naturally assimilate nitrogen though sedimentation, uptake, and atmospheric losses. Nitrogen is taken up by the plants, incorporated into their biomass, and returned back to the wetlands and water column as an organic nitrogen source.

This is then modified through bacterial activity to ammonia and is further modified and transformed to nitrites and nitrates by bacterial processes, all of which takes place in an aerobic water column (hence the importance of adequate dissolved oxygen). But the largest removal process is really denitrification. Wetlands aren't always accruing nitrogen from the amount being added; they're also losing a lot to the atmosphere as part of the natural cycle. And this is one of the sustainable processes that make wetlands work for a large-scale nitrogen improvement (Bays 2004).

It is possible to make some broad conclusions about treatment wetland performance (Bays 2004; France 2003). The general range of performance from wetlands is in the order of 50 to 90 percent removal for BOD, total suspended solids, and the nutrients nitrogen and phosphorus. But importantly, in terms of public health concerns and considering wetlands as part of a treatment train for potable water supply or for wastewater treatment, they're especially good at removing pathogenic bacteria to background levels. Other constituents such as metals are also removed well by wetlands, essentially down to detection limits.

One example to show how treatment wetlands work is a project by CH2MHILL near Indio in southern California (Bays 2004). This is an area with an extremely low level of rainfall of about 8 cm per year, and with evaporation rates in the order of about 250 cm per year—in other words, essentially the same kind of climate as that in southern Iraq (see chapter 21 for other CH2MHILL projects in California). This project involved construction of a small wetland, 6 ha in size, that receives about 3,000 m^3 per day of flow. What is special about this case study is that the inflow to the wetland is treated primary effluent, not secondary treatment quality effluent. As a result, the inflowing water is characterized by very high BOD. And with a long residence time, in the order of thirty days or so, there's plenty of opportunity for water to evaporate from the system, something that needed to be factored into the design (Bays 2004). A curve was fitted to the system which mapped the removal rate of BOD through time and along the treatment train, which consisted of three separate cells. In the first cell, there was an almost straight-line reduction, showing a very high removal rate from about 140 to 40 mg/L BOD. In the second cell, the removal rate leveled off, ending in about 30 mg/L, and in the last cell there was a bit more removal, with a final outflow concentration of about 10 mg/L. This sort of analysis gives insight into how the system actually works. This first treatment cell takes out those constituents within BOD that are readily decomposable and degradable, and as a result, does most of the "heavy lifting" (Bays 2004). By the time water moves into the second cell, it has been conditioned somewhat. The BOD has changed and been transformed and it takes longer for the organic compounds to decompose and be degraded. And by the third cell, the BOD has been removed and transformed so now it's even easier for the wetland to remove that organic load. For the case of wetlands in Iraq that are receiving a high BOD load, they could easily be calibrated to this model (Bays 2004). In this respect, removal rate constants could be developed that would be specific to the climate and the amount and type of wastewater being generated for each locale in the country.

HYDROLOGY AND HYDRAULICS

Contaminant removal processes in wetlands are intimately connected to issues concerning water movement and residence within the treatment cells. It is important to define a few terms in this respect (Bays 2004). "Water regime" refers to the spatial and temporal distribution of water elevations in a wetland; that is, how deep it is and how depth is distributed throughout the system. "Hydro-period" refers to the number of days per year or percentage of time that surface water is present in a given wetland location. Because for treatment wetlands the strategy is to take advantage of the natural process that occurs in the sediments in a flooded condition, an objective becomes to keep the system flooded year-round. But that may not necessarily be a sustainable water regime for a given area, and it is possible that a desirable state may actually be to cycle the hydro-period by varying it over time. "Flooding frequency" considers how often the wetland floods. This is critically important for any natural system, particularly those like the Iraqi wetlands that are (or rather, once were) dependent on rivers carrying rain runoff. Today, the water regime in the Iraqi marshes has been highly modified, with the once major spring freshet pulses being replaced by a steady year-round flow of greatly diminished magnitude (France 2011).

The critically important part of wetland design, particularly in southern (hot) regions, is a preliminary water balance calculated in order to ensure that an adequate supply of water is present to maintain the restored wetlands. For wetlands in arid regions such as those in southern Iraq, the bottleneck for system sustainability is of course the absence of rain during the summer. Enough water, therefore, has to be present or be renewed frequently enough in order to sustain the system, and usually this is determined by what inflow–outflow and evaporation rates can be expected, as well as the leakiness of the soil in terms of its infiltration rate. The hydraulic loading rates for wetlands are normally in the order of 1 to 10 cm per day. This is not a large amount of water to apply to a system, which means when these numbers are extended out into large-scale applications, a great deal of land is needed to treat the water. Given that land in southern Iraq is not in short supply, Bays (2004) believes that prospects there look good from this respect. The residence time for a system ranges anywhere from ten to thirty days based upon the relationship of the volume to the flow. But in fact the issue is that flow is not uniform in wetlands and because of this water balances are often difficult to calculate, so much so that this topic represents a large part of the current research efforts in terms of trying to understand how wetlands operate (Bays 2004).

The Valley Sanitation District Wetland in Indio, southern California, is, as mentioned, in the same kind of climatic regime as that in southern Iraq and in consequence experiences an amazing loss of over 200 cm of evaporation per year which absolutely controls the water balance (this particular artificial system is lined so there are no groundwater losses). Indeed, with temperatures in excess of 35°C, these wetlands can lose over 30 cm of water a month during the summer, such that it is actually possible to measure the water levels receding on a daily basis. This type of information can be used to estimate infiltration rates and evaporation rates in other wetlands without bottom liners.

Some design suggestions can be offered about how to distribute flow in created or restored wetlands (Bays 2004). It is critically important to do everything possible to promote an even sheet flow by making maximal use and efficiency of the area available in order to maximize remedial treatment by bacteria and plants through minimizing the potential for short circuiting. It is also important to maintain low flow velocities to minimize erosion and sediment resuspension. Sheet flow can be maintained through the use of multiple outlets which have the effect of reducing corner dead zones by ensuring a uniform pattern in movement or by installing a deep zone collector system at the end of the wetland.

Water movement in wetlands is not uniform due to the fact that with increasing depth in a system, drag increases in proximity to the sediments and bottom vegetation. In larger open-water wetlands such as those in Iraq, wind-induced mixing occurs which differentially circulates water through the wetlands depending on bottom topography. Because deeper areas get preferential flow, this turns into a major design issue since there is need to ensure adequate flow to all regions of the wetland. Work is currently underway on developing a hydrodynamic model (Lightbody, S., pers. comm.) that will be a major step forward in how to approach managing surface-flow wetlands by generating models to predict water movement patterns through these systems under different design scenarios. For example, one approach to improving hydraulic efficiency in wetlands is to create deep zones that are akin, for example, to the kind of pathways seen in pictures (France 2007) of how the marsh Arabs used their boats to travel around the islands and along the canals within the system. These different deep zones, if oriented perpendicular to flow, turn into opportunities for re-areated water to slow down and respread out through the system as well as providing habitat areas for fish and other wildlife and also opportunities for solid settling and removal in the system (Bays 2004).

Such wetland design components are typically brought into a wetland to respond to these hydraulic issues, as, for example, at the Tres Rios wetlands in Arizona, a water treatment project built by CH2MHILL in an arid region receiving scant rainfall. The demonstration project, which has provided great insight into how to operate wetlands in this arid region, is characterized by alternating deep and shallow zones as well as habitat islands and outlet control structures. Hydraulic analysis has shown that adding deep zones served to increase the hydraulic efficiency and led to improvements in water quality performance (Bays 2004).

In terms of distributing water to wetlands, solutions can be simple up-flow stand pipes, initial spreader swales, or the more costly grated pipes. Shallow marsh zones are required for the best water quality treatment (France 2003). Water collection following transport through wetlands can be simple open pools, ponds, or collector swales.

PLANTING DESIGN

Plants can be installed or allowed to colonize naturally. In the United States, it is extremely rare to adopt a completely natural colonization approach because many plants that would become established may actually be somewhat deleterious to the cover and development of other more desirable plant communities (Bays 2004).

Typically these plants are installed as seedlings obtained from nurseries. Such an approach is of course impractical in situations such as restoring the large-scale wetlands in southern Iraq. Here, however, the reflooded marshes are being colonized again with Phragmities, demonstrating that a regional seed bank still exists upon which to plan future restoration efforts. In this case, simply getting the water to the new wetlands and retaining it there provide an opportunity for growth and reestablishment through colonization of natural plants.

Plants typically installed in treatment wetlands include cattails (*Typha* spp.), bulrushes (*Scirpus* spp.), and the tall reed grass endemic to the Iraq marshlands (*Phragmities* spp.), all of which have similar treatment performances, with the ultimate choice being a matter of personal and regional interest, to some extent consistent with the local ecology and expectations of regulators and wetland owners (Bays 2004). Increasingly CH2MHILL is using mixed native species to produce a resulting flora that may be just as diverse as a nearby natural system as well as optimally functional in terms of contaminant removal.

PUBLIC USE AND OPERATIONS AND MAINTENANCE

Accounts by British adventurers visiting the marshlands of southern Iraq are filled with anecdotes about widespread infection among the resident marsh Arabs (Thesiger 1964). Because of the overall rarity of surface water in Mesopotamia, the poor quality of the water of the Euphrates and Tigris Rivers is even today not a detriment to recreational activity (Figure 18.2). Nevertheless, in situations where wastewater effluent is a major water source to wetlands, as for example in the Las Vegas Wash (chapters 8 and 13), public movement should be restricted to avoid contact. In Iraq, wetland designs should try to create a series of zones of wetland treatment

FIGURE 18.2 Playing in a branching channel of the Tigris River in Syria near the Turkish and Iraqi borders.

systems where the first cells doing most of the work would be an area of reduced human activity, and as one moved down through the wetland system increasingly more human contact would be permitted and facilitated.

Treatment wetlands are essentially simple to operate but do require monitoring. In terms of the water quality, it is necessary to gauge the performance of the system by measuring the changes in contaminant concentrations as the water moves through the system (Bays 2002). If the wetland is not working up to expectations, it's usually an indication that there are issues, perhaps hydraulic ones, that are impeding performance. Hydraulically, the main issues to ensure performance are to keep the water moving and to maintain a constant depth or targeted depth range to avoid desiccation, a critical concern in a place of high evaporation and aridity such as Iraq.

In terms of vegetation management, in the United States it is quite common to encounter problems with herbivory. When wetlands are restored or created, in almost no time at all "the word gets out" and animals begin to arrive, sometimes to undesirable levels such that they completely eat their way through the vegetation that has been installed. Managing nuisance herbivory can therefore often be an issue in the long-term survival of shoreline or shallow-water plantings (Pouder and France 2002). In the restored Iraqi marshlands, however, we may actually want to encourage herbivory. The water buffalo in Iraq are part of the ecosystem that's been established there (Maxwell 1966; France 2007), and we would want to factor this in for the sustainable operation of the system (Bays 2004).

Wildlife management also becomes a very significant issue in treatment wetlands. Many of these wetlands function as secure oases free from disturbance and provide a number of habitat types, important for maximizing the diversity of wildlife use (France 2003). But most importantly, because nutrient concentrations are high in the treated wastewater, the wetlands offer a rich availability of the food resources for wildlife, which, as a result, may be there in abundance, sometimes at levels up to ten times what would be expected to occur in natural wetlands (Bays 2004). Building these treatment marshes therefore creates attractive habitat that is colonized almost immediately, particularly in arid regions. For example, one year after their construction, the Tres Rios wetlands were fully colonized and producing high densities of breeding birds (numbers increased from less than fifty birds to over eight hundred birds per unit area within a single year; Bays 2004).

Design considerations to enhance wildlife habitat include a variety of morphological, hydrological, and ecological attributes (France 2003; Bays 2004). Variable water levels—making sure water fluctuates over time—are important in enabling wading birds to have access to the wetland. Alternatively, deep zones provide habitat for waterfowl. A diversity of plants, including certain species with known qualities that are beneficial to wildlife, can substantially increase biodiversity by building vertical structure through adding perimeter trees for birds to perch upon, as well as by building horizontal structure through making sure a gradient of habitat types or riparian ecotone exists. Building islands to provide refugia from nest predators is perhaps the most fundamentally useful and simple thing to do in a wetland construction project toward fostering avian diversity and abundance. And adding snags and platforms, of course, is an easy and simple way in which to build hall-out habitat for birds and turtles.

Treatment wetland parks (*sensu* Bays et al. 2002) create aesthetic, educational, and recreational opportunities. CH2MHILL has learned that public recreational use of these created treatment wetland systems can exceed all expectations (Bays 2004). In the Wakodahatchee wetlands project in Florida, for example (Bays 2002), the boardwalk intersects all the zones within the wetland system, providing access to the public from the inlet to the outlet of the system, and creating an element of exploration and adventure for visitors at the same time as an opportunity for them to go into the heart of the wetland that they would never have normally encountered. This exceptional wetland project, which has been featured in local and national media, has become extremely popular with visitors, the number of people on tours increasing from 165 individuals in all of 1997 to over two thousand on just a single day in 2004. As a result, the parking lot was recently expanded in size by a factor of fivefold. Treatment wetlands which provide habitat for wildlife create a destination for wildlife watchers which can actually be an economic attraction. And this linking of tourism with water treatment is a truism in every wetland park (France 2003); in other words, once you create these opportunities, people will use them, and they will bring money with them (Bays 2004).

CASE STUDIES IN WATER TREATMENT AND RESTORATION BENEFITS

The largest constructed treatment wetlands in the world are the 45,000 acres of surface-flow wetlands that have been built over the last decade at a cost of over a billion dollars to help restore the Everglades in southern Florida. The Everglades Nutrient Removal Project is a thirty-eight-thousand-acre wetland constructed as a demonstration project and sized and located to treat the phosphorous-rich agricultural runoff before it flows into what's left of the historic Everglades. Importantly, this project provides a great model to learn from with respect to restoring and managing the Iraqi marshes in that all the new treatment wetlands were built where the former Everglades marshes had once existed (Bays 2004). This represents a good example in which the remediation of water quality actually becomes part of the restoration of the site. The treatment wetland is composed of four separate cells receiving over 160 million gallons a day of inflow with phosphorous concentrations of 100 µg/L and with a target outflow concentration of 25 µg/L. This project is the kind of scale that needs to be thought about in Iraq, where wetlands there have already been really reflooded due to dike breaching. Bays (2004) believes that if redevelopment projects in Iraq are too small in scale, some of the larger scale phenomena necessary for sustainable restoration will be missed. And it is critically important to include residents in the decision making. The award-winning Wakodahatchee wetlands project in one of the most densely populated areas of southeast Florida, for example (Bays 2002), shows that people can cohabitate beside and make public use of such treatment wetland marshes. Such wetlands therefore operate as "green infrastructure" in regenerative landscape design (France 2008).

Other projects, though much smaller in scale than the Florida example, are useful to examine given their location in arid regions. One such case study in the city of

Laguna Niguel, southern California, concerns irrigation overflow from residential neighborhoods that was adding a significant coliform bacteria load to the stream which, because of extremely low rainfall and high evaporation, reached high concentrations. The system built by CH2MHILL was a simple series of successive ponds down the drainage gradient that within a year was completely vegetated and within three years had also developed a bordering riparian shrub forest. Importantly, treatment performance has been spectacular in these wetlands with coliform reductions of over 99 percent, thereby achieving levels that are far below the public health standards. This case study provides wonderful proof that wetlands can indeed be used to disinfect and improve water quality, an issue of obvious significance to the situation in southern Iraq (Bays 2004).

Another small project that CH2MHILL is presently working on in Oxnard, southern California, addresses the salinity question, another issue of pressing concern in the restoration of the southern Iraqi marshes (France 2011). The city Water Division was expanding its water supplies with a new reverse-osmosis membrane treatment facility and although outfall to the nearby ocean was possible, local interest in wetland restoration prompted the question of suitability for treatment restoration. The project takes reverse-osmosis membrane and treatment concentrate and applies it to a series of constructed marshes in which six different technologies are examined side by side. The idea was to create a new water source of about 4,000 mg/L TDS and to select those brackish and natural saltmarsh plants tolerant of brine concentrate and then see how well they performed. The six different technologies included a high marsh that was shallowly flooded, a low marsh that was flooded to a foot or two or so in depth, a gravel-based subsurface flow wetland, and a peat-based vertical-flow system. Inflow nitrate concentrations were very high (40–50 mg/L) due to the source of water initially being groundwater with a consequent long legacy of agricultural contamination. Both the surface-flow low marsh and the vertical-flow systems significantly reduced nitrates down below the World Health Organization standard of 10 mg/L due to having the right kind of habitat and low-oxygen conditions to encourage denitrification. The chemical constituent of most concern with known ecotoxicological properties is selenium, also an issue in the Iraqi marshes. Only the vertical-flow wetland was found to do a good job in removing selenium due to the anaerobic environment which reduced it to an inorganic form that was essentially bound in sediments in the system, thereby being unavailable to wildlife (Bays 2004).

Another CH2MHILL project with useful comparisons for Iraq is the Sweetwater Wetlands in the desert city of Tuscon, Arizona (Bays 2002). High-strength wastewater, not unlike what would be expected in some parts of Iraq, is received from a traditional wastewater treatment plant and the quality improved by movement through the wetland prior to its use for recharging the local groundwater. This system is also designed to function as a park that provides and sustains wildlife habitat at the same time as providing educational and recreational amenities through the construction of numerous landscape features to encourage use of the site. The project is also a good example of where progress toward vector control has been a significant issue (Bays 2004). High densities of mosquito larvae were initially present (up to eight thousand individuals caught per trap night) because biological predators had not been integrated into the design (mosquito fish, a common vector control tactic in the United States, were not introduced by administrators because they are a nonnative species for this area). Managers therefore replaced the predatory role of mosquito fish by applying biological larvicides in concentrations sufficient to reduce the larval abundance down to only several hundred individuals per trap night.

A final project in India that CH2MHILL has worked on with the Canadian International Development Agency involved building a treatment wetland in a small town in Punjab, India, a situation not dissimilar to the kind of local site-based wastewater system that might be considered in Iraq (see chapters 19 and 20). This wetland system treats a mixture of very-high-strength wastewater not unlike that characteristic of urban situations in southern Iraq. The system has now become a centerpiece for the entire town and was a product of extensive community involvement in the planning and design of the project. Such engagement is recognized to be important to the long-term success of many landscape regeneration projects (chapter 2; France 2008). Further, involvement of the local community in the creation of this wetland system served to foster understanding and acceptance of the new wastewater technology and helped to define appropriate building practices that were implemented in construction (Bays 2004).

CONCLUSION

There is a systematic process needed for engineering natural treatment systems (Table 18.1). There are planning, design, construction, and management steps, all of which vary somewhat from project to project, but almost all of which are always considered for any single project (Bays 2004). Initially important is a clear setting of project goals (France 2003). What is the need that the wetlands could meet? How do we plan to achieve this goal through efficient construction? And how can this be accomplished in a way that has the acceptance of the public and the stakeholders? This last question is of critical importance in that both the public and the owner must accept the project and someone must be willing to maintain or manage it. This is really the only way these projects can be sustained over time.

One last thought concerns the importance of technology transfer approaches to help the Iraqi marsh restoration. Demonstration facilities, important for facilitating watershed stewardship and management (France 2005), can be built that model the approach to restoration using treatment wetlands as well as providing opportunities for instruction and research (Bays 2004). Good examples of this approach in the United States include the Olentangy River Wetland Reserve in Columbus, Ohio, designed and supervised by staff from the local university. This project is a good example of learning by doing, and has been the source of numerous theses, dissertations, and publications. And this is exactly what Iraqis should do (i.e., build a series of such treatment wetland learning centers in different regions and educate themselves in how to go about undertaking this kind of work) (Bays 2004). This is the best and most efficient way, ultimately, to build acceptance and application of natural treatment systems in Iraq. Such demonstration wetlands provide internal research and an education platform in which to train Iraqi scientists and engineers. The point is that no matter what the scale of magnitude of the final project, whether

TABLE 18.1

**Systematic Process Needed for Engineering
Natural Treatment Systems**

Planning

Envision the project and charter a team.

Characterize the water source.

Target the desired water regime.

Determine the area needed for calculated performance.

Collect data.

Compare sites, concepts, and preliminary cost estimates.

Inform the public and media.

Select the development site and design concept.

Design

Develop a detailed design.

Estimate costs.

Reconcile the design with budget (or vice versa).

Construction

Publish plans and specs.

Select a bid.

Construct the project.

Perform a quality check and final checklist.

Management

Initiate start-up carefully.

Operate within determined specs.

Monitor

Adjust in response to change.

Keep the public informed.

Source: Bays (2004).

it is the Everglades or small arid-region wetlands in the West, all started with a relatively humble pilot system or demonstration project. And this is often the only way to get acceptance for large-scale, large-expenditure projects such as those being contemplated for Iraq (Bays 2004).

The external training of Iraqi leaders will be important for building morale and understanding. The goal is to make the future water resource leaders in Iraq accept this technology and have it trickle down to those in the trenches undertaking the restorative redevelopment of the marshlands. Bays (2004) therefore believes that the directors of the various environmental ministries in Iraq need to take a tour to visit the projects described in this chapter so that they can become educated and conversant in the technology and then become effective champions for implementing such natural treatment projects back home.

ACKNOWLEDGMENT

Adapted from Bays (2004).

REFERENCES

Apfelbaum, S. I. 2005. Stormwater management; A primer and guidelines for future programming and innovative demonstration projects. In *Facilitating watershed management: Fostering awareness and stewardship*, ed. R. L. France. Lanham, MD: Rowman & Littlefield.

Bays, J. S. 2002. Principles and applications of wetland park design. In *Handbook of water sensitive planning and design*, ed. R. L. France. Boca Raton, FL: CRC Press.

———. 2004. Wastewater and salinity treatment in wetlands. Paper presented at the Mesopotamian Marshes and Modern Development: Practical Approaches for Sustaining Restored Ecological and Cultural Landscapes conference, Cambridge, MA, October.

Bays, J., J. Cormier, N. Pouder, B. Bear, and R. France. 2002. Moving from single-purpose treatment wetlands toward multifunction designed wetland parks. In *Handbook of water sensitive planning and design*, ed. R. L. France. Boca Raton, FL: CRC Press.

France, R. L. 2003. *Wetland design: Principles and practices for landscape architects and land-use planners*. New York: Norton.

———, ed. 2005. *Facilitating watershed management: Fostering awareness and stewardship*. Lanham, MD: Rowman & Littlefield.

———, ed. 2007. *Wetlands of mass destruction: Ancient presage for contemporary ecocide in southern Iraq*. Winnipeg, MB: Green Frigate Books.

———, ed. 2008. *Handbook of regenerative landscape design*. Boca Raton, FL: CRC Press.

———. 2011. *Restoring the Iraqi marshlands: Potentials, perspectives, practices*. Sussex, UK: Sussex Academic Press.

Kadlec, R. H., and S. Wallace. 2004. *Treatment wetlands*, 2nd ed. Boca Raton, FL: CRC Press.

Maxwell, G. 1966. *People of the reeds*. New York: Pyramid.

Ochsenschlager, E. L. 2004. *Iraq's marsh Arabs in the Garden of Eden*. Philadelphia: University of Pennsylvania Museum of Archaeology and Anthropology.

Pouder, N., and R. France. 2002. Restoring and protecting a small urban lake (Boston, Massachusetts). In *Handbook of water sensitive planning and design*, ed. R. L. France. Boca Raton, FL: CRC Press.

Steinfeld, C., and D. Del Porto. 2007. *Reusing the resource: Adventures in ecological wastewater recycling*. New Bedford, MA: Ecowaters.

Thesiger, W. 1964. *The marsh Arabs*. New York: Penguin.

Vymazal, J., H. Brix, P. F. Cooper, and M. B. Green. 1998. *Constructed wetlands for wastewater treatment in Europe*. Leiden: Backhuys.

Water Environmental Federation. 2005. *Natural systems for waste water treatment: WEF manual of practice*. Alexandria, VA: Water Environmental Federation.

19 Advanced Ecotechnology for Decentralized Wastewater Treatment and Reuse

David Austin

CONTENTS

Introduction 391
Culture and Wastewater Reuse 392
Wastewater Reuse: Why Do It? 397
What Is Decentralized Reuse? 397
Public Health and Environmental Concerns 399
Overview of Wastewater Ecotechnology 402
Conclusions 408
References 409

INTRODUCTION

This chapter is written for the lay reader interested in decentralized wastewater reuse and ecotechnology for wastewater treatment. These topics are considered within the context of this book: the future of the Iraqi marshes. Because comprehensibility is inversely proportional to technical rigor, and there is much intellectual territory to cover, technical detail is set aside to convey key concepts.

Stepping back from detail has the advantage of bringing context into focus. Technologies exist within, not separate from, the opportunities and limitations of the societies that produce them, the economies and institutions that support them, and the cultures that accept them. Substantial consideration is therefore given throughout this chapter to these contexts.

Ecological engineering has emerged since the 1990s as a viable engineering discipline (Mitsch and Jorgensen 2004; Kangas 2004; Kadlec and Knight 1996). In its simplest embodiment, ecological engineering and the ecotechnology it produces are related to environmental engineering. It is distinguished by a broader view of engineering that incorporates ecological science into design philosophies and methodologies to achieve engineering goals (Kadlec and Knight 1996). More broadly, ecological engineers seek to design and manage infrastructure and ecosystems from a systems ecology perspective to benefit both society and nature (van Bohemen 2005; Mitsch and Jorgensen 2004). The ecological engineer's view of infrastructure embraces concepts of ecosystem function (e.g., the role of wetlands in flood control and water quality improvement) as foundational elements of engineering design (see chapter 19).

A parallel development within environmental engineering has been the evolution of viable alternatives to centralized wastewater treatment and reuse (Crites and Tchobanoglous 1998). There is now a widespread realization in the technical and regulatory communities in the Unites States that continued expansion of centralized collection, treatment, and reuse (or discharge) is neither economically feasible nor technically necessary in many areas of the country (U.S. Environmental Protection Agency [EPA] 1997). Some developing countries are turning to decentralized wastewater treatment infrastructure even in urban environments to escape the huge capital costs associated with centralized infrastructure (Viet Ahn 2004). As will be discussed in this chapter, ecotechnology is well suited to many decentralized wastewater treatment applications.

Why is this topic relevant to Iraq? Others have pointed out that dams on the Euphrates and Tigris Rivers have deprived the Iraqi marshlands of the abundant water enjoyed since the dawn of civilization. Both Iraqi and Turkish dams (Kazem and Osman 1998) have radically altered the ancient patterns of spring floods that flushed and renewed the marshes. Expanding populations and severely curtailed water volumes induce water stress in terms of both demand and pollution loading. Land-based communities within or near restored marshlands can ill afford to pollute their own water supplies. Moreover, smaller communities cannot necessarily expect to be served by centralized wastewater infrastructure. It is likely, therefore, that a decentralized treatment and reuse model, as adapted by the Iraqi engineering community to local needs and culture, will prove useful in this region.

CULTURE AND WASTEWATER REUSE

Wastewater reuse is not entirely a technical matter. Engineering, microbiology, and epidemiology necessarily inform choices of technology and management for wastewater reuse, but they alone do not *sufficiently* inform these choices. Culture plays a decisive and varied role. Wastewater reuse is not viable if its intended beneficiaries are appalled at a proposed use of water initially containing human excreta, no matter how clean the treated water may be. Senior design and planning professionals in the field of wastewater reuse must either consider the sociocultural aspects of reuse or risk certain failure of the project (Dingfelder 2004; Wegner-Gwidt 1998; World Health Organization [WHO] 1989).

Culture and religion are intimately related. In areas of the world where civil law is influenced by religious law, how local religious authorities regard proposed wastewater reuse can be every bit as decisive as secular regulations or public opinion. The emergence of powerful Islamic political movements in Iraq since the fall of the Ba'athist regime clearly reveals the strength of Islam in contemporary Iraqi culture. How will the influence of Islam affect wastewater reuse? As shall be seen below,

Islam is admirably well equipped to provide a sound foundation of rational decision making for reuse.

Before investigating Islam and wastewater reuse, it is helpful to put the broader issues of reuse and culture in context by seeing how powerfully culture can influence reuse in a secular, technocratic society such as the United States. There is also a technical context that is needed. What is meant by wastewater reuse, and what types of reuse are culturally sensitive?

A wide variety of applications may use treated wastewater. Most often, treated wastewater is used for irrigation of landscapes, fodder crops, and human food crops. Direct recycle of highly treated wastewater for potable reuse is very rare. Indirect potable reuse is more common, such as discharge of highly treated wastewater to aquifers and rivers that supply raw potable water. Any category of reuse that involves human contact with, or ingestion of, clean water of sewage origin can arouse hostility from intended beneficiaries that are deeply emotional in nature (Dingfelder 2004). (This chapter does not consider reuse without scientifically justified levels of treatment. It is worth noting, however, that farmers appropriate raw sewage in many areas of the world by custom or desperation [Austin and Asano 1996; Ghosh 1991; WHO 1989].)

It may surprise many readers that the United States has no uniform standards for wastewater reuse (EPA 1992). These regulations are left to individual states and vary widely among them. The states of California and Florida have the most experience with wastewater reuse. Both states experience significant water stress. Both have extensive wastewater reuse infrastructure. The technical standards adopted by these states for reuse are strict and essentially the same. They entail the highest possible levels of treatment and disinfection for effluent exposed to direct public contact (Crook 1998). These standards are achieved with redundant disinfection and filtration systems that are expensive to build and operate. Reuse standards based on the California–Florida model try to minimize risks to public health by essentially removing all pathogens from treated effluent, which is sometimes, but erroneously, referred to as a "no-risk" philosophy. (The "no-risk" label is technically a misnomer because there are measurable risks in water reuse that drafters of the California regulations carefully considered [Crook 1998].)

In contrast, the World Health Organization recommends reuse standards based on a different public health philosophy. Reuse is a strong need in regions of water stress, but many societies cannot afford "no-risk" reuse standards. The WHO standards are based on an epidemiological assessment of the risk of acquiring illness from ingesting food crops irrigated with reused wastewater (WHO 1989). In general and highly simplified terms, a disinfection standard is considered sufficient if the risks of additional illness are statistically indistinguishable from the background rate of illness caused by pathogens of fecal origin. Some consider the WHO standards to be less rigorous than the California–Florida model because of the potential insensitivity of epidemiological analyses to detecting actual infections caused by contact with treated wastewater (Crook 1998).

California and Florida regulators would consider the WHO standards to be completely unacceptable for agricultural reuse. Yet both the California–Florida model and WHO reuse standards are scientifically justified. Why is there such a large difference?

Economics is reasonably given as the principal raison d'être for the contrasting standards (Crook 1998; WHO 1989). The WHO (1989) endorses low-cost wastewater treatment technologies for reuse in food crop irrigation and in aquaculture. Economics is certainly important, but does not explain why the WHO standards for reuse have been adopted, with modifications, by France and Spain (Bontoux 1998), which undoubtedly can afford the California–Florida-type standards. Moreover, why would any country or state in the United States spend fortunes on "no-risk"-level treatment for reuse if the WHO standards can be attained at far lower cost and are based on sound science?

Culture better explains these differences than science. The United States provides a good example of these cultural influences. Consider the differences between the reuse standards of the states of California and New Mexico.

California produced the first comprehensive set of wastewater reuse regulations in the United States. These regulations have been influential worldwide (Crook 1998). The cultural background of the California reuse regulations was that of a wealthy, populous, technically sophisticated, and litigious society anticipating, but not yet suffering, substantial water stress.

A litigious society without a universal perception of water scarcity presents daunting policy challenges to those considering a treatment standard based on epidemiological models. These models tacitly admit a certain rate of illness caused by reuse, however lost in statistical noise. There is therefore an unacceptable level of government legal exposure to lawsuits seeking damages for illnesses that could be reasonably attributed to exposure to reused wastewater. The "no-risk" philosophy of reuse treatment standards is an appropriate response to this cultural reality because it obviates a great deal of litigation by effectively removing pathogens from reused wastewater exposed to direct public contact.

Fear of litigation does not fully explain the prevalence of "no-risk" reuse standards in the Unites States and elsewhere. As stated earlier, both reuse philosophies stand on firm scientific ground. How well a given engineer or scientist is persuaded by one philosophy or the other probably has more to do with the culture of the technical community in a given country than anything else. Reuse regulations reflect the consensus of the relevant technical community. Nevertheless, litigation over reuse has been important in California, despite mature and extraordinarily rigorous reuse standards.

California has been an arena of epic legal and political battles involving large municipal reuse projects in San Diego and Los Angeles (Lee 2005; Waldie 2002). "Toilet to tap" has been a rallying slogan of those in California opposed to vital reuse projects characterized by indirect connection to public water supplies. This vociferous opposition arises despite treatment of wastewater to drinking water standards, extreme dilution of discharged wastewater with the receiving waters, and then further treatment and disinfection of (receiving) water prior to public use (Hartling and Nellor 1998; Mills et al. 1998; Olivieri, Eisenberg, and Cooper 1998).

Public perception sometimes has little in common with rationality, as can be seen in the spectacle of (former) Los Angeles Mayor James K. Hahn crying, "Lips that touch reclaimed water must never touch mine" (Waldie 2002). Perhaps the honorable mayor is unaware that kissing is commonly, indeed enthusiastically, practiced in cities located along the Mississippi River. These cities draw raw water into drinking water treatment facilities downstream of the previous cities' sewage discharges. Sewage discharging into the Mississippi River from those cities comes nowhere near to meeting the treatment standards of the California reuse project that so alarmed Mayor Hahn.

In comparison to California, New Mexico is a poor, sparsely populated, desert state where water scarcity is ingrained into the public psyche. Drought often prevents irrigation of cherished urban amenities, such as sports fields and city parks. New Mexico society does not appear to be intrinsically less litigious than that of California. A crucial difference is that water scarcity is obvious to all, and its effects are widely perceived as both economic and cultural losses.

A key mandate for the New Mexico Environment Department (NMED) Reuse Workgroup (of which the author was a member) was to create wastewater reuse guidelines that would both protect public health and provide economically viable means of irrigating urban amenities. Municipalities across New Mexico had already rejected the California–Florida model in a previously proposed reuse rule. Public parks and ball fields were turning to sun-baked dust for want of irrigation water, and few municipalities could afford to build or operate reuse facilities treating to the proposed rule.

The NMED chose members of the Reuse Workgroup to represent a cross-section of state government, industry, municipal, and public stakeholders. Although discussions were mostly technical, the cultural realities and needs of the state were never off the table. The Reuse Work Group was able to craft a middle ground between the California–Florida model and WHO standards that allowed affordable irrigation of public urban amenities and protected public health to rigorous scientific standards. Given the absence of agricultural demand for reuse, the guidelines prohibited spray irrigation of food crops to avoid unnecessary cultural and technical contention (NMED 2003). Even though WHO standards were probably sufficient to protect public health from incidental contact with reused wastewater in parks and sports facilities, they were not considered to be viable policy because of likely resistance from technical professionals. The reuse guidelines have been met with widespread acceptance in New Mexico.

As in the United States, there is every reason to believe that wastewater reuse across predominantly Muslim societies will also be culturally nuanced. How can it be any other way given the sectarian, geographic, ethnic, economic, and cultural diversity of Muslim countries? Moreover, the authority of clerics over civil society varies between these countries (cf. Iran vs. Turkey), and rulings on religious law from ranking clerics in one sect of Islam are not binding upon Muslims of another sect.

Despite these differences, the unifying power of scripture in Islam is great. The Qu'ran and Hadith have a great deal to say about water, cleanliness, and the environment that informs both cultural attitudes and Islamic law regarding water management and wastewater reuse (Al-Jayyousi 2001; Amery 2001; Faruqi 2001). The Qu'ran is abounding in imagery of water and its sacred nature. Indeed, paradise is described as "Gardens beneath which running waters flow" (47:12). Culturally and legally, therefore, it is meaningful and important to consider wastewater reuse from the perspective of Islam.

Islamic law allows beneficial and healthy reuse of wastewater. A pioneering fatwa (authoritative ruling on Islamic law) on the subject issued by the Council of Leading Islamic Scholars (CLIS) of Saudi Arabia in 1978 clearly lays out the operating principles (Abderrahman 2001).

Impure wastewater can be considered as pure water and similar to the original pure water, if its treatment using advanced technical procedures is capable of removing its impurities with regard to taste, color, and smell, as witnessed by honest, specialized, and knowledgeable experts. Then it can be used to remove body impurities and for purifying, even for drinking. If there are negative impacts from its direct use on human health, then it is better to avoid its use, not because it is impure but to avoid harming human beings. The CLIS prefers to avoid using it for drinking (as possible) to protect health and not to contradict with human habits.

This fatwa reflects careful attention to Islamic law from learned clerics in a conservative society where sharia (legal code based on the Qu'ran and Hadith) is the law of the land. Other Islamic authorities may be likely to arrive at similar conclusions regarding water reuse. It is important to note that this fatwa explicitly rejects any notion of the spiritual impurity of wastewater; rather, the central issue is the protection of human health, just as in WHO or California reuse standards.

The conclusions of the CLIS may also reflect popular attitudes toward reuse in predominantly Muslim societies. In Palestine, polling data indicate widespread acceptance of the concept of reuse for irrigation of food crops among the general public and farmers provided that the level of treatment protects public health (Al Khateeb 2001).

Other elements of Islamic law may influence the structure of reuse projects beyond technical standards of treatment. According to Faruqui (2001), the priority of water use rights within Islamic law is well established: first, there is the right of humans to drink or quench their thirst; second, the right of cattle and household animals; and third, the right of irrigation. The environment has clear and unmistakable rights in Islam, as can be seen in the Hadith (Al Bukhari 5505): "He who digs a well in the desert … cannot prevent the animals from slaking their thirst at this well." Furthermore, ownership of water is structured in Islamic law. Full cost recovery of water supply, treatment, storage, and distribution is permitted, as are similar costs with wastewater (Faruqui 2001). Developers of water resources, however, have restricted rights within Islamic law to ensure just distribution of water and that the fundamental right of access to water is not violated (Caponera 2001; Faruqui 2001; Kadouri, Djebbar, and Nehdi 2001).

Islamic scholars have ruled that the status of reused water is "similar to the original pure water" if it is sufficiently treated. The complex and varied features of Islamic law regarding water rights logically apply to reuse of treated wastewater and will be influential factors in shaping reuse projects in Islamic societies. Also, the spiritual station of water in Islamic scripture informs the cultural attitudes of all Muslims toward water.

Through the ancient Muslim practice of *ijtihad* (inquiry, interpretation, and innovation by Muslim scholars), Islamic societies clearly have the tools to rationally consider reuse of wastewater in light of modern technology. It remains to be seen how reuse will play out in Iraqi society given the current warfare and political turmoil, but it does seem clear that the influence of Islamic authorities on civil affairs, such as reuse, will be important. Evidence from scripture, actual practice, and attitudes in neighboring societies leads one to conclude that the influence of Islam will be to promote wastewater reuse.

WASTEWATER REUSE: WHY DO IT?

The overall rationale for wastewater reuse is to control water pollution, to augment existing water supplies, and/or to create an alternative water resource to preserve higher quality water sources (Asano and Levine 1998). Water reuse can be an important part of overall water management because only a small percentage of daily water use within a city actually requires potable quality water. Wastewater treated to reuse standards is also a consistent source of water supply because of the relatively constant use of water by people. As such, reused wastewater is well suited to applications with consistent demands or a structured succession of uses that come and go with the seasons.

Each potential reuse application is a function of local needs, opportunities, and constraints. For instance, landscape irrigation is most meaningful where cultural values prize green parks or common public green space, agricultural irrigation may not be well suited to urban areas, and so on. Except for direct potable use, highly treated wastewater is best considered like any other water resource in terms of appropriateness of applications, but is an "extra" source of water in terms of satisfying overall water demand. Like any valuable resource, reuse demand can exceed supply.

Control of pollution provides a rationale for wastewater reuse. The high treatment standards required for reuse also protect the environment. Removal of pathogens, either entirely (California–Florida model) or below set concentrations (WHO model), entails a high-level treatment for other wastewater constituents that are harmful to the environment. Thus, a well-designed and managed reuse system is environmentally beneficial.

WHAT IS DECENTRALIZED REUSE?

Large sewer networks that bring wastewater to a central treatment facility typically serve cities. If these central facilities treat wastewater to reuse standards, a large network of pipes is needed to redistribute the treated effluent to users. In contrast, decentralized wastewater reuse systems operate over smaller areas and with much smaller flows. The savings in pipe capital costs can be large.

In terms of flows, the border between large, central treatment systems and small, decentralized treatment systems is somewhat arbitrary. Crites and Tchobanoglous (1998) set it at 3,750 cubic meters per day (1.0 million gallons per day). Above this flow, large-scale technologies, such as activated sludge, are as cost-effective and reliable. Below this flow, other technologies tend to be more cost-effective and reliable than scaled-down versions of large-scale technologies. In particular, small communities in the United States often struggle with the high operational costs and inconsistent performance of activated sludge systems under this size (EPA 2000).

The scale of decentralized reuse in the United States probably falls within the range of one hundred to two thousand homes. A smaller population base typically cannot support the technology and operational requirements to treat wastewater to minimum reuse standards. A larger population base will support a more centralized treatment and reuse model.

Few things are as unglamorous and prosaic as sewers. Nevertheless, the presence of working sewers is fundamental to civilization. They merit attention here because the type of collection systems (sewers) often defines decentralized treatment.

Typically, the largest expense for a centralized reuse system is not treatment; it is collection. Traditional gravity sewer systems are expensive to construct and maintain because of deep excavations, large pipes, and the need for lift stations to periodically pump sewage up to higher elevations once sewer pipes get too deep. Decentralized treatment systems use smaller diameter pipes in shallower excavations, thereby realizing large cost savings.

Small-gravity sewers are cost-effective for buildings in close proximity to one another, such as a small town or cluster development. In these situations, a simple collection net of 200 mm pipe may convey wastewater to a local treatment system without additional pumping. Where frost is not an issue, excavations for pipe may be shallow, cutting costs even more. This type of sewer must be laid at a constant slope to maintain a scouring velocity to prevent buildup of solids in the pipe. Some sites are ill suited for this type of construction. Conventional sewers also will convey all sewage solids directly to the local treatment system. Because decentralized treatment is done near the people producing wastewater, fresh sewage solids present odor problems that must be carefully managed to avoid adverse public reactions.

Septic tank effluent gravity sewers (STEGs) are common in decentralized applications (Figure 19.1). In this system, interceptor (septic) tanks take waste from a building and provide primary treatment. The liquid fraction of the sewage then discharges out of the septic tank in a small-diameter pipe.

Sewer pipes in STEG systems need only convey the liquid fraction of wastewater because large particles settle out in the interceptor tanks. Several benefits immediately follow. Pipes can be small diameter (50–100 mm) because there are no large solids to plug flow. Pipes can be laid with a variable (inflective) grade in which no elevation of the sewer pipe is above a discharge elevation from an interceptor tank. Inflective grading can substantially facilitate sewer construction. There is no need for large volumes of flushing water to keep pipes open as in conventional gravity sewers because there are no large solids to settle in pipes. This feature promotes lower water use.

Additionally, the interceptor tanks that make these hydraulic advantages possible also provide primary treatment of wastewater by settling and anaerobic digestion. Primary treatment lowers the cost of the treatment system.

Other collection system types are more energy intensive, but are often popular in the United States. Pressure sewers use grinder pumps or pumps in septic tanks (STEP system) to convey sewage. The advantage of pressure sewers is that they may follow virtually any gradient and can be made from small-diameter, flexible pipe that is cheap and easy to place in the ground in comparison to conventional gravity sewers. Vacuum sewers are also used in decentralized applications because of the

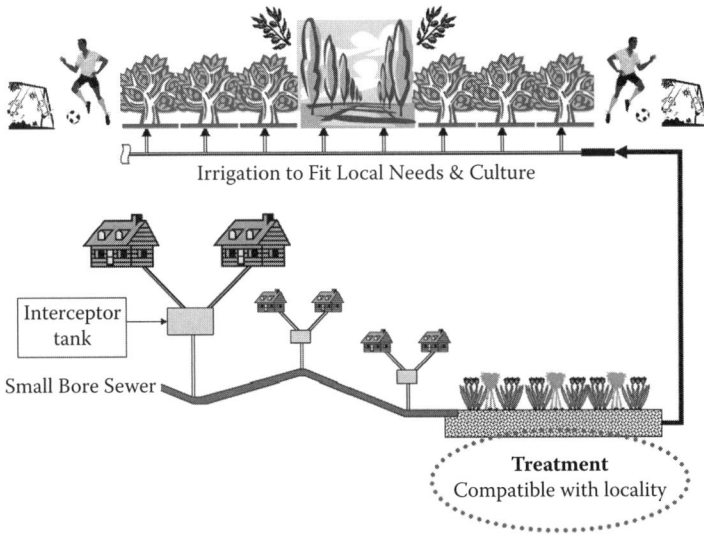

FIGURE 19.1   Decentralized collection, treatment, and reuse schematic. The rising and falling slope of the small-bore sewer depicts the inflective sewer grading possible with STEG collection systems. The local treatment facility is depicted as a constructed wetland, but could be other technologies as well.

advantages of variable-grade piping. The additional mechanical complexity of pressure or vacuum sewers requires periodic maintenance supported by local manufacturing and service industries or they will fail.

A decentralized collection system delivers wastewater to treatment systems near the sources of wastewater, including homes, schools, business, and others. Decentralized treatment must be a good neighbor to that community. Careful attention to odor control in design and operation is therefore crucial. Passing exhaust gases from wastewater treatment through earth or compost filters is one simple and effective way of controlling odor. Mechanical–chemical odor-scrubbing systems are too expensive to consider for most decentralized treatment applications.

To be visually acceptable, decentralized treatment systems should be hidden, either out of sight or in plain view. The former category is represented by underground treatment systems (Figure 19.2) and systems hidden in buildings or behind landscaping. The latter category is represented by wetland treatment systems (Figure 19.3). Certain wastewater treatment systems (e.g., package plants) present greater challenges for visual integration into the local community (Figure 19.4).

PUBLIC HEALTH AND ENVIRONMENTAL CONCERNS

Wastewater reuse must protect human and environmental health. Both of these topics are vast and can only be discussed briefly here because of limitations of space. Reuse water must not put people at risk of becoming ill from pathogens. Environmental health takes into consideration the impact of reuse water on surface water, groundwater,

FIGURE 19.2 Underground decentralized wastewater treatment system in the United States. Hatches at the surface provide access for operations and maintenance. Minimal landscape screening could hide this system. Photo courtesy of North American Wetland Engineering, LLC.

FIGURE 19.3 Vertical-flow wastewater treatment wetland in the United States. Photo courtesy of North American Wetland Engineering, LLC.

FIGURE 19.4   A package treatment system located in a U.S. residential neighborhood. There is an aesthetic clash with the neighborhood. Landscaped berms or a thick screen of trees may be able to visually remove this facility from adjacent sightlines. Photo courtesy of North American Wetland Engineering, LLC.

and soils. All of these issues can be successfully addressed by competent engineering design informed by science and knowledge of local conditions.

Obligate internal pathogens cause the diseases of concern to wastewater reuse. For these organisms, direct human-to-human infection is not the mode of transmission. For most waterborne pathogens, human waste must leave the body of one person and be ingested by another. Infection depends upon ingestion of a life cycle stage that lives outside of the human and is shed to the environment in urine or feces (Figure 19.5). Additionally, there are other diseases, such as schistosomiasis, that enter the human host through the skin via direct contact with wastewater contaminated with human urine or feces (WHO 1993).

Wastewater treatment for reuse must interrupt the pathogen or parasite life cycle to eliminate it from its host population in the intermediate to long term. To achieve that goal with irrigation reuse, wastewater must meet recognized standards of treatment. Irrigation reuse places treated wastewater in soils. A high degree of disinfection occurs as wastewater percolates through soils (EPA 1981). A well-managed irrigation reuse program also can interrupt the potential flow of pathogens into local surface waters, resulting in additional public health benefits. Additionally, there are effective management practices to minimize potential contamination of food crops irrigated with treated wastewater (WHO 1989).

One potential disadvantage of irrigation reuse comes from salt. Wastewater is more saline than the original freshwater sources. Some soils with high clay content may be vulnerable to irrigation with slightly saline water without periodic flushing

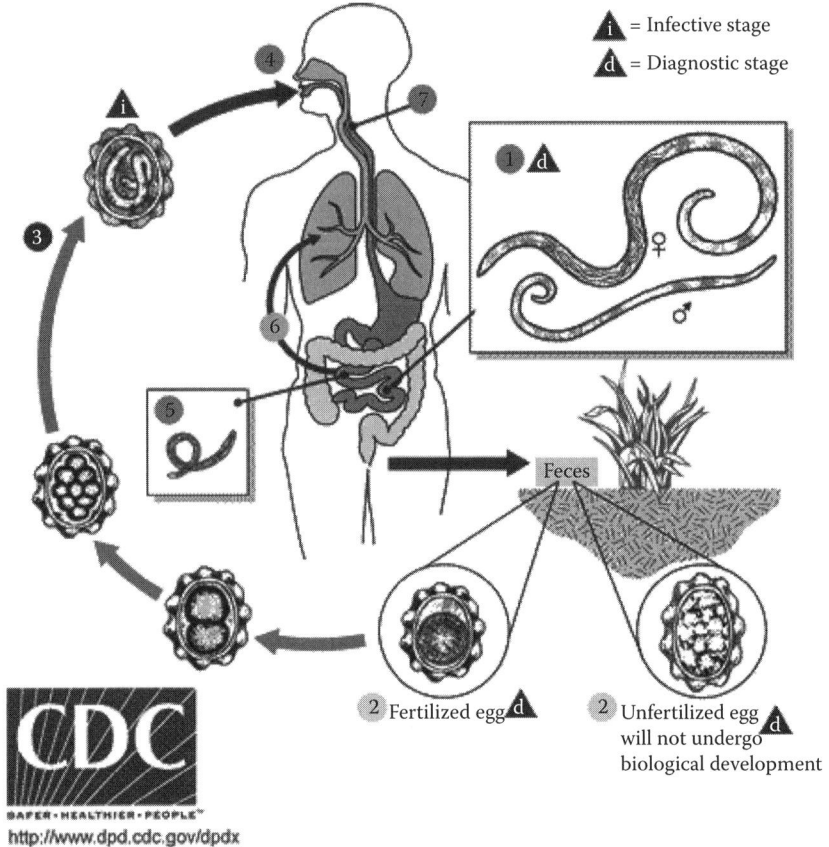

i = Infective stage

d = Diagnostic stage

Fertilized egg

Unfertilized egg will not undergo biological development

Feces

CDC
SAFER · HEALTHIER · PEOPLE™
http://www.dpd.cdc.gov/dpdx

FIGURE 19.5 *Ascaris lumbricoides* life cycle. This organism is of special concern to the World Health Organization because of a low infective dose. Reuse technology must stop loading of eggs to the environment. Reuse management practices, such as no irrigation of crops eaten raw or unpeeled, interrupt the infective stage pathway. Preferably, both appropriate technology and management practices are used together. This concept applies to protecting public health from any waterborne pathogen. Treatment and management strategies take into account the life cycle biology of pathogens.

with fresher water. In some cases, soil structure can collapse, resulting in an impermeable hardpan. The level of vulnerability of a given soil is readily assessed by qualified technicians and can be incorporated into the irrigation management plan.

OVERVIEW OF WASTEWATER ECOTECHNOLOGY

Wastewater treatment systems for reuse provide safe water for nonpotable uses. Ecotechnology seeks to do so with simplified civil works and by reducing chemical use, operating costs, energy consumption, and wastewater biosolids production. These goals are desirable in any engineering design. Ecotechnology cannot be used everywhere because it often requires more land than conventional technologies. Sites with small areas that have to process large flows of wastewater require more energy-intensive, conventional treatment systems.

Ecotechnology for wastewater treatment is typically composed of constructed pond or wetland systems. The discussion will start with ponds and then move on to wetlands (see also chapters 17 and 19).

As described earlier, pathogens are a primary concern in any type of reuse system. Each pathogen in wastewater is a biological particle with a known size and predictable physical characteristics. Large particles, such as *Ascaris* eggs (diameter [d] = 70 microns [μm]), will settle to the bottom of large, quiescent ponds. Any pathogen can stick to larger particles and settle. This latter mechanism is especially important for viruses (d < 0.1 μm), which have no tendency to settle in water as individual particles.

Classic engineering theory accounts for the removal of pathogens in ponds by means of settling, but it misses the key ecological observation that pathogens are biological particles that are food for other microorganisms. Particles in suspension of d < 10 μm, a size class covering most pathogens, have little or no tendency to settle and are removed principally by means of predation by zooplankton in pond systems (Figure 19.6). Plankton ecology provides mechanistic insight into the pathogen removal properties of wastewater treatment lagoons.

Although seldom recognized as such, the lagoon systems favored by WHO (Figure 19.7) for agricultural reuse are a de facto ecotechnology because of pathogen removal by zooplankton. Their suitability for decentralized reuse is limited by large

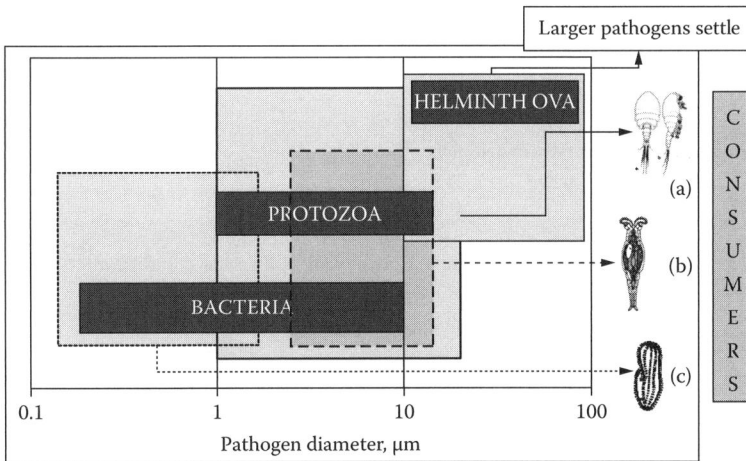

FIGURE 19.6 Zooplankton predation of biological particles by size class. Each pathogen class has a distinct size range. Representative consumers graze a particular size class of biological particles: Adult microcrustaceans, such as *Daphnia* or copepods (a), consume particles in the approximate range of 1 to 25 μm (Dodson and Frey 1991), rotifers (b) in the 4 to 17 μm size range (Wallace and Snell 1991), and protozoa (c) in the 0.1 to 2.0 μm size range (Taylor and Saunders 1991). Viruses are left out of this figure but are also subject to predation by smaller classes of protozoa (Kim and Unno 1996). Images (a) and (b) courtesy of the National Oceanic and Atmospheric Administration.

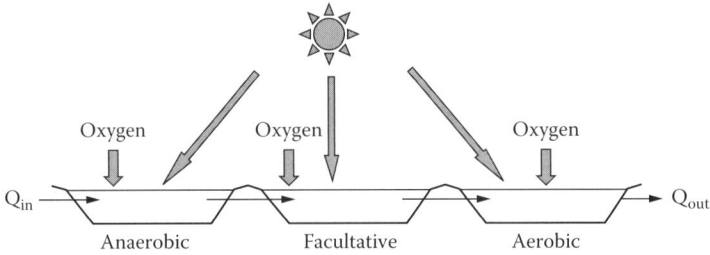

FIGURE 19.7 Schematic of lagoons in series. The actual number of lagoons may vary. Passive lagoons in series meet World Health Organization (1989) treatment requirements for irrigation reuse. Wastewater (Q) entering the system consumes available oxygen in the first lagoon, making the first lagoon anaerobic. The second lagoon shifts back and forth between aerobic and anaerobic conditions. The final, or polishing, lagoon should be aerobic. Zooplankton (consumers of pathogens) need aerobic conditions. All lagoons have large algae populations because of sunlight and ample nutrients. The algae produce oxygen during the day, but consume it at night. Water entering the first lagoon should take several weeks to pass through all lagoons (Crites, Reed, and Middlebrooks 2006), providing ample time for zooplankton consumption of pathogens.

TABLE 19.1
Energy Use Comparison

System	Area Footprint, m^2/m^3 of Wastewater
Lagoon	150
Surface-flow wetland	94
Horizontal subsurface-flow wetland	30
Aerated or tidal-flow wetland	16
Activated sludge	< 1

Note: The comparison was done for 200 m^3/d (1,000 people at 200 liters per capita use per day) systems, all designed to achieve the same reuse and nitrogen removal standards.

land requirements (Table 19.1) and odor. The first stage of the lagoon system is typically anaerobic and may produce foul odors.

Constructed wetlands are a class of ecotechnology well suited to decentralized applications (see chapter 17). They are much smaller than lagoons (Table 19.1), and they have features that control odor.

Surface-flow wetlands (Figure 19.8) are similar to wetlands in nature. Shallow water facilitates thick growth of wetland plants. Wastewater is treated as it flows through plant thatch. As with all treatment wetlands, plants themselves do not substantially contribute directly to treatment, but create an environment where a microbial ecosystem can utilize wastewater as a growth substrate and thereby treat it to design standards.

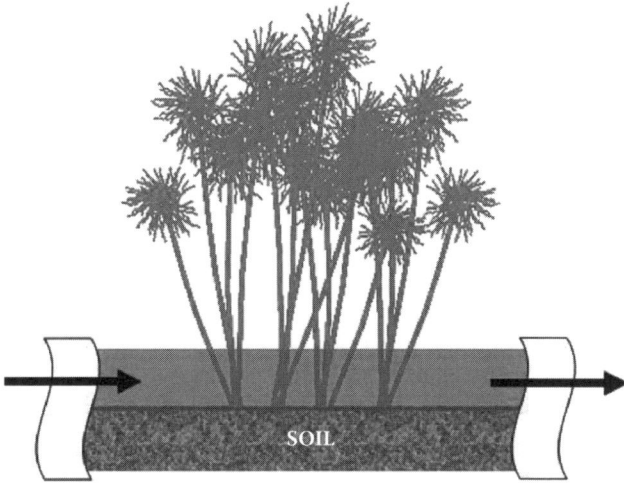

FIGURE 19.8   Surface-flow wetland section schematic. Depth is typically 0.3 m or less. The underwater plant thatch surface is large and biologically active. Both settling and predation mechanisms for pathogen removal are strongly present.

Pathogen removal is well documented in surface-flow systems (Kadlec and Knight 1996). Removal mechanisms are essentially the same as in lagoons. Based on settling theory and known pathogen removal rates, a surface-flow wetland system could meet WHO reuse standards as a final treatment stage.

Surface-flow wetlands are best suited to a final treatment stage characterized by aerobic conditions. If loaded with raw wastewater, anaerobic conditions will become established, creating an ideal mosquito (disease vector)-breeding habitat. A positive dissolved oxygen concentration favors predators of mosquito larvae. Suitable pretreatment, such as a mechanical system or a subsurface-flow wetland, must be part of the overall design for a surface-flow wetland to be part of a reuse system.

The most common type of wetland treatment system is a subsurface-flow wetland treating effluent from a septic tank (Figure 19.1) (Wallace and Knight 2006). There are many advantages to this type of treatment. Septic tanks provide primary treatment for raw wastewater. Wastewater flows through aggregate underneath the surface of the wetland (Figure 19.9), preventing the creation of mosquito-breeding habitat. The treatment area requirements are much smaller than those of surface wetlands (Table 19.1). Treatment is passive, requiring no energy inputs. Most importantly, subsurface-flow wetlands are effective at removing *Ascaris* and other parasitic nematode eggs (Stott, May, and Mara 2003). Removal of bacteria pathogens to WHO standards (1,000 colonies of fecal coliforms/100 ml effluent) by subsurface-flow wetlands has been well documented (Kadlec and Knight 1996).

Wastewater lagoons and surface- and subsurface-flow wetlands can all be passive technologies working on gravity alone. The advantage of a purely passive system is simple and low-cost operation. A need for more advanced treatment or smaller treatment areas may not allow a purely passive system (Table 19.1). Local topography may not support a purely passive system and may require some pumping to

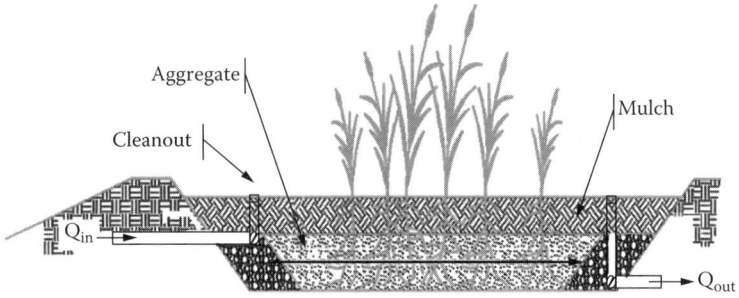

FIGURE 19.9   Subsurface-flow wetland schematic. The mulch layer is used in cold climates to prevent freezing in winter. In warm climates, a thin layer of mulch prevents odors from escaping the wetland bed.

transport wastewater. Control of nutrients or cultural values also can create demand for advanced treatment.

Ecotechnology is not restricted to passive systems. Supplemental mechanical power inputs to semipassive systems are a common feature of ecotechnology to achieve higher levels of treatment performance (Austin, Lohan, and Verson 2003; Wallace 2001b; Behrends et al. 2000) or to reduce the area required for treatment (Table 19.1). These power inputs include aeration and flood and drain pumping.

Aerated wetlands are subsurface-flow systems supplied with air distribution tubing underneath the aggregate and a blower to push air through the tubing into the wetland bed (Figure 19.10). For domestic sewage, these systems probably best serve fifty to one thousand individuals to realize economies of scale in construction and operation. Both smaller and larger systems have been successfully used. The largest

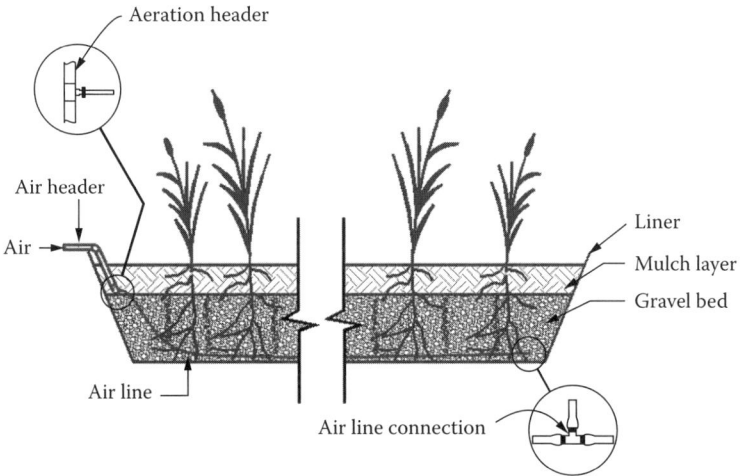

FIGURE 19.10   Aerated wetland schematic. System depicted is known as Forced Bed Aeration™ (Wallace 2001b). The wetland bed alternates between aerobic aerated zones and anaerobic passive zones.

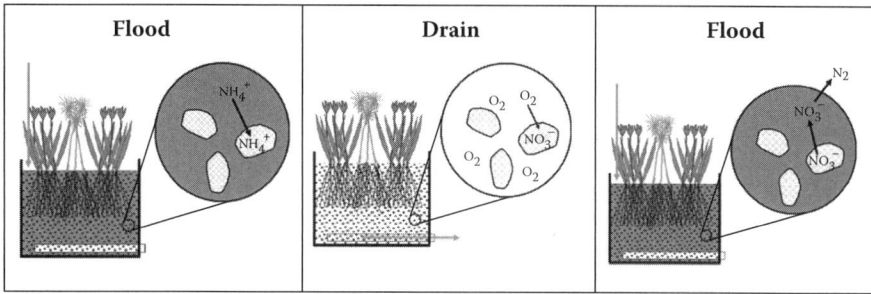

FIGURE 19.11 Nitrification and denitrification schematic for a tidal flow wetland. In the flooded phase, ammonium cations (NH_4^+) adsorb to negatively charged surfaces of aggregate and organic material. In the drained phase, air enters pore spaces and oxygen saturates the thin films of water remaining in the wetland bed. Bacteria rapidly nitrify adsorbed ammonium anions to nitrate (NO_3^-). In the next flooded phase, negatively charged nitrate ions desorb into bulk water, where bacteria convert them to atmospheric nitrogen (N_2).

is a 6,000 m^3/d system in operation since 2003 for groundwater remediation (Wallace and Kadlec 2005).

Flood and drain wetlands, known as tidal flow systems, also use limited power to improve treatment. Oxygen transfer to the wetland in flood and drain cycles takes advantage of the mechanical efficiency of pumps and cation exchange chemistry to remove nitrogen (Figure 19.11).

Several types of wetland systems use some kind of mechanical power inputs to boost treatment performance. All of them are semipassive technologies, using limited power inputs to improve treatment. Introduction of oxygen into the wetland system boosts already effective, passive, anaerobic treatment mechanisms to a more energetic state. Anaerobic treatment still occurs, but is combined with aerobic treatment.

For domestic wastewater, the advantage of this thermodynamic boost is manifested in the removal of excess nitrogen. Urine quickly breaks down into ammonia, the principle form of nitrogen in wastewater. Ammonia is stable under anaerobic conditions. Passive subsurface-flow wetlands are incapable of nitrification (bacterial conversion of ammonia to nitrate) because they are purely anaerobic. Nitrification requires oxygen. If nitrification occurs, bacteria can then convert nitrate into nitrogen gas under subsequent anaerobic conditions (denitrification). Removal of nitrogen from wastewater can be desirable for reuse applications. Although nitrogen is a vital plant nutrient, excess nitrogen can harm plants. Moreover, nitrate contamination of groundwater is not desirable. Aeration of the wetland provides a means to nitrify wastewater ammonia which, when combined with subsequent denitrification, is the crucial step for nitrogen removal in wastewater destined for reuse.

Conventional wastewater treatment consumes a lot of energy. Purely passive treatment systems consume no process energy. The semipassive nature of certain advanced ecotechnologies occupies a middle ground of energy use (Table 19.2). Semipassive systems can be designed to treat wastewater when power is turned off, albeit to a lesser standard, provided that sewage still flows. Most conventional technologies,

TABLE 19.2
Energy Use Comparison

System	Energy Requirement (kW hours/day)
Activated sludge	125
Aerated wetland	43
Tidal wetland system	27

Note: The comparison is for 200 m³/d systems (approximately 1,000 people at 200 liters per person per day). All systems are designed to achieve the same reuse and nitrogen removal standards.

such as activated sludge, fail completely without power. Semipassive ecotechnology therefore may be a good choice for regions where power supply is unreliable.

The cost of ecotechnology is favorable when compared to other wastewater technologies. A rigorous cost analysis is out the scope of this chapter, but consideration of construction techniques is instructive. The technologies described in this chapter use shallow, lined excavations, and local soils or aggregates. Designs must adapt to local materials and construction methods. Operational power requirements are much less than conventional technology. Land must be available for the treatment system, typically 0.5 to 5.0 m² per person, depending on the type of wetland and design goals. Assuming land is available at a reasonable price, the capital costs of ecotechnologies are competitive with conventional technologies. Lower power requirements mean that ecotechnology costs less to operate than conventional technologies.

CONCLUSIONS

Wastewater reuse and ecotechnology cover broad intellectual territory. Subjects germane to this chapter have included Islamic water law, culture, epidemiology, public policy, parasitology, plankton ecology, and engineering design. Such a range of relevant subjects is not surprising when considering wastewater reuse in an ecological context, given that water is an integrating factor for all life and that ecology is a study of the relationship between living things, their environment, and each other.

We have seen that cultures react to wastewater reuse both rationally and irrationally. How or if wastewater reuse projects proceed may have little to do with need or technical merit. Cultural values rule.

How culture will affect choices for wastewater reuse in small towns and villages near the Iraqi marshes remains to be seen. It does seem clear that the influence of Islamic law will not hinder development of decentralized reuse projects given the clear language of the Saudi fatwa on reuse. Rather, if local imams can be persuaded that human health is safeguarded, then they actually may assist in developing favorable public opinion toward reuse projects. The local politics of who actually owns and gets the benefit of reused water may also be influenced by local religious leaders.

In terms of science and engineering, ecotechnology is suitable for wastewater reuse projects. It can meet the WHO standards for agricultural irrigation. Ecotechnology may also have a special role to play in decentralized reuse applications. Decentralized treatment for reuse does not have the visual and olfactory luxury of being far away and out of mind that most centralized treatment systems enjoy. Few people enjoy seeing the stark industrial features of conventional wastewater treatment in their neighborhood. Wetland treatment systems have the ability to blend into the landscape. The beauty of plant-based systems is a benefit to public acceptance of wastewater treatment.

Choice of ecotechnology will depend heavily on geography and local resources. Climate is an essential consideration. Designs change a great deal between warm and cold or moist and dry climates. Local availability of materials is fundamental to design. Economics cannot be ignored. Just as vital is the question of institutional and cultural support for a technology under consideration. The engineered ecology of a wetland or pond treatment system for wastewater reuse is intimately bound to the human ecology of place, culture, and economics.

REFERENCES

Abderrahman, Walid. 2001. Water demand management in Saudi Arabia. In *Water management in Islam*, ed. N. I. Faruqui, A. K. Biswas, and M. J. Bino. Tokyo: United Nations University Press.

Al-Jayyousi, Odeh. 2001. Islamic water management and the Dublin Statement. In *Water management in Islam*, ed. N. I. Faruqui, A. K. Biswas, and M. J. Bino. Tokyo: United Nations University Press.

Al Khateeb, Nader. 2001. Sociocultural acceptability of wastewater reuse in Palestine. In *Water management in Islam*, ed. N. I. Faruqui, A. K. Biswas, and M. J. Bino. Tokyo: United Nations University Press.

Amery, Hussein. 2001. Islam and the Environment. In *Water management in Islam*, ed. N. I. Faruqui, A. K. Biswas, and M. J. Bino. Tokyo: United Nations University Press.

Austin, David, and Takashi Asano. 1996. Appropriate technologies for agricultural use of wastewater. *New World Water* 6:103–5.

Austin, David, Eric Lohan, and Elizabeth Verson. 2003. Nitrification and denitrification in a tidal vertical flow wetland pilot. In *Proceedings of the Water Environment Federation 76th annual technical conference*, Los Angeles, October.

Asano, Takashi, and Audrey Levine. 1998. Wastewater reclamation, recycling, and reuse: An introduction. In *Wastewater reclamation and reuse*, ed. T. Asano. Boca Raton, FL: CRC Press.

Behrends, Lesley, L. Houke, E. Bailey, P. Jansen, and D. Brown. 2001. Reciprocating constructed wetlands for treating industrial, municipal, and agricultural wastewater. *Water Science and Technology* 44(11–12):399–405.

Bontoux, Laurent. 1998. The regulatory status of wastewater reuse in the European Union. In *Wastewater reclamation and reuse*, ed. T. Asano. Boca Raton, FL: CRC Press

Caponera, Dante. 2001. Ownership and transfer of water and land in Islam. In *Water management in Islam*, ed. N. I. Faruqui, A. K. Biswas, and M. J. Bino. Tokyo: United Nations University Press.

Crites, Ronald, and George Tchobanoglous. 1998. *Small and decentralized wastewater management systems*. San Francisco, WCB/McGraw-Hill.

Crites, Ronald, Sherwood Reed, and E. Joe Middlebrooks. 2006. *Natural wastewater treatment systems*. Boca Raton, FL: CRC Press.

Crook, James. 1998. Water reclamation and reuse criteria. In *Wastewater reclamation and reuse*, ed. T. Asano. Boca Raton, FL: CRC Press

Dingfelder, Sadie. 2004. From toilet to tap: Psychologists lend their expertise to overcoming the public's aversion to reclaimed water. *Monitor on Psychology* 35(8):26–28.

Dodson, Stanley, and David Frey. 1991. Cladocera and other branchiopods. In *Ecology and classification of North American freshwater invertebrates*, ed. J. H. Thorp and A. P. Covich. San Diego, CA: Academic Press.

Faruqui, Naser. 2001. Islam and water management: Overview and principles. In *Water management in Islam*, ed. N. I. Faruqui, A. K. Biswas, and M. J. Bino. Tokyo: United Nations University Press.

Ghosh, Dhrubajyoti. 1991. Ecosystems approach to low-cost sanitation in India: Where the people know better. In *Ecological engineering for wastewater treatment: Proceedings of the International Conference at Stensund Folk College*, ed. C. Etnier and B. Guterstam, Stensund, Sweden, March 24–28. Gothenburg, Sweden: Bokskogen.

Hartling, Earle, and Margaret Nellor. 1998. Water recycling In Los Angeles County. In *Wastewater reclamation and reuse*, ed. T. Asano. Boca Raton, FL: CRC Press.

Kim, Tae Dong, and Hajime Unno. 1996. The roles of microbes in the removal and inactivation of viruses in a biological wastewater treatment system. *Water Science & Technology* 33(10–11):243–50.

Kadlec, Robert, and Robert Knight. 1996. *Treatment wetlands*. Boca Raton, FL: CRC Press.

Kadouri, M., Y. Djebbar, and M. Nehdi. 2001. Water rights and water trade: An Islamic perspective. In *Water management in Islam*, ed. N. I. Faruqui, A. K. Biswas, and M. J. Bino. Tokyo: United Nations University Press.

Kangas, Patrick. 2004. *Ecological engineering: Principles and practice*. Boca Raton, FL: Lewis.

Kazem, Mohamed, and Khalil Osman. 1998. Conflicting claims to Euphrates water muddy Syrian-Turkish relations. *Muslimedia*, July 16–31. www.muslimedia.com.

Lee, Mike. 2005. Perceptions of purity still cloud city's push to reuse wastewater. *San Diego Union-Tribune*, July 12.

Mills, William, Susan Bradford, Martin Rigby, and Michael Wehner. 1998. Groundwater recharge at the Orange County Water District. In *Wastewater reclamation and reuse*, ed. T. Asano. Boca Raton, FL: CRC Press.

Mitsch, William, and Sven Jorgensen. 2004. *Ecological engineering and ecosystem restoration*. New York: John Wiley.

New Mexico Environment Department (NMED). 2003. *New Mexico Environment Department (NMED) policy for the above ground use of reclaimed domestic wastewater*. August 7. Santa Fe: New Mexico Environment Department.

Olivieri, Adam, Don Eisenberg, and Robert Cooper. 1998. City of San Diego health effects study on potable water reuse. In *Wastewater reclamation and reuse*, ed. T. Asano. Boca Raton, FL: CRC Press.

Stott, R., E. May, and D. Mara. 2003. Parasite removal by natural wastewater treatment systems: Performance of waste stabilization ponds and constructed wetlands. *Water Science and Technology* 48(2):97–104.

Taylor, William, and Robert Saunders. 1991. Protozoa. In *Ecology and classification of North American freshwater invertebrates*, ed. J. H. Thorp and A. P. Covich. San Diego, CA: Academic Press.

U.S. Environmental Protection Agency (EPA). 1981. Process design manual for land treatment of municipal wastewater. EPA 625/1-81-013. Washington, DC: U.S. Environmental Protection Agency.

————. 1992. Guidelines for water reuse. EPA/626/R-92/004. Washington, DC: U.S. Environmental Protection Agency.

————. 1997. Response to Congress on the use of decentralized wastewater treatment systems. EPA 832-R-97-001b. Washington, DC: U.S. Environmental Protection Agency.

————. 2000. Constructed wetland treatment of municipal wastewaters. EPA/625/R-99/010, September. Washington, DC: U.S. Environmental Protection Agency.

Viet Anh, Nygen, subproject super. 2004. Appropriate decentralised sanitation options for urban and peri-urban areas in Vietnam. Subproject of Capacity Building Project in Environmental Science & Technology in Northern Vietnam, Hanoi, Vietnam.

Waldie, D. J. 2002. Los Angeles' toilet-to-tap fear factor. *Los Angeles Times*, December 1.

Wallace, Robert, and Terry Snell. 1991. Rotifera. In *Ecology and classification of North American freshwater invertebrates*, ed. J. H. Thorp and A. P. Covich. San Diego, CA: Academic Press.

Wallace, Scott. 2001a. System for removing pollutants from water. U.S. Patent 6,200,469. March 13.

————. 2001b. *Treatment of cheese-processing waste using subsurface flow wetlands.* Columbus, OH: Battelle Institute, 2001.

Wallace, Scott, and Robert Kadlec. 2005. BTEX degradation in a cold-climate wetland system. *Water Science & Technology* 51(9):165–71.

Wallace, Scott, and Robert Knight. 2006. *Feasibility, design criteria, and O&M requirements for small scale constructed wetland wastewater treatment systems.* Alexandria, VA: Water Environment Research Foundation.

Wegner-Gwidt, Joyce. 1998. Public support and education for water reuse. In *Wastewater reclamation and reuse*, ed. T. Asano. Boca Raton, FL: CRC Press.

World Health Organization. 1989. Health guidelines for the use of wastewater in agriculture and aquaculture. Technical Report Series 778. Geneva: World Health Organization.

————. 1993. The control of schistosomiasis: Second report of the WHO Expert Committee. Technical Report Series 830. Geneva: World Health Organization.

20 Infrastructure Models for Ecological Wastewater Management

Scott Wallace

CONTENTS

Introduction... 413
Creating Successful Wastewater Infrastructure....................................... 414
The Effect of Scale in Wastewater Infrastructure 414
Single-Home Infrastructure 417
    Single-Home Wastewater Management Options............................... 417
    Conventional Technologies 418
    Development of New Ecotechnology.. 419
    Water as a Carrier Medium 419
Village-Scale (Cluster System) Infrastructure 420
    Alternative Collection Systems.. 420
Cluster Treatment Systems 422
Moving Forward in Iraq.. 424
References.. 424

INTRODUCTION

This chapter summarizes the planning and management challenges of maintaining a successful wastewater infrastructure. Every culture places its own set of religious, technical, and economic constraints on the development of societal infrastructure. Within Islamic law, for example, there is sufficient foundation through fatwa to support water reuse (Abderrahman 2001), as discussed in chapter 18.

There are ecotechnologies available today that are capable of treating sewage to targeted reuse standards. Ponds (Mara 1976) and various forms of constructed wetlands (Kadlec and Knight 1996) are leading candidates of ecotechnology selection, as discussed in chapters 17 and 18. However, the choice of technology is not nearly as important as the cultural fit of the technology. Successful wastewater projects incorporate technologies that are culturally acceptable and that can be designed, constructed, operated, and maintained within the local economy.

The world population has increased from approximately 2.5 billion people in 1950 to 6.3 billion people in 2003 (United Nations 2003). Approximately 81 percent of the world population lives in developing countries (Population Reference Bureau 2002). As the world's population continues to grow, it is shifting from rural to urban areas. Total urban population was projected to be 2.3 billion in 1990 and is projected to increase to 4.6 billion by 2020, with 93 percent of this growth occurring in less developed countries.

The shift from rural to urban areas continues to create dense population centers that lack basic sanitation facilities, representing about 38 percent of the current world population (Postel 2003). Approximately 40 percent of the world's population is experiencing water stress (Glieck 1993). Continued exploitation of water reserves is likely to have detrimental long-term consequences on the availability of fresh water for human communities and native ecosystems (Kivaisi 2001). As water scarcity becomes more and more common, reuse of treated wastewater becomes increasingly important.

In the context of the Mesopotamian marshes, Iraq is undergoing a transformation from a water-rich country to a water-poor country. The construction of dams along the Tigris and Euphrates Rivers (including the massive Ataturk Dam in Turkey) has significantly altered the natural hydrology of the marshes, and water diversions for irrigation are severely limiting the quantity and quality of water available for marsh restoration. Current estimates surmise that only 15 to 20 percent of the drained marshes can be restored (Richardson et al. 2005) given the limitations on water availability now, and the uncertainty of water availability in the future. In a water-scarce future, wastewater reuse will play an important role in meeting Iraq's water needs.

CREATING SUCCESSFUL WASTEWATER INFRASTRUCTURE

Creating an infrastructure that successfully meets the needs of a society depends on implementing the "correct" technology. In this context, "correctness" can be defined as how well the selected technology meets the needs of society. In Iraq, there is clearly a large, unmet need for wastewater treatment, and as water scarcity becomes more common, effluent reuse will become a preferred management option. So, if the need is apparent, and water is scarce, why is technology selection difficult?

The societal "fit" of a technology is critical in technology selection, and is dependent on many factors. The technology needs to be compatible with cultural values and traditions, but it also has to make sense economically. In most cases, this means that the wastewater system can be built using locally available materials and is easy to operate and repair. The system must provide effective treatment, yet be capable of being locally operated and maintained. Finally, there must be societal support for the system, which requires that the benefits of the treatment system (in terms of public health, environmental protection, and economic return) are apparent to the community. However, technologies that can be offered as a solution are highly dependent on the scale of the infrastructure, as discussed in the next section.

THE EFFECT OF SCALE IN WASTEWATER INFRASTRUCTURE

Wastewater infrastructure around the world generally falls into one of three models (Hallahan and Wallace 2001). These three infrastructure models are categorized by

TABLE 20.1

Wastewater Infrastructure Models

Typical Population Density	Planning and Organization Capacity	Most Suitable Infrastructure Scale
High	Regional	Centralized systems ("The Big Pipe")
Medium	Local	Clustered systems (decentralized)
Low	Family	Single-home systems

the population density and/or organizational capacity of a community, as summarized in Table 20.1.

In many areas of the world, a high population density is not synonymous with the ability to plan and organize infrastructure on a regional scale. In these cases, the ability of governments to plan and organize is a much more important determinant in the selection of appropriate wastewater infrastructure than population density. The existence of government agencies does not ensure that effective planning and organization do, in fact, occur. For the purpose of this chapter, it is important to define the concept of institutional capacity. "Institutional capacity" is defined as the ability for an organization to develop a plan *and to provide effective implementation of that plan*. The importance of organizational and planning capacity in technology selection is key in understanding why so many wastewater technology transfers between developed and developing countries fail.

Determining the scale at which the wastewater infrastructure will operate is significant because each of the three infrastructure models is embodied by unique technologies, and there is very limited technology crossover between the three models. By selecting the scale level, planners determine the range of technology options that will be available to the engineers who construct the infrastructure (Wallace et al. 2005).

A theoretical framework for wastewater management has been developed for Zimbabwe (Nhapi, Gijzen, and Siebel 2003), but is applicable to many regions of the globe, including Iraq. This framework summarizes the effects of population density, effluent reuse, level of treatment, and appropriate technology options, as shown in Figure 20.1.

In reviewing Figure 20.1, it can be seen that as the treatment level increases, there is a cascade from property level (single-home) systems up to large-scale centralized treatment systems (lower right quadrant). As the scale level increases, there is generally an increase in population density, with an upward cascade from low-density to high-density areas. The technologies used for wastewater management (lower left quadrant) cascade upward from simple graywater/blackwater systems (implemented at the single-home level) up to large-scale advanced treatment systems (employing biological nutrient removal [BNR]). As the scope and complexity of the infrastructure change, the end use of treated effluent also changes (upper left quadrant), cas-

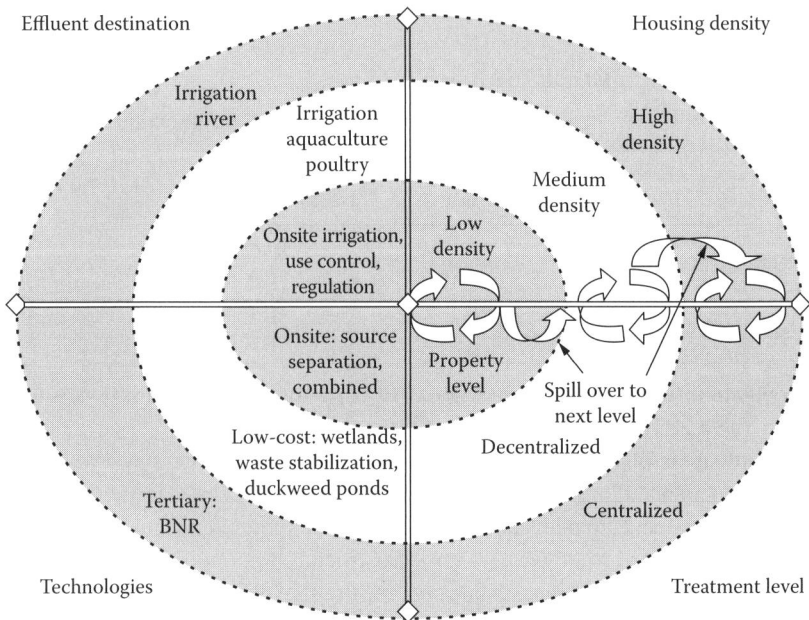

FIGURE 20.1 Theoretical approach to decentralized wastewater management. Reprinted with permission from Nhapi, I., H. J. Gijzen, and M. A. Siebel. 2003. A conceptual framework for the sustainable management of wastewater in Harare, Zimbabwe. *Water Science and Technology* 47(7–8):11–18.

cading upward from simple graywater irrigation systems (at the single-home level) up to large-scale treatment works that primarily discharge to rivers.

Moving along the scale gradient produces a set of technical solutions that are distinctly different from the technology options offered at a previous scale. For instance, sewage treatment for a large city could be provided at any of the scale levels. If the decision is to centrally manage the wastewater, technologies most likely to be the most cost-effective at this level include conventional sewers, pumping stations, and one large mechanical treatment plant, operated by a full-time specialist staff supported by the government. This is a reasonable approach, provided that there is planning and organizational capacity capable of managing the wastewater infrastructure from a regional perspective.

Another equally valid approach would be to divide the city into sections and build a smaller treatment plant for each section of the city. This offers advantages because less sewer pipe is needed, the treated effluent is physically closer to the site(s) of water reuse, and the treatment facilities, which are smaller and less complicated than large treatment plants, can be managed by local people. Technologies likely to be cost-effective at this scale level include small-diameter sewers, ponds, and water reuse. This is a reasonable approach if wastewater is managed at the local level.

Finally, the city could opt to not construct any sewer system at all. Then by default, responsibility to manage wastewater falls on individual families at the single-home level. Technologies likely to be effective at this level include gray-water separation, settling tanks, subsurface-flow constructed wetlands, and local irrigation.

For a small town near the edge of the marshes, the wastewater system will not be large enough to allow many of the conventional "big-pipe" technologies to be cost-effective, so the decision matrix narrows to cluster and single-home options. For scattered rural dwellings, the choice is clear: the only option is to manage wastewater at the single-home level.

SINGLE-HOME INFRASTRUCTURE

In the absence of any other planning and organization capacity, the single-home system is the default infrastructure model, and technology selected for wastewater management should be consistent with this model.

SINGLE-HOME WASTEWATER MANAGEMENT OPTIONS

The wastewater management options available to Iraqis at the single-home level have a range of pros and cons. These management options include direct discharge to the marshes, soak-away pits and latrines, and onsite treatment and reuse.

Direct discharge to the marshes has one overriding advantage, because it is the lowest up-front capital cost option. It is also current status quo. However, the direct discharge of pathogens, especially trematode worms (*Schistosomosa spp.*), creates long-term health problems for people living in and near the marshes. Potentially valuable water and nutrients are thrown away, and this model generates no economic return.

Soak-away pits or latrines allow wastewater to soak into the groundwater (or into the marshes), but due to the filtering effect of the soil and/or sand used in these devices, pathogens are generally removed from the wastewater (especially larger particle-sized pathogens such as geohelminths, tapeworms, and trematode worms). These systems achieve the goal of pathogen removal but do not recover potentially valuable water and nutrients.

Onsite treatment and reuse systems are advantageous in that they not only control pathogens but also recover water and nutrients for irrigation, thereby creating the opportunity to generate an economic return for the citizens of the community. Treated effluent can be reused to irrigate gardens or other agricultural production areas. These benefits will likely be the greatest to residents living along the perimeter of the marshes. A drawback to onsite treatment and reuse, however, is that there is a cost involved in constructing the infrastructure and management required for such a treatment and reuse system (even if it is managed at the single-home level). The key to "bottom-up" (e.g., not supported by taxes) infrastructure development is to introduce a system that is culturally acceptable, and it is apparent to the community that the long-term benefits of the treatment system outweigh its up-front capital cost.

CONVENTIONAL TECHNOLOGIES

Even in countries that enjoy a high level of infrastructure support, like the United States, single-home (onsite) treatment systems are very common. Onsite systems serve approximately 25 percent of the U.S. population and 33 percent of new home construction (U.S. Census Bureau 1990). Because the United States has been a relatively water-rich country, the primary technical focus has been on below-ground dispersal of wastewater (U.S. Environmental Protection Agency [EPA] 1980). These systems typically employ a settling tank (septic tank) and underground distribution pipes. This technology has essentially remained unchanged since the early part of the nineteenth century, as illustrated in Figure 20.2.

The use of "old" wastewater technology is of little benefit to Iraq because a primary focus of implementing sewage treatment will be to produce water of sufficient quality for reuse. In the last 20 years, there has been considerable interest in decentralized wastewater management in the United States (EPA 1997) that has led to the development of a variety of compact treatment systems (EPA 2002). While the majority of these devices use pumps and other mechanical support systems, some systems, such as subsurface-flow constructed wetlands, are suitable for widespread adoption around the world.

FIGURE 20.2   1927 Sears Roebuck advertisement for a single-home septic system. Reprinted with permission from Crown Publishers, Inc.

DEVELOPMENT OF NEW ECOTECHNOLOGY

Subsurface-flow wetland technology was initially developed in Germany in the 1950s (Börner et al. 1998) and has spread worldwide, with increasing use and interest over the last 25 years (IWA Specialist Group 2000; Kadlec and Knight 1996; Steiner and Watson 1993; EPA 2000; Wallace and Knight 2006).

Single-home subsurface-flow wetland systems are the most widely adopted form of ecotechnology in the world (Wallace and Knight 2004). There are estimated to be over fifty thousand systems worldwide (Kadlec 2004). A typical subsurface-flow constructed wetland is shown in Figure 20.3.

Subsurface-flow constructed wetland treatment systems are compact, provide effective treatment of pathogens, and do not expose wastewater during the treatment process. They are capable of producing effluents that meet World Health Organization (WHO; 1989) reuse guidelines, and up-to-date design criteria are readily available (Wallace and Knight 2006). Because these wetlands can be constructed with local materials and labor (often by the homeowners themselves), they have become very popular in some developing areas. A typical small-scale subsurface-flow wetland system is shown in Figure 20.4.

WATER AS A CARRIER MEDIUM

In the initial stage of technology development, wetlands were used in relatively water-rich regions of Europe and North America. In many developing countries, water is not available in sufficient quantity (or it is too expensive) to be used in conventional flush toilets. Instead, pit latrines, pour-flush toilets, or composting toilet systems may provide for more effective management of fecal material. This does not necessarily preclude ecological wastewater management, as indicated in Figure 20.5.

In the absence of conventional flush toilets, separation of fecal material (black-water) and graywater (urine) is commonly employed. Because graywater-urine is

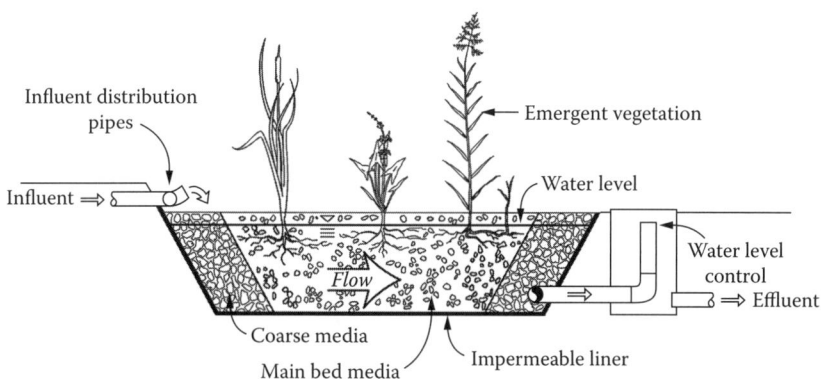

FIGURE 20.3 Schematic of a subsurface-flow wetland. Reprinted with permission from Wallace, S. D., and R. L. Knight. 2006. Feasibility, design criteria, and O&M requirements for small-scale constructed wetland wastewater treatment systems. Project 01-CTS-5. Alexandria, Virginia: Water Environment Research Foundation.

FIGURE 20.4 Single-home subsurface-flow constructed wetland in Akumal, Mexico. Photo courtesy C. Sparks.

much less polluted, it usually requires less treatment to produce reuse-quality water. In these instances, graywater can and should be viewed as a resource as opposed to waste product, since it has such a high potential for reuse. A single-home graywater system is shown in Figure 20.6.

VILLAGE-SCALE (CLUSTER SYSTEM) INFRASTRUCTURE

Wastewater management at the village scale is characterized by the use of a collection network that collects the sewage from two or more homes and routes it to a treatment facility that is owned by the community. *This requires a much higher degree of institutional support than the single-home model. For the cluster-system infrastructure model to work effectively, a management entity must be in place to operate and maintain the shared collection, treatment, and reuse system.*

ALTERNATIVE COLLECTION SYSTEMS

Collection systems can consist of conventional gravity sewer and manholes, although alternative systems that remove the majority of the solids from the wastewater before transport are likely to be much more cost-effective for small communities (Mara 1996). A typical settled-sewage system is shown in Figure 20.7.

In many instances, these collection systems can be constructed of "new" piping materials like high-density polyethylene (HDPE). This type of piping system is tough and flexible (minimizing the number of fittings). Because sections of pipe are melted together (fusion welded), the result is a watertight piping system. These factors are contributing to the popularity of HDPE pipe in developing countries. Figure 20.8 shows a typical network of small-diameter HDPE sewer pipes.

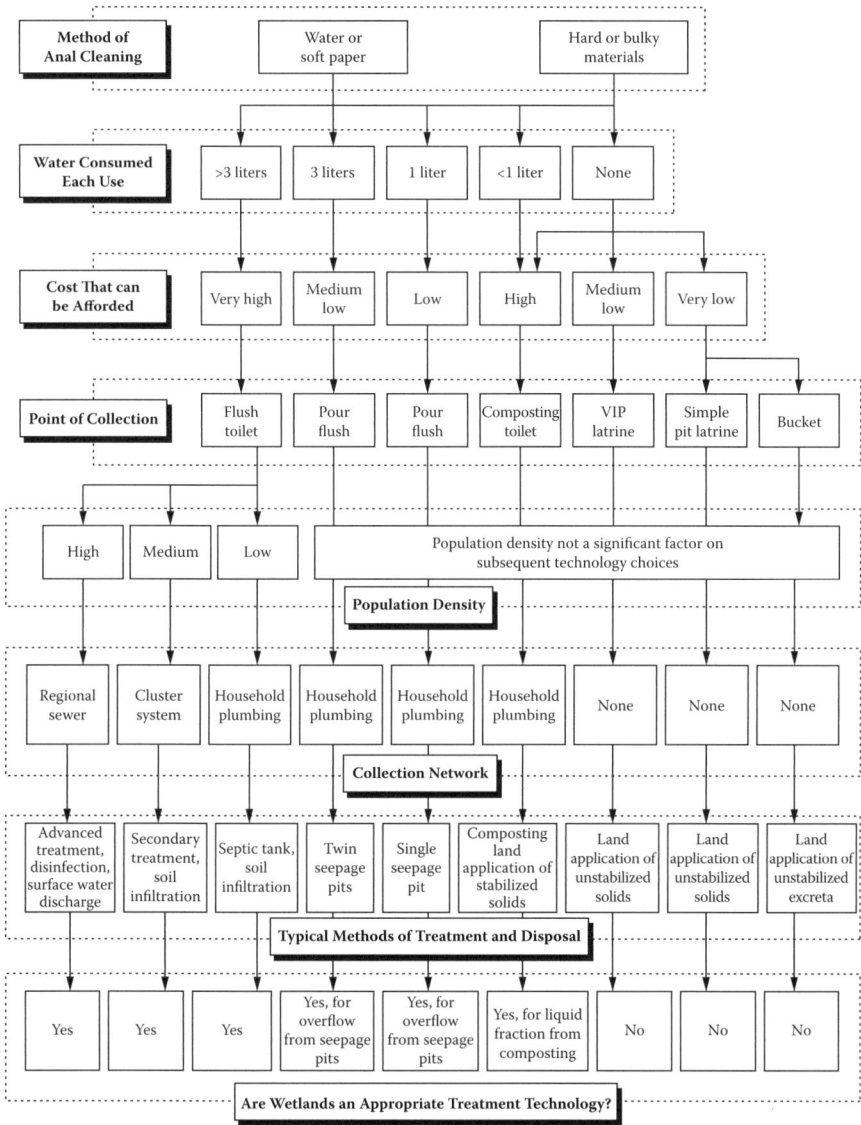

FIGURE 20.5 Use of single-home ecotechnology for wastewater management. Reprinted with permission from Wallace, S. D., and R. L. Knight. 2006. Feasibility, design criteria, and O&M requirements for small-scale constructed wetland wastewater treatment systems. Project 01-CTS-5. Alexandria, Virginia: Water Environment Research Foundation.

FIGURE 20.6   Single-home graywater system.

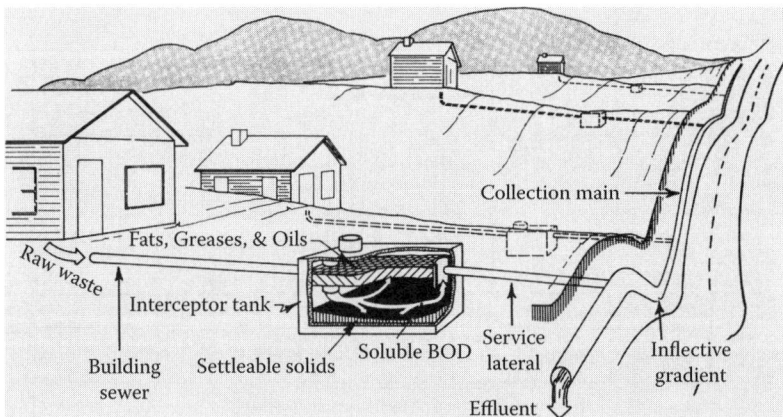

FIGURE 20.7   Settled-sewage collection system. Reprinted from EPA (1991).

CLUSTER TREATMENT SYSTEMS

While subsurface-flow constructed wetlands are suitable for smaller cluster systems, other forms of ecological wastewater management become more cost-effective as systems grow larger. These include surface-flow wetlands and pond systems, as discussed in chapter 18. A typical surface-flow wetland is shown in Figure 20.9.

Because these passive ecological systems are land intensive (see Table 18.1), semipassive systems (wetlands or ponds that rely on some degree of mechanical assistance) are often used. The advantage of semipassive systems is that they occupy less land area, retain some of the resiliency of passive ecological systems (not all

FIGURE 20.8   Small-diameter pipes for settled-sewage system. Inset: Machine for fusion welding. Photos courtesy North American Wetland Engineering LLC.

FIGURE 20.9   Surface-flow wetland near Pensacola, Florida. Photo courtesy S. Wallace.

treatment is lost if the mechanical components break down), and have relatively low energy inputs compared to conventional mechanical treatment processes (see Table 18.2).

The larger challenge with semipassive treatment systems lies in management of the system. While simple to operate, they are not maintenance-free. Through the introduction of mechanical treatment components, a host of maintenance and repair tasks is added to the responsibilities of the management entity. Communities *must* be aware of the increased complexity involved in managing systems that include mechanical components if these technologies are going to be effectively employed.

MOVING FORWARD IN IRAQ

This chapter has discussed ecological wastewater management and the infrastructure models that determine the range of technology choices available to communities. As Iraq transitions from a water-rich to a water-poor country, increasing priority will have to be given to recovery and reuse of treated effluents.

The Mesopotamian marshes are a unique landscape. The infrastructure that will be developed to serve the communities in and around the marshes will be similarly unique and will depend heavily on the creativity and resourcefulness of Iraqi planners and engineers, and most of all on the marsh Arabs themselves.

While the template for infrastructure development has yet to be created for the communities that depend on the Mesopotamian marshes, much can be learned from analogous ecological technologies employed in other parts of the world, including ponds, surface-flow constructed wetlands, and subsurface-flow constructed wetlands. All of these are mechanically simple, biologically complex technologies that are capable of producing effluents that meet WHO guidelines for reuse. Technology selection is likely to be the easiest component of infrastructure development in Iraq.

However, no technology transfer will be successful unless there is the institutional capacity to support the selected scale of infrastructure. This chapter has explored the single-home and village-scale (cluster) infrastructure models in detail, based on the assessment that these are the infrastructure models most likely to be successful in the Mesopotamian marsh region. In the absence of clearly defined and effective management entities, the default level of management is at the single-home level. As a result, planners and engineers should base their technology choices accordingly.

REFERENCES

Abderrahman, W. 2001. Water demand management in Saudi Arabia. In *Water management in Islam*. Tokyo: United Nations University Press.

Börner, T., K. Von Felde, E. Gschlössl, T. Gschlössl, S. Kunst, and F. W. Wissing. 1998. Germany. In *Constructed wetlands for wastewater treatment in Europe*. Leiden, the Netherlands: Backhuys.

Glieck, P. H. 1993. *Water crisis: A guide to the world's freshwater resources*. Washington, DC: Oxford University Press.

Hallahan, D., and S. D. Wallace. 2001. Wastewater system options: Providing solutions for small communities. *CE News*, October.

IWA Specialist Group on Use of Macrophytes in Water Pollution Control. 2000. *Constructed wetlands for pollution control: Processes, performance, design and operation*. London: International Water Association.

Kadlec, R. H. 2004. *Constructed wetlands: State of the art (oral presentation)*, vol. 3, ed. A. Liénard and H. Burnett. Avignon, France: IWA.

Kadlec, R. H., and R. L. Knight. 1996. *Treatment wetlands*. Boca Raton, FL: CRC Press.

Kivaisi, A. M. 2001. The potential for constructed wetlands for wastewater treatment and reuse in developing countries: A review. *Ecological Engineering* 16(4):545–60.

Mara, D. 1976. *Sewage treatment in hot climates*. London: John Wiley.

————. 1996. *Low-cost urban sanitation.* West Sussex, UK: John Wiley.

Nhapi, I., H. J. Gijzen, and M. A. Siebel. 2003. A conceptual framework for the sustainable management of wastewater in Harare, Zimbabwe. *Water Science and Technology* 47(7–8):11–18.

Population Reference Bureau. 2002. *World population data sheet.* Washington, DC: Population Reference Bureau.

Postel, S. 2003. From Rio to Johannesburg: Securing water for people, crops and ecosystems. World Summit Policy Brief no. 8. Washington, DC: Worldwatch Institute.

Richardson, C. J., P. Reiss, N. A. Hussain, A. J. Alwash, and D. J. Pool. 2005. The restoration potential of the Mesopotamian marshes of Iraq. *Science* 307:1307–11.

Steiner, G. R., and J. T. Watson. 1993. General design, construction, and operation guidelines: Constructed wetlands wastewater treatment systems for small users including individual residences, 2nd ed. TVA/WM-93/10. Chattanooga: Tennessee Valley Authority Resource Group Water Management.

U.S. Census Bureau. 1990. Historical census of housing tables: Sewage disposal. www.census.gov/hhes/www/housing/census/historic/sewage.html.

U.S. Environmental Protection Agency (EPA). 1980. Design manual: Onsite wastewater treatment and disposal systems. EPA 625/1-80/012. Washington, DC: U.S. EPA Office of Research and Development.

————. 1997. Response to congress on use of decentralized wastewater treatment systems. EPA 832/R-97/001b. Washington, DC: U.S. EPA Office of Water.

————. 2000. Constructed wetlands treatment of municipal wastewaters. EPA 625/R-99/010. Washington, DC: U.S. EPA Office of Research and Development.

————. 2002. Onsite wastewater treatment systems manual. EPA 625/R-00/008. Washington, DC: U.S. EPA Office of Research and Development.

United Nations. 2003. *World population prospects: The 2002 revision highlights.* New York: United Nations.

————. 2004. *Water Environment Research Foundation (WERF) small-scale treatment wetland database,* vol. 3, ed. A. Liénard and H. Burnett. Avignon, France: IWA.

Wallace, S. D., and R. L. Knight. 2006. Feasibility, design criteria, and O&M requirements for small-scale constructed wetland wastewater treatment systems. Project 01-CTS-5. Alexandria, VA: Water Environment Research Foundation.

Wallace, S. D., G. F. Parkin, B. Ballavance, and R. C. Brandt. 2005. Ecological wastewater management in Iowa: Hope for Iowa's unsewered communities. Report prepared for the Iowa Policy Project. Mount Vernon: Iowa Policy Project.

World Health Organization. 1989. Health guidelines for the use of wastewater in agriculture and aquaculture. Technical Report Series 778. Geneva: World Health Organization.

Agriculture and Water Management

21 Salinity Management in Arid Regions

Lessons for Iraq from the Western United States*

CONTENTS

Introduction 429
Scientific and Engineering Issues 434
Relevant Applications 437
Application in the Iraq Marsh Context 441
Acknowledgment 448
References 449

INTRODUCTION

Growing plants in arid regions is a major challenge, relying upon careful management of the scarce water resources (Figure 21.1). As water moves across rocks and through soils, it dissolves salt which in turn becomes something that also needs careful management, particularly in desert regions such as southern Iraq. Salt is seldom managed as a solid unless there is an economic benefit to it or unless it's a last resort as, for example, in evaporation ponds where some farmers make salt because they have no other place to put it.

Water meets a range of needs in Iraq (Naff and Hanna 2002), including supporting not only the marshes and their functions like fisheries but also rice and other agricultural production, as well as supporting urban and industrial activities throughout the country. With all these competing uses, John Dickey and Mark Madison (2004) of CH2MHILL believe that water will almost certainly be the limiting factor for marsh restoration, particularly in relation to its role in salinity management. Some fairly sophisticated water budget and management models have been developed for both the macro and regional scales in Iraq (France 2011). Today, while natural flood peaks are greatly diminished by the dams, there are opportunities for artificial annual flooding to create flood storage capacity which may be feasible in

---

* Adapted by Robert L. France from Dickey, J., and M. Madison. 2004. Moving salt and water in managed ecosystems: Case studies from history and the western United States. Paper presented at the Mesopotamian Marshes and Modern Development: Practical Approaches for Sustaining Ecological and Cultural Landscapes conference, Cambridge, MA, October.

FIGURE 21.1   Irrigation practices in the United Arab Emirates (old farm well head, restored channels in an ancient oasis, drip apparatus sustaining a tree, and a small farm growing fodder for camels) and Syria (diesel pumps and metal and plastic piping sustaining large-scale grain fields).   **(Continued)**

the January to February period, and which would allow opportunity for moving salt with water.

Fish production is critical to the restoration of the Iraqi marshes because of cultural and economic reasons (France 2011, 2012). Some preliminary data from the marshes suggest that populations of such valued species drop off precipitously at high levels of salinity.

FIGURE 21.1 (*Continued*) Irrigation practices in the United Arab Emirates (old farm well head, restored channels in an ancient oasis, drip apparatus sustaining a tree, and a small farm growing fodder for camels) and Syria (diesel pumps and metal and plastic piping sustaining large-scale grain fields).

None of this is new. Marsh and land salinization have been recurring problems in southern Iraq, evidence for which extends back thousands of years. Sumerian and Akkadian lamentations recorded as cuneiform inscriptions (France 2007) bemoan the problem, as the following three examples show:

> Below, the water from the depths is bolted, it does not flow/ That is why the dark ploughland has whitened/ That is why in the pastureland grains do not sprout. (Atahasis II)

FIGURE 21.1 (*Continued*) Irrigation practices in the United Arab Emirates (old farm well head, restored channels in an ancient oasis, drip apparatus sustaining a tree, and a small farm growing fodder for camels) and Syria (diesel pumps and metal and plastic piping sustaining large-scale grain fields).

> *The dark pastureland was bleached/ The broad countryside filled with alkali.* (Lugalbanda and the Thunderbird)
>
> *Its central plain…/ grew reeds of lament…/ the water flowing sweet/ flowed now as saline waters.* (The Cursing of Akkade)

Indeed, soil salinization was so severe that tablets from many thousands of years ago now in Harvard's Semitic Museum (see photos in France 2007) often show overt whitening as salts emerge from the baked clay and fill in the depressions left by ancient scribes' styluses (J. Greene, personal communication). Even

FIGURE 21.1 (*Continued*) Irrigation practices in the United Arab Emirates (old farm well head, restored channels in an ancient oasis, drip apparatus sustaining a tree, and a small farm growing fodder for camels) and Syria (diesel pumps and metal and plastic piping sustaining large-scale grain fields).

the geographical northward migration of sequential Mesopotamian civilizations from Sumeria (south of present-day Basra) to Akkadia to Babylonia to Assyria (north of present-day Baghdad) can be linked to ruined land due to overwhelming problems in salinization. And it is possible that the gradual shift in crops from wheat to barley might be a result of the latter's higher salinity tolerance. Today, salt is very evident on the ground around the ancient ruins of Mari, an important ancient Mesopotamian city near the Euphrates River and beside the modern border with Iraq (Figure 21.2).

Like wetlands ecology and design, salinity management is a broad area which will be significantly condensed in this chapter. Sulfate is one of the components of salinity, and there are often problems with sulfide toxicity in wetlands—which can, however, be managed by manipulating water levels to provide redox conditions for mitigation. This strategy can also be applied to broader marsh areas where water levels can be regulated, and certainly there are many such water control facilities in the marshland region at present (indeed, that is one of the problems). It is also possible to recycle salt in the tissues of tolerant plants such as *Spartina* for beneficial use such as for feeding cattle (Dickey and Madison 2004).

After discussing the background to salt management, a series of relevant applications from elsewhere will be presented. This will be followed by potential application of these ideas to the Iraqi marshlands, including refining some of the recommendations already made in terms of regional restoration that are specifically targeted to address salinity concerns.

FIGURE 21.2 Resembling snow, extensive surface deposits of salt at the ancient Mesopotamian city of Mari, located near the Syria–Iraq border.   (*Continued*)

SCIENTIFIC AND ENGINEERING ISSUES

It is important to first establish some definitions (Dickey and Madison 2004). Salinity is composed of common ions dissolved in water ranging from sodium to nutrients and trace elements like selenium, molybdenum, and boron, noting that at higher concentrations some of these latter elements can become toxic. "Applied water" is the water applied as irrigation or flowing into a marsh system. "Consumptive use" is water actually taken up by plants. "Drainage" is the remainder not taken up by plants that's leaving the field or the marsh system. This is drainage for beneficial purpose. On the

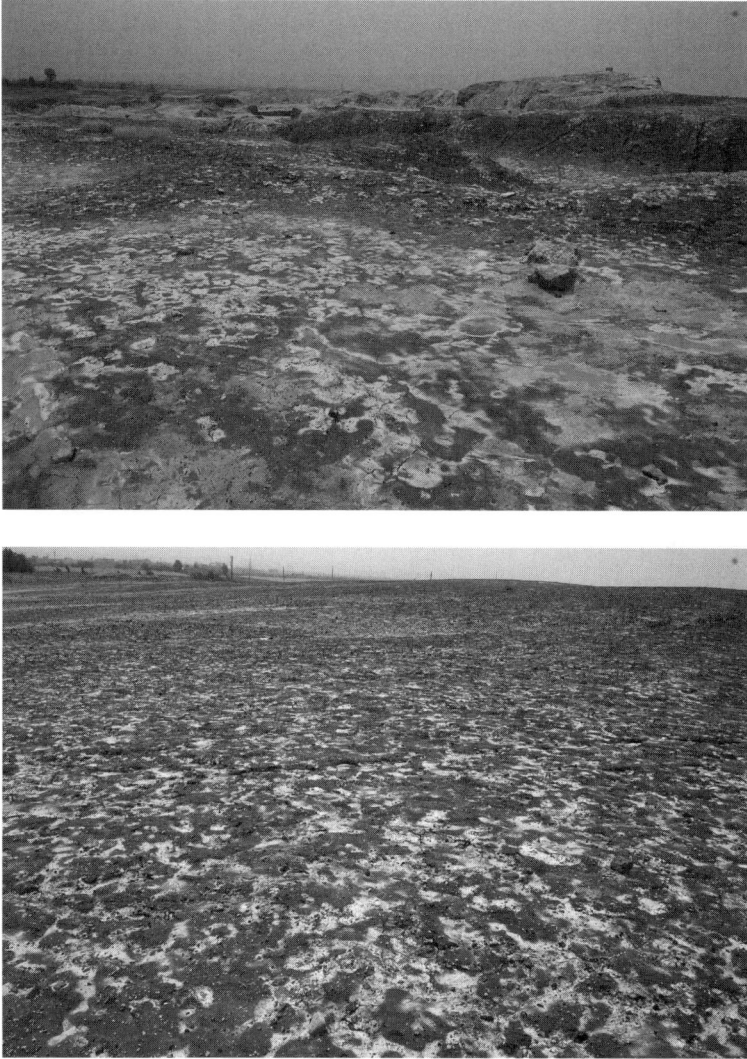

FIGURE 21.2 (*Continued*) Resembling snow, extensive surface deposits of salt at the ancient Mesopotamian city of Mari, located near the Syria–Iraq border.

other hand, "overdrainage" is the removal of water in excess of that required to maintain ecologically suitable and productive soil or marsh hydrology and water quality.

To effectively manage an ecological system, it is necessary to have some sort of target or criteria to aim for, and salinity is no different in this regard (Dickey and Madison 2004). One way to do this is to set criteria in terms of primary productivity in relation to the sensitivity of various species in the habitat to salinity. Graphs of plant productivity versus salinity show a threshold at which the impact begins to be felt followed by productivity decreases, with the more tolerant species having higher

FIGURE 21.2 (*Continued*) Resembling snow, extensive surface deposits of salt at the ancient Mesopotamian city of Mari, located near the Syria–Iraq border.

thresholds and descending more slowly. Similar relationships of yield and productivity exist for fish species, and it will be necessary to establish these criteria if the goal of successfully creating and maintaining marsh habitat is to be met.

Leaching and flushing water comprise one effective way to manage salt (Dickey and Madison 2004). If the water supply quality can be managed, the same threshold goal can be achieved simply by running more water through the system. But the problem of course is that more water needs to be available; for example, if the target and inflow salinity approach one another, it is necessary to use vast amounts of water to maintain the targets. So the ideal situation occurs with relatively low supply salinity relative to the goal or the existence of a more tolerant and thus more water-efficient system that operates along a different flushing curve. Success therefore depends on the applied water quality and the species salinity tolerance. The water and salt balances need to be integrated to effectively manage the system to maintain habitat by being a guide toward various best management practices or strategies. If it is possible to reduce or eliminate various salt inputs such as those from industry or cities, it then becomes much easier to manage the entire system. Depending on what targets are set, it may be possible to flush a large amount of water through the system to create outflows that are suitable at least for the more tolerant species. This is, of course, dependent on the goals and amount of available water but does present another opportunity.

On the watershed scale, the strategies for managing salt range from source controls such as reducing consumptive use, fallowing land, changing crops (all designed to send more freshwater down the system and less up into the atmosphere), or simply running a less efficient upstream water system which will allow more to flow downstream into the marshes and adjourning fields (Dickey and Madison 2004). Other options include addressing the timing and seasonal release of surface and subsurface water, or reducing salt load through avoiding adding new salt in industry or with saline groundwater. For example, sometimes saline aquifers exist that are over-charged by inefficient irrigation and then end up recharging rivers with extremely saline water. There are good programs in the Colorado River Basin that have been implemented to avoid that sort of salt loading just by shifting to a more efficient sys-tem of irrigation. Sequestration of salt upstream in soil and groundwater or surface waters is another option. Here, salt is simply stored in a salt pan without any water running through.

In-marsh solutions include recycling and the encouragement of tolerant ecosys-tems that nevertheless have value. The high value attributes of the historic marsh ecosystem (France 2012) will no doubt be very important parts of the future of the marshes (France 2011), but if there are also opportunities to recycle water and create more saline habitat that has value of other sorts, that may be a good way to expand the marsh opportunities (Dickey and Madison 2004). Sequestration, flushing, leach-ing, and disposal are all valid management options. The ample drainage opportuni-ties that now exist in the Iraqi marshes are going to be needed, perhaps not in their entirety, but certainly for some strategic use to route salt out to the ocean.

RELEVANT APPLICATIONS

California has many similarities with Iraq: it is arid, it has a desert agriculture that is very developed, it has many water control structures, and it has areas where water once was but now is not. California has also lost most of its wetlands to a degree even greater than that occurring in Iraq (France *in* Reed 2005), and it is struggling to try to restore some of these.

In most of the American West and other places around the world where ground-water is harvested, total dissolved solids (TDS) increase as irrigation utilizes groundwater (Dickey and Madison 2004). Pumping the water out to irrigate a crop evapotranspires water through the crop and leaves behind a salt residue. The result of this is that in many arid areas where the groundwater is deep, the salts will eventually leach back to groundwater. Over time, if large amounts of groundwater are pumped out and the freshwater is evaporated, the amount of salt that goes back to ground-water will increase its TDS, occasionally to levels such that the system becomes no longer usable for crop irrigation. Such a scenario can be prevented through the adop-tion of more efficient irrigation or by reducing the area irrigated. The goal here is to balance the amount of water extracted by crop evapotranspiration with the amount that comes back into the system through either rainfall or groundwater recharge, toward an end of maintaining the freshwater in the long-term with a constant con-centration of salt getting cycled through the system. Such systems don't really make salt, but they also don't really destroy it, though they can concentrate it if crops are grown and freshwater is used.

One hypothetical example (Dickey and Madison 2004) illustrates how, for an eighty-acre farm in California, salt accumulation becomes a severe problem rep-resentative of issues developing in dry-land agriculture around the world (see also chapter 22). The rainfall in the area is about five inches, amounting to about 10 mil-lion gallons per year of freshwater almost free of salt. At only five inches per year, however, this is not enough to grow a crop and so groundwater must be extracted in an amount of about eighty-five inches per year. Although this groundwater has a relatively low TDS concentration of 997 mg/L, it will nevertheless generate about a million and a half pounds of salts per year applied to this eighty-acre field. The plants transpire sixty-five inches per year, consuming freshwater and leaving behind all the salts. The salts concentrate in the root zone, and twenty-five inches of water have to be applied as overirrigation to keep this upper five feet of soil fresh so crops can continue to be grown by moving the salt to below the root zone.

That salt—the million and a half pounds that came out of the groundwater—gets moved with that deep percolation water back to the groundwater. But now it is at 3,600 mg/L due to being concentrated by plant extraction of freshwater that exceeds the rainfall. Groundwater concentrations have consequently increased over time, going from 1,000 mg/L to 1,400 mg/L in eight years. This farm needs to develop a more efficient form of irrigation so that sixty-five inches per year of supplemental irrigation are not required. Ways need to be explored either to use less water to grow almost the same crops in a sustainable water balance or to develop an alternative, supplemental water source from somewhere other than the groundwater (Dickey and Madison 2004).

One potential solution occurs through seasonally reclaiming soils and exporting salts. Seasonal drainage occurs in many agricultural areas where groundwater exists close to the surface or when a low rate of irrigation during the summer accumulates salts in the root zone. The plants are still extracting freshwater and leaving behind salt, so whatever salt is in the water source can be stored in the soil during the growing season. If the region receives winter rainfall, this can be used with some additional irrigation to leach the salts into the drainage system so that they can be discharged with the winter floods when the system is full of freshwater from rainfall and it thus can tolerate the salt present without causing downstream damage. This is a valuable strategy if one is concerned about upstream drainage water affecting the quality of the downstream water resource. Therefore, controlled irrigation upstream is required so salts can be stored in the system during the growing season until such time during the winter storms when they can be safely flushed out. At this time, tolerance is increased because the plants are growing more slowly and are not affected as seriously by the salt, and of course there is more water in the basin so salt concentrations are diluted.

An example of applying such a seasonal salt management strategy is the Oremet titanium mine in Albany, Oregon (Dickey and Madison 2004). Titanium is an ore found in titanium tetrachloride liquid and whose extraction leaves behind a high amount of chloride. The chloride is leached with nitric, sulfuric, and phosphoric acids in a very polluted waste stream which, however, can be neutralized and put into the soil to grow trees and grasses. About 2–3 tons per ha per year of waterborne salt is applied to these crops in this way. To make the process sustainable, salts are accumulated during the growing season (to about 1,500 mg/L in the surface-applied irrigation water) until the winter, when rainfall and rising streams are used to flush the salts out of the system and into the drains to collect the salinity and discharge it downstream at a very high flow rate so it will not impact biota (i.e., concentrations of below 500 mg/L).

In another example, Dickey and Madison (2004) worked on a comprehensive salt management plan for Owens Lake, located just west of Death Valley in California, a system they refer to as "an environmental problem of epic proportion," and thus of relevance to that which has transpired in southern Iraq. Owens Lake is the body of water that was drained when the city of Los Angeles diverted its river into the L.A. Aqueduct to use for drinking water. It is now a huge environmental problem: about $260 \, km^2$ of dusty saline desert lake bed without water. The EPA has issued a mandate to restore the habitat to bring a halt to the dust storms that plague the area from the mobilization of the soil that was once the bottom of the lake (this soil is not covered by vegetation due to being very saline and powdery and is thus easily airborne).

CH2MHILL installed new infrastructure to restore the habitat. The first phase was to plant $52 \, km^2$ of vegetation for dust control by flooding saline and fresher habitats. Meadows were therefore created over a portion of the total barren area which with time will grow toward complete cover. The second phase involved reflooding. CH2MHILL diverted water back out of the L.A. Aqueduct and into the former lake in order to reestablish this habitat. This approach required a high level of infrastructure in the form of pipelines and hydrological control systems as Dickey and Madison were trying to minimize the amount of water they used to accomplish the task. If they had had access to all the water from the aqueduct, the hydrological management demands would have been much less. In other words, the more habitat that one attempts to create with the least amount of water, the more management that will be required.

This case study has potential relevance to the Iraqi marshes. The water that is diverted into the drained Owens Lake bed is always fresh, but the further out from the edge where the water is being inputted the greater the evaporation and consequently the higher the salinity of the remaining water. Far out in the middle of the former lake, the environment is more like seawater and at the far edge most distant from the inflow there, it is actually a salt sink where CH2MHILL is building a containment area in which deposited salts can be concentrated by being flushed there from the rest of the area. Significantly, there are great wildlife habitat benefits in the entire area, even those of saltwater (Dickey and Madison 2004). And in the area where salt is being accumulated, U.S. Borax has built a very large salt-mining operation to remove minerals from the lake bed.

The soils around the Owens Lake area look very similar to the soils in the Iraqi marshes: windblown sands and silts. The soil in the lakebed is an anoxic black clay with a strong sulfur smell, and when it dries out it cracks into large blocks of clay which fill in with windblown sands and silts, a very poor soil structure for surface-irrigated agriculture (this is similar to the situation described in chapter 6 for the Hula Swamp area in Galilee). When water is placed on such soil, it almost immediately runs off the nearly watertight clay into the porous sand structure around it, resulting in a very poor distribution of water and causing problems if one wanted to establish agriculture (Dickey and Madison 2004).

Green salt grass is used for revegetation in the Owens Lake project due to its high salt tolerance. About 850 ha are being grown with fairly saline water through subsurface drip irrigation which is very efficient in that there is almost no surface evaporation. This technology was used because if the water is not put back into the lake,

Los Angeles sells it as drinking water and so it has a huge value. And like the Iraqi marshlands, Owens Lake has more desiccated land than available water such that the desire to establish more habitat must be balanced against a very water-limited environment (Dickey and Madison 2004). One of the big management trade-offs of not having enough water is the development of farming in the bottom of a lakebed in an area that historically had a high water table. But simply putting water back into such areas and creating open-water marsh areas and wet playa regions would create new problems. The already high water table would be raised higher such that a subsurface drainage system would be needed in order to remove the groundwater from under the salt grasses or they would also turn white with salt during evaporation. So the subsurface drip infrastructure helps to keep salt from the surface, and the deep drainage infrastructure allows percolation to collect that saltwater and move it to other areas for alternative habitat creation (Dickey and Madison 2004).

The Sacramento Valley Basin-wide Management Plan is another relevant example (Dickey and Madison 2004). As in southern Iraq, the valley is bordered by mountains that provide a rain-on-snow hydrology system. There are huge silt loads and big floods in the river, the presence of many dams, problems with salinization, and widespread destruction of wetlands. Thus, there are many parallels with the marshlands of Iraq. The Sacramento Valley is primarily used for irrigated rice production and so has to remain fairly fresh, and again like the Iraqi marshes, it is not very tolerant of salinity.

In the 1960s, background measures of soil salinity showed a good deal of flushing taking place and the resulting presence of a high-quality and abundant water supply. This was compromised by drought and regulatory pressures in the 1980s and 1990s, and by 1994 salinity was found to have increased fourfold over the previous three decades. Impacts were starting to be noticed on rice yields due to less water flushing through the system. This is very similar to the Iraqi marshes and other locations where problems are felt hardest in areas that have the most significant natural drainage limitations. Field irrigation practices had not changed over this period, so the problem was due to having a drainage and irrigation system that was set up for flushing but with no flushing actually taking place. As a result, there is a new diminished water supply that has become mismatched with the unchanged facilities and is now causing problems. The take-home message is that if the water supply is going to be changed, the operations and facilities have to be changed as well (Dickey and Madison 2004). So this means that the operations of the Iraq marshes will also have to change given the drastic alteration in water supply that has occurred there over the last twenty years.

The chain of cause and effect in the Sacramento example is that the drought has resulted in a restricted water supply. Water detention in fields is a matter of course in rice production, and this means that in order to continue agriculture more and more of the water needs to be recycled. And as more water is recycled, it becomes more saline and rice production begins to be impaired. The basin then shifts to a more saline equilibrium condition with more and more areas becoming hotspots in which salinity thresholds are exceeded. Criteria therefore aren't being met, and there is a need for altering facilities management to avoid the continued and increasingly precipitous trend toward reduced crop production.

APPLICATION IN THE IRAQ MARSH CONTEXT

The first thing that needs to be attended to in dryland agriculture is the distribution of water in time and in space in order to keep salinities in an acceptable range (see also chapter 22). Those living and farming between the Tigris and Euphrates Rivers (Figure 21.3) have known this for millennia, as, for example, shown in an ancient Sumerian farmer's first instruction to his son (circa 1700 BCE): "When you have to

FIGURE 21.3 The unchanging landscape of the Fertile Crescent, the birthplace of agriculture. Vast fields of grain located beside the 3,000–4,000-year-old archaeological tells of Halaf, Harmoukar, and Sheikh Hamad between the Tigris and Euphrates Rivers in what was once ancient Mesopotamia.   **(Continued)**

FIGURE 21.3 (*Continued*) The unchanging landscape of the Fertile Crescent, the birthplace of agriculture. Vast fields of grain located beside the 3,000–4,000-year-old archaeological tells of Halaf, Harmoukar, and Sheikh Hamad between the Tigris and Euphrates Rivers in what was once ancient Mesopotamia.

prepare a field, inspect the levees, canals and mounds that have to be opened. When you let the flood water into the field, this water should not rise too high in it. At the time that the field emerges from the water, watch its area with standing water, it should be fenced. Do not let cattle herds trample there." The instructions go on to discuss tillage and to promote drainage and subsequent irrigation.

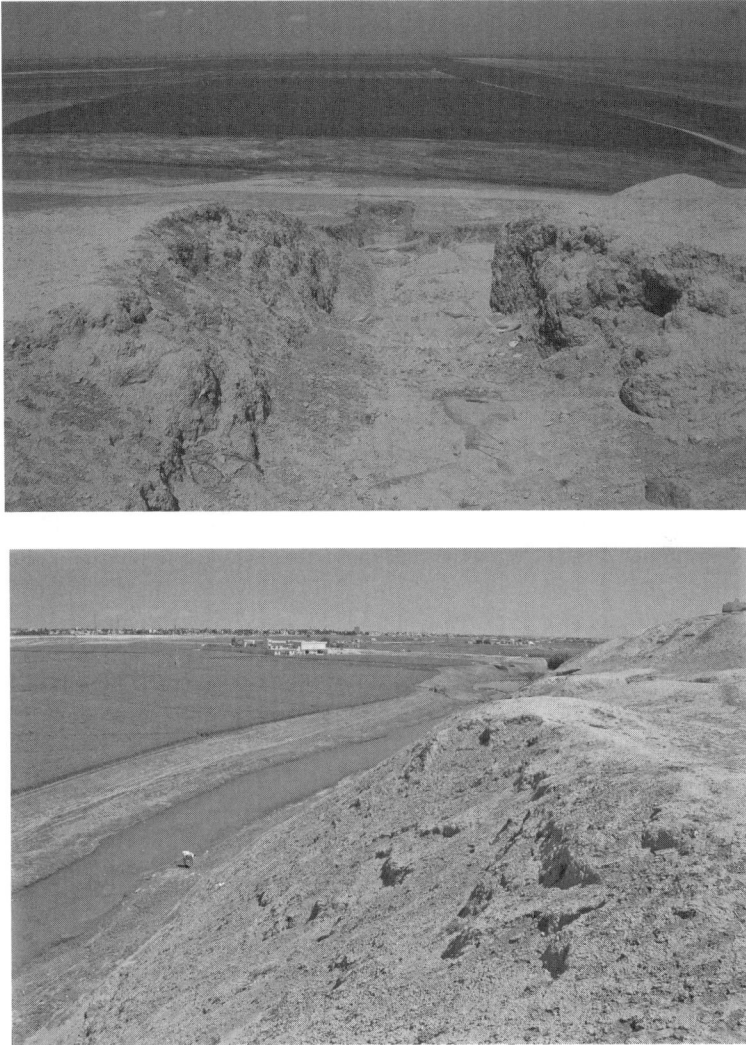

FIGURE 21.3 (*Continued*) The unchanging landscape of the Fertile Crescent, the birth-place of agriculture. Vast fields of grain located beside the 3,000–4,000-year-old archaeological tells of Halaf, Harmoukar, and Sheikh Hamad between the Tigris and Euphrates Rivers in what was once ancient Mesopotamia.

Before the current situation in southern Iraq is considered, it is useful to review some selected historic landmarks in Mesopotamian agriculture and water management (Dickey and Madison 2004):

- Domestication of plants in about 8000 BCE.
- Irrigation beginning by 6500 BCE and becoming widespread by 3500 BCE as far as 200 km from the coastline.

- Irrigators and civilization battle floods and huge silt load in rivers and canals until the thirteenth century.
- Mongolian invasions, the Black Death, and Tamerlane put an end to the maintenance of irrigation and drainage works in the Middle Ages.
- The twentieth century has some rehabilitation of infrastructure:
  - 1950s: beginning of large impoundments, flood reduction, and drainage construction to address declining water quality.
  - 1960s: Gradual decline in water quality (i.e., increased salinity) due to reception of concentrated return flows from upstream urban, industrial, and agricultural water users.
  - 1970s: shift from irrigation and drainage to overdrainage to "reclaim" (i.e., dry out) the marshes.
  - 1980s: marshes are embroiled in the Iran–Iraq war.
  - 1990s: radical acceleration of overdrainage and associated oppression.
- 2000s: regime change, unrest, and movement to restore marshes.

The end result has been less water of poorer quality due to increasing salinity, less silt and organic matter, a shift from irrigation and drainage to overdrainage facilities, and a reduced indigenous population and knowledge base (Dickey and Madison 2004).

When one examines a salinity map of Iraq, it is possible to trace the water and salt coming down from the mountains and tending to concentrate in the lower watershed (this happens in any watershed) with drainage management affecting the regional distribution of the salt.

There are salt risks present if no marsh restoration occurs:

- Loss of marsh cultural, ecosystem, and economic values
- Loss of critical time, knowledge, and ecosystem resiliency
- Loss of outflow water quality polishing with consequent potential impacts on the marshes and the Gulf
- Increased dust emissions from desiccated land surfaces
- Increased vulnerability to future flooding
- Acceleration of seawater intrusion due to lack of recharge to the delta

And there are salt considerations to be aware of during the restoration process:

- Avoid ecotoxicity by restricting exposure to highly concentrated waters that contain specific elements like selenium that are associated with the salt ions.
- Gather and use indigenous knowledge of salt management in terms of salt-tolerant plant communities, land variability, and flow regimes, and subsistence agriculture based on historical wisdom.
- Identify and protect critical salt management infrastructure.
- Change operation or configuration of overdrainage facilities to keep water present or to help return it to the marshes.
- Avoid new salt loads such as those from flood storage in saline sinks prior to discharge into the marshes.

- Balance water spreading with supply and salinity goals in terms of sizing the restored marsh for the projected flow (e.g., if the marsh is too big, the water will be spread too thinly and problems will ensue).
- Protect renewed water flow with durable agreements.

The goal in all this is to blend efficiency, reuse drainage, encourage tolerant plant communities, and develop and capitalize on new and indigenous knowledge to promote salt management for the creation of desired values (Dickey and Madison 2004).

It will probably never be possible to restore the freshwater Iraqi marshes to their full areal grandeur given the projections of water that will be available based on upstream hydrodevelopment. But it will be possible to have a large region of the marsh preserved with the same historic water quality, and with negotiations of instream flows it is possible that this area might be expanded over time. Probably the biggest opportunity is to enlarge the marsh area with low-quality water from agricultural drainage or the water that currently comes out of the marshes that can be placed in another area to be managed separately and that might be restored with halophytic plants collected from further downstream in the estuary where the water is saltier (Dickey and Madison 2004). The strategy here is to bring those plants and fish northward to create another ecosystem rather than draining that poor-quality water out of the wetland system entirely. And this could be repeated again with plants brought up all the way from the delta to create inland seawater marshes in which perhaps shrimp farming could take place. The idea here is to create a mosaic of marshes of ranging salinity depending on water sources and quality. So though the full area could become wet again, it would be a different kind of wet, with different types of vegetation, aquatic life, and value. The bottom line is that one can accommodate a high degree of salinity if one is accepting of creating different habitats. Even highly concentrated salt sinks, it is important to note, have habitat value with lots of bird use. One particularly informative example in this regard is the Al Wathba Nature Reserve outside of Abu Dhabi City in the United Arab Emirates (Figure 21.4). Here water from a domestic sewage treatment plant is reclaimed to provide wildlife habitat. The problem, however, is that the area is underlain by a unique form of encrusted salt deposits called *sabkha* (Figure 21.5), which has contributed to very high salinity levels of the water. Rather than attempt to reduce the salt concentrations, managers actively manipulate the water levels to allow for the development of a healthy brine shrimp population (Anonymous 2006), which in turn attracts flamingos (Figure 21.6), the first breeding population of such on the Arabian Peninsula in almost a century (Anonymous 2004).

The United Nations Environment Programme's recommendations for restoring the marshlands (Patrow 2001) include the following:

- International agreements on sharing the Tigris and Euphrates waters
- Mitigation of dam impacts on downstream ecosystems
- Reestablishing the flood regime
- Protecting water quality
- Reevaluating the role of engineering works
- Designation of protected areas

FIGURE 21.4   Al Wathba Nature Reserve, located in the desert outside of Abu Dhabi City.

- Assistance and repatriation of refugees
- Monitoring and long-term capacity building

With regard to salinity, there are two big areas in which concentrations can be modulated (Dickey and Madison 2004). One is by modifying the water supply and the other is by specifically working to manage the water quality. It will be necessary to allocate and define the water flow as the basis for planning a balanced and productive marshland system. This will help to calculate the salt balance needed in order to deliver the right type of water to the right places in the marshes through a basin-wide

FIGURE 21.5   Sabkha salt flats located ten kilometers inland along the gulf coast of the United Arab Emirates.

management plan that deals with water and salinity, not just water, through both national and international accords. Specifically, these opportunities for controlling salinity could include the following:

- Avoid routing salts from saline sinks to the marshes.
- Minimize saline groundwater recharge to streams located higher up in the watershed.
- Minimize industrial loading.
- Reuse water in tolerant plant communities.
- Provide needed routes of salt removal and export.
- Use water efficiently to flush and leach salt.
- Balance water resources and salt management.
- Sequester and import salt where appropriate.
- Monitor specific elements to avoid ecotoxicity.

One obvious major management option (Dickey and Madison 2004) is to eliminate overdrainage where possible by

- routing all peak flows to depressions and/or to the Gulf,
- intercepting surface and subsurface flow to wetlands,
- empoldering with drainage or evaporation, and
- balancing legitimate and competing land and water use as applicable.

Adequate water quality can be achieved by getting salinity right in the protected areas through recognizing site-specific opportunities and constraints for salt management such as defining the maximum tolerable salinity for each plant community,

FIGURE 21.6 The only breeding population of flamingos on the Arabian Peninsula, attracted to the saline waters and resident brine shrimp of Al Wathba.

and allocating adequate water of the appropriate quality in order to meet those goals. In the end, the goal is to create adaptive marsh systems which must be managed adaptively.

ACKNOWLEDGMENT

Adapted from Dickey and Madison (2004).

REFERENCES

Anonymous. 2004. *Management options for the improvement of flamingo breeding at Al Wathba Wetland Reserve*. Abu Dhabi: Terrestrial Environmental Research Center, Environmental Research and Wildlife Development Agency.

———. 2006. *Water quality and* Artemia *monitoring at Al Wathba*. Abu Dhabi: Terrestrial Environmental Research Center, Environmental Research and Wildlife Development Agency.

Dickey, J., and M. Madison. 2004. Moving salt and water in managed ecosystems: Case studies from history and the western United States. Paper presented at the Mesopotamian Marshes and Modern Development: Practical Approaches for Sustaining Ecological and Cultural Landscapes conference, Cambridge, MA, October.

France, R. L. (Ed.) 2007. *Wetlands of mass destruction: Ancient presage for contemporary ecocide in southern Iraq*. Winnipeg, MB: Green Frigate Books.

———. 2011. *Restoring in Iraqi marshlands: Potentials, perspectives, practices*. Sussex, UK: Sussex Academic Press.

———. 2012. *Back to the Garden: Searching for Eden in the Mesopotamian marshes*. Cambridge, MA: Harvard University Press.

Naff, T., and G. Hanna. 2002. The marshes of southern Iraq: A hydro-engineering and political profile. In *The Iraqi marshlands: A human and environmental study*, ed. E. Nicholson and P. Clark. London: Politico's.

Patrow, H. 2001. *The Mesopotamian marshlands: Demise of an ecosystem*. Nairobi, Kenya: United Nations Environment Programme.

Reed, C. 2005. Paradise lost? *Harvard Magazine* 107:3.

22 Managing Scarce Water Resources in Agriculture in West Asia and North Africa

Theib Y. Oweis and Ahmed Y. Hachum

CONTENTS

Extent of Water Scarcity in Dry Areas 452
Declining Agricultural Share of Water.. 455
Options for Coping with Water Scarcity.. 456
    Developing New Water Supplies.. 456
    Utilizing Marginal-Quality Water 457
    Regional Water Transfer.. 458
    Rainwater Harvesting 459
    Import Virtual Water... 459
    Demand Management, Water Pricing 460
Increasing Agricultural Water Productivity... 460
    Evolution of the Water Productivity Concept 461
    From Efficiency to Productivity 461
    The Water Accounting Framework 463
    Water Productivity Performance Indicators 464
    Economic Water Productivity.. 464
Responses to Improve Water Productivity 465
    Deficit Irrigation.. 465
    Changing Current Land Use 466
    Precision Irrigation.. 466
Case Studies from the Middle East.. 468
    Supplemental Irrigation Improves Rainfed Agriculture.................................. 468
    Improved Water Productivity in Irrigated Agriculture 476
    Water Harvesting for the Drier Environments—Steppe.................................. 478
    WP Improvement and the Environment... 481
    Policies and Institutions 482
Conclusions.. 482
References... 483

EXTENT OF WATER SCARCITY IN DRY AREAS

Water is life, and is absolutely essential to human development and prosperity. It is estimated that nearly one billion people live in the dry areas, where about half of the workforce earns its living from agriculture. The world data on water resources indicates that the West Asia and North Africa (WANA) region, mostly Middle Eastern countries, faces the most serious threat of water shortages (Figure 22.1). Water in this region has shaped up some of the greatest civilizations in the history of mankind along the Tigris, the Euphrates, and the Nile. Since then water has always played a crucial role in the development, stability, and conflicts of this region.

Population growth rates in WANA of up to 3.6 percent are among the highest in the world. An estimated 690 million people in these areas presently have an income of less than two dollars a day; of these, 142 million earn less than one dollar a day. Rural women and children suffer the most from poverty and its social and physical deprivations. Degradation of land and other natural resources is both the cause and the effect of poverty. Population pressure and the lack of adequate agricultural technologies are major forces driving the poor to make desperate choices. But the worst enemies of the poor in the region are water scarcity and drought.

The dry areas in general and the WANA region in particular are characterized by low and variable precipitation and high evaporation rates. Drought is an inherent characteristic of climates with pronounced rainfall variability. The large internal deserts in the WANA region result in average annual precipitation rates of less than 100 mm for one-third of the region. The average precipitation varies from 85 mm in the Arabian Peninsula, 106 mm in North Africa, and 318 mm in West Asia (Table 22.1).

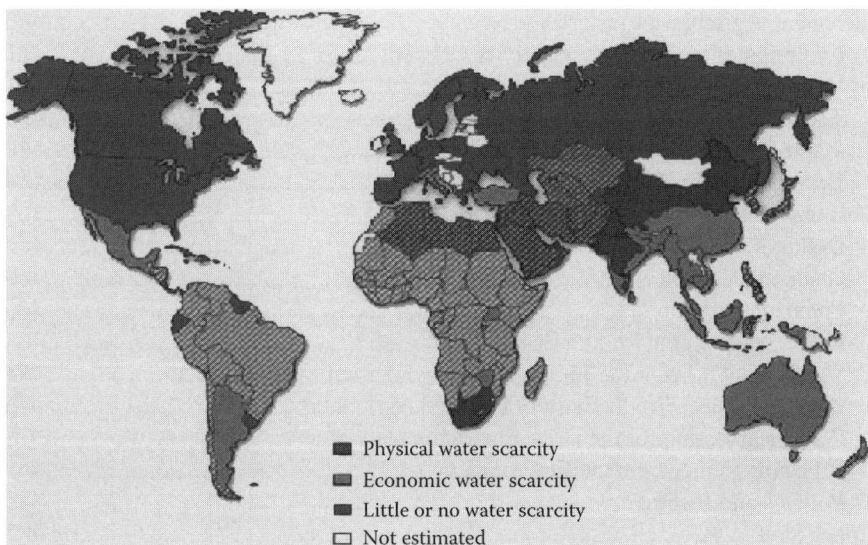

FIGURE 22.1   Global outlook on water scarcity. *Source*: International Water Management Institute. N.d. [Home page]. www.iwmi.cgiar.org.

TABLE 22.1
Water Resources in Countries in the WANA Region

Country	Population[a] 2002 10⁶	Area Total 10⁹ m²	Precipitation[b] Average/Year mm	Precipitation[b] Average/Year 10⁹ m³	Renewable Water Resources Internal 10⁹ m³ yr⁻¹	Renewable Water Resources Total[c] 10⁹ m³ yr⁻¹	Total/ Cap.[d] m³yr⁻¹ cap⁻¹	Water Use[e] m³yr⁻¹ cap⁻¹
			West Asia					
Bahrain	0.7	0.7	83	0.1	0.004	0.12	181	469
Iraq	24	438	216	95	35	75	3111	1861
Iran	72	1,648	228	376	129	138	1900	1036
Jordan	5.2	89	111	9.9	0.68	0.88	169	208
Kuwait	2.0	18	121	2.2	0.00	0.02	10	235
Lebanon	3.6	10	661	6.9	4.8	4.4	1220	392
Oman	2.7	310	125	39	1.0	1.00	365	532
Palestine	1.1	6	316	0.1	0.8	0.8	587	—
Qatar	0.6	11	74	0.8	0.05	0.05	86	513
Saudi Arabia	21.7	2,150	59	127	2.4	2.4	111	851
Syria	17	185	318	59	7	26	1541	1232
Turkey	69	775	593	460	227	229	3344	563
United Arab Emirates	2.7	84	78	6.5	0.2	0.2	56	886
Yemen	20	528	167	88	4	4	206	361
Total/Average	242	6253	225	1271	412	482	1990	619
			North Africa					
Algeria	31	2,382	89	212	14	14	456	200
Egypt	70	1,001	51	51	1.8	58	830	1011
Libya	5.5	1,760	56	99	0.6	0.6	109	909
Mauritania	2.8	1,026	92	95	0.4	11.4	4028	638
Morocco	31	447	346	155	29	29	936	427
Tunisia	10	164	313	51	4.2	4.6	472	289
Total/Average	151	6780	*158*	662	50	118	*785*	*579*

[a] Total population in 2002 (WRI 2003); data for Palestine are from 2000 (FAO 2003a).

[b] Precipitation is the annual average for the years 1961–1990, as used by the IPCC. Adjustments were made in case of large discrepancies with national data (FAO 2003a). Precipitation averages are aerial averages.

[c] Total renewable resources (actual) are internal plus external resources, considering formal and informal agreements (FAO 2003a).

[d] Computed from total water resources (FAO 2003a) and 2002 population (WRI 2003).

[e] Total water use per capita data are from 2000 (FAO 2003a).

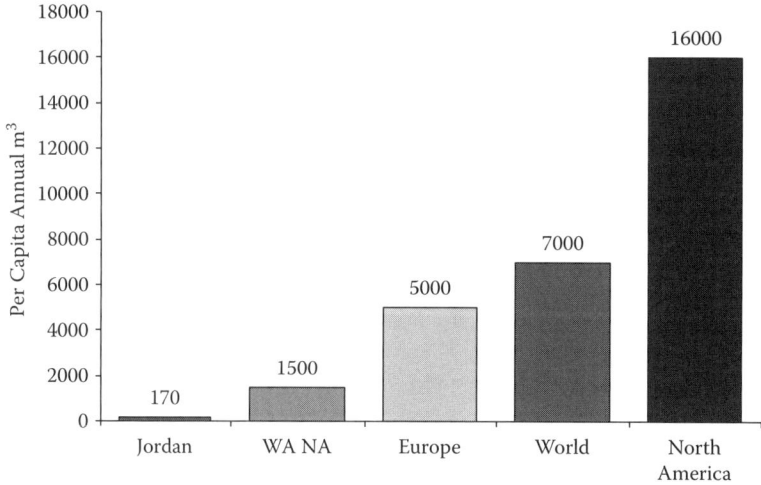

FIGURE 22.2 Annual renewable water share.

The average annual per capita renewable supplies of water in WANA countries is now about 1,500 m³, which is well below the world average of about 7,000 m³ (Figure 22.2). This level has fallen from 3,500 m³ in 1960 and is expected to fall to less than 700 m³ by the year 2025. In 1990 only one-third of the WANAs countries had per capita water availability of more than 1,000 m³, the threshold for water poverty level. In Jordan, the annual per capita share has dropped to less than 200 m³. Mining groundwater is now common in the region risking both water reserves and quality. In many countries securing basic human water needs for domestic use is becoming an issue, not to mention the needs for agriculture, industry and environment.

The water scarcity situation in WANA is getting worse every day. It is projected that the vast majority of the WANA countries will reach the severe water poverty level by the year 2025; one-third of them are already below that level. Over the coming years this situation will worsen with increasing demand, given the fact that the possibility of new supplies is limited (Seckler et al. 1999). The increasing pressure on this resource will, unless seriously tackled, escalate conflicts and seriously damage the already fragile environment in the region. This is particularly obvious between countries in the Middle East with shared water resources.

River flows generated in the mountains of Africa and Turkey are providing critical water resources to some of the water-scarce countries downstream. Egypt, Sudan, Iraq, Syria, and many other countries in the region obtain more than 50 percent of their renewable water resources from neighboring countries. Syria, Algeria, and Lebanon are increasingly affected as scarcity grows year after year. Nonconventional water resources, which are contributed by rainwater harvesting, desalination of seawater, physical transport of water, trading of virtual water through import of grain and other agricultural commodities, saline drainage water, and wastewater generated by household, industrial, and municipal activities, also provide an important, but limited, share of the water demand of the water-scarce countries.

DECLINING AGRICULTURAL SHARE OF WATER

The world contains an estimated 1,400 million cubic km of water. Only 0.003 percent of this vast amount, about 45,000 cubic km, is what we call "freshwater resources," which is water that theoretically can be used for drinking, hygiene, agriculture, and industry. In fact, only about 9,000 to 14,000 cubic km is economically available for human use—a small amount compared to the total amount of water on earth. With the population increasing by two billion by 2030, the question is: will there be enough water to sustain human life?

Agriculture is by far the largest user of water, accounting for about 70 percent of all withdrawals from rivers, lakes, and aquifers, and up to 95 percent in many developing countries. Currently, agriculture consumes about 80 percent of water resources in the WANA region. This share is decreasing not only because of fast population growth, but also due to diversions to domestic and industrial sectors and to maintain ecosystems. In North African countries it is projected that agricultural share will drop to about 70 percent and 50 percent by the years 2025 and 2050, respectively (Figure 22.3).

The amount of water used in food production is significant, and most of it is supplied directly by rainfall. Based on current average food intake per capita, approximately 1,000 m^3 water is needed to produce the food per person per year (FAO 2003b). Thus, with a world population of 6 billion, net water needed to produce enough food for this population is 6,000 km^3. Most of the water used in agriculture comes from rainwater stored in the soil profile (green water) and only 15 percent of water for crop production is provided through irrigation (blue water). Thus, irrigation needs 900 km^3 of water per year for crop production. On average, 40 percent of water withdrawn from rivers, lakes, and aquifers effectively contributes to crop production. The remainder (60 percent) is lost through nonbeneficial evaporation, deep percolation, and through growth of weeds. Therefore, the current global water withdrawal for irrigation is in the order of 2,000 to 2,500 km^3 per year. Thus, there is enough freshwater to feed the people of the globe, but the problem it is not uniformly distributed in space and time and WANA receives the least.

The water needed for crops amounts to 1,000 to 3,000 cubic meters per tonne of cereal harvested. It takes 1 to 3 tonnes of water to produce 1 kg of grain. Furthermore, it is estimated that only 45 percent of the water used in agriculture is effectively used by crops (UN/WWAP 2003). The other 55 percent is partially lost by either evaporation or by losing quality while joining salt sinks. The other part recharges aquifers, or flows downstream to be reused. Therefore, agriculture is not seen as the most efficient water user. The ever-growing competition among water-using sectors is certainly forcing agriculture to give up part of its share to higher priority uses, especially the domestic and industrial sectors. Meanwhile, agriculture must cope with the increasing demand for food, feed, and fiber, but with less water.

In many countries of the world, the challenge, as set by the CGIAR Challenge Program on Water for Food (CPWF), is to retain diversions of water to agriculture not higher than the year 2000 level. In WANA the question is how fast it drops below that target (Figure 22.3).

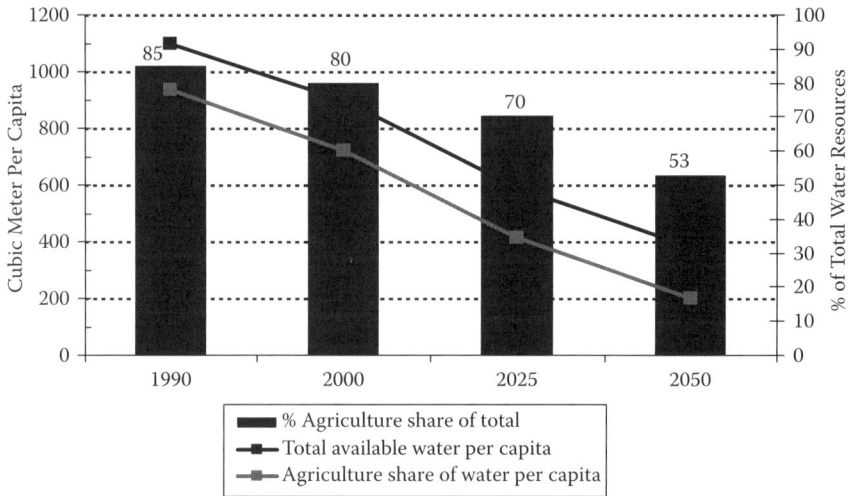

FIGURE 22.3   Declining agricultural water in the WANA region. *Source*: Oweis (1997).

Water is a vital element in the continued economic development of the dry areas for improved living conditions of the people. The current water supplies will not be sufficient for economic growth in all the countries of the region except Turkey and Iran. Water scarcity in this region has already hampered the development in all countries. It is, therefore, essential that substantial changes be made in the way water is valued and managed to help overcome water shortages.

OPTIONS FOR COPING WITH WATER SCARCITY

Potential consequences of severe water shortages may include: abandonment of agricultural land, water quality deterioration, waterborne diseases, and conflicts in transboundary basins. Major investments are necessary to develop and implement new water policies, monitoring systems, water-saving technologies, and management practices to sustain agricultural production and communities. A number of promising options to cope with the dilemma of water scarcity in the region will be elaborated.

DEVELOPING NEW WATER SUPPLIES

Although very limited, new waters in the dry areas mainly come from nonconventional sources that include desalination, marginal quality water, water transfers, and water harvesting.

Desalinization is an expensive process, and hence is currently mainly used in areas where an affordable energy source is available such as the Gulf countries where part of the desalinized water is used for irrigation. The total production of fresh water may reach up to 913 billion m^3 annually for some 18 million inhabitants. The seawater desalinization cost ranges between US$1.00 to 1.80 per m^3. Some reported lower costs but with subsidized energy. Costs may become feasible for agricultural use, particularly with the development of new technologies, possibly making use of natural gas as a source of energy. Breakthroughs in the cost of desalination would open real opportunities for several countries of the region. However, potential breakthrough in the desalinization technology is hampered by lack of funds to support research in this field.

Marginal-quality water development and use offers some promise. Potential sources include natural brackish water, agricultural drainage water, and treated effluent. Research shows that substantial amounts of brackish water exist in dry areas that can either be utilized directly in agriculture or desalinated at low cost for human and industrial consumption (see chapter 21). The treated effluent is an important source of water for agriculture in areas of extreme scarcity, such as Jordan and Tunisia where it counts for about 25 percent of total water resources in the country. Egypt is currently producing about 1.2 billion m^3 per year of recycled water from the city of Cairo and by 2010 it is expected that 4.9 billion m^3 year will be in the country (El-Beltagy et al. 1997). There is, however, several health and environmental issues associated with its use in agriculture.

UTILIZING MARGINAL-QUALITY WATER

It is projected that the water-scarce countries will have to increasingly rely on marginal-quality water resources. Contingent upon the use of appropriate approaches, marginal-quality water resources have the potential to narrow the gap between water availability and its demand in water-scarce regions. Whether beneficially used or wasted, marginal-quality water needs appropriate treatment and disposal in an environmentally feasible manner (see chapter 18).

Estimates show that at least 20 percent of global irrigated areas fall under irrigation with saline and/or sodic waters or irrigation-induced saline and sodic soils. In case of other major category of marginal-quality waters, one-tenth of the global population eats food produced using wastewater. As populations in different countries and regions continue to grow and more freshwater is diverted to cities for domestic, commercial, and industrial uses—about 70 percent of water used in domestic, commercial, and industrial activities is returned as wastewater—the use of wastewater is certain to increase, both in terms of volumes and areas irrigated (Scott et al. 2004). The quality of wastewater used for irrigation varies greatly, both within and between countries. There is no precise account of the global extent of wastewater irrigation. Although overestimated, the tentative figure suggests about 20 million hectares (about 8 percent of global irrigated area) are under irrigation with treated, partially treated, and diluted wastewater (Scott et al. 2004).

In the future, marginal-quality waters are expected to become an integral part of the agricultural production systems in water-scarce countries (Qadir and Oster 2004). Different types of marginal-quality waters need suitable pre-use treatment and/or appropriate management during the course of their application. However, a significant part of wastewater is used by the farmers for agricultural production in untreated or partly treated form in many developing countries in an unregulated manner. The protection of public health and the environment are the main concerns associated with such wastewater reuse (see chapter 19). The use of saline and/or sodic drainage and groundwater resources is also expected to increase and warrants attention in order to cope with the inevitable increases in salinity and sodicity that will occur (as described in chapter 21).

Agricultural drainage water is becoming an appealing option to many countries not only to protect natural resources from deterioration, but also to make a new water resource available for agriculture. In the last two decades, the reuse of drainage water in agriculture and its impacts on the environment have become the focus of research scientists in many parts of the world, particularly in dry areas. In Egypt, reuse of drainage water increased from 2.6 billion m^3 per year in the 1980s to about 4.2 billion m^3 per year in the early 1990s. Now two new projects will bring the total reused drainage water to approximately 7.2 billion m^3 per year, some 12 percent of total water resources available to Egypt. Treating these drainage waters as a "*resource*" rather than as a "*waste*" contributes to the alleviation of water scarcity, environment protection, and sustainability of agricultural production systems. The following aspects have become high priority in West Asia and North Africa:

1. Qualitative and quantitative assessment of marginal-quality water resources and extent of its use in agriculture
2. Short- and long-term implications of using marginal-quality waters for crop production systems in an unregulated manner; crop quality, health and environmental implications (land and water degradation)
3. Evaluation of productivity constraints under existing scenarios of marginal-quality water use
4. Identification and promotion of the community-based and farmer-led development of low-cost and environmentally feasible crop, irrigation, and soil management strategies of using marginal-quality waters
5. Promotion of social, financial, and environmental benefits for using marginal-quality waters and salt-affected soils for crop production systems
6. Policy and institutional aspects of using marginal-quality waters in agriculture; analysis of the present scenarios to overcome policy and institutional constraints
7. Capacity-building options for the national agricultural and extension systems and awareness among the farmers for greater understanding of the potentials of plants, soil, and water for agricultural produce from marginal-quality waters

REGIONAL WATER TRANSFER

Transboundary water transfers between basins and across national borders have been extensively discussed in the region over the last two decades (Kally 1994). Importation of water is under active considerations in the Middle East. The two options most relevant are to involve transportation by pipeline (Turkey's proposed peace pipeline) and by ship or barrage (big tanks or "*Medusa*" bags). Both suggestions are subjected to economical, political, and environmental measures, which are yet to be examined. In the WANA region, attempts on transferring water by balloons and tankers have been made but the cost is still high for agricultural purposes. The project to transfer water by pipelines from Turkey to the Middle East countries was unsuccessful because of financial and political consideration. Potential for such projects can only be realized with good regional cooperation and trust building. As water scarcity in the region grows, the issues associated with cross-boundary water resources become urgent and require solutions. Internationally agreed laws and codes of ethics need to be developed to insure water rights and to open the way for innovative projects in the region.

RAINWATER HARVESTING

Water harvesting provides opportunities for decentralized community-based management of water resources. Hundreds of billions of cubic meters of rainwater in the drier environments is lost every year. This loss occurs mostly in the marginal lands, which occupy a major part of WANA, due to lack of proper management. The development of water harvesting systems in these areas can save substantial amounts of water that is otherwise lost and become irrecoverable. The International Center for Agricultural Research in the Dry Areas (ICARDA) has demonstrated that over 50 percent of this water can be captured and utilized for agricultural production if integrated on-farm water use techniques are implemented properly (Oweis et al. 1999). However, issues of policies and socioeconomic aspects require special attention for achieving greater success.

IMPORT VIRTUAL WATER

Virtual water refers to the water used abroad to produce imported food. It is seen as one of the economical solutions to water scarcity in WANA. It can be shown that, in water-scarce areas, it is sometimes cheaper to import a commodity than to produce it locally. Due to higher water productivity, it has been claimed that cereal international trade reduces global crop water use by 164 km^3 (from effective rain and irrigation) and 112 164 km^3 of net irrigation water. This represents a savings of 6 percent in global water use by crop and 11 percent in irrigation water depletion. The problem with the concept of virtual water is that the production value is only measured by market prices. The value of agricultural production, particularly in poor rural areas, is not only in the market value of the product but more as social and environmental and food security returns. All these dimensions and aspects have great values that should be considered when including the virtual water in the national water resources budget.

Trade of virtual water has always been necessary, but what is important is that rural agricultural communities continue to practice agriculture and no policies on virtual water should phase them out as a result. Relying on virtual water to feed a nation may negatively effect the scientific advancements on improvements in the water sector of the country. Extremely water-short and dry countries have no choice but to import food. Other countries with less water stress have to balance increasing pressure on water resources and dependency on imports. A certain degree of food sufficiency is still a national priority, despite of the political and economic considerations. The dilemma is when the country suffers acute water scarcity and cannot afford virtual water.

DEMAND MANAGEMENT, WATER PRICING

Water demand management in agriculture is the management of water through influencing consumer behavior by introducing incentives to use water more efficiently. This will involve many elements such as legislative measures, including pricing mechanism and financial incentive as well as penalties. It also involves direct technical measures to control and ration water by flow regulating devices. However, effective public awareness programs should come at the top of the action list to arrive at fruitful management for water demand. It requires that all users recognize and accept that water supplied to them has a value that varies depending on the purpose of its use. Farmers should understand that the opportunity cost of this water is very high and what they are paying for it is a small fraction of its real value. Media and extension services can play a role in achieving this awareness. Water pricing is difficult to implement in most WANA countries. Reasons include economic but also cultural and sociopolitical one. Water is seen as a gift from God to humankind and should be accessible to all. If a pricing mechanism is to be implemented, care must be taken in considering the limited capacity of the resource-poor farmers in addition to other constraints. Effective alternatives to water pricing as a means of demand management is yet to be developed in WANA.

INCREASING AGRICULTURAL WATER PRODUCTIVITY

As agricultural water is declining in most of WANA countries and the need for food in increasing, the region is facing the challenge of producing "more food with less water." The only way to achieve this is by increasing the production and/ or return per unit of available water, usually termed as "water use efficiency" or "water productivity." The understanding of how water is acquired, used, and managed is the key to the solution. This is a cross-discipline solution, which certainly requires the concerted action of all players: farmers, water managers, hydrologists, agronomists, water resources specialists, engineers, socioeconomists, and policy makers.

Until recently, the focus of water management in agriculture was to maximize marketable yield per unit area of land. With the growing crises of water scarcity, malnutrition, and water-driven ecosystem degradation, the focus had to be shifted towards maximizing return per unit of water used. The drive behind this shift is more obvious in the dry regions of the world, because water, not land, is the most limiting constraint to increased food production and sustainability of the system. In the humid regions, this shift is somewhat controversial, because water is not necessarily the most limiting constraint. In the humid industrialized countries, other water-related issues, such as pollution, sanitation, and ecosystem degradation, have higher priority.

Evolution of the Water Productivity Concept

Water's abundance and continuous renewal through the global hydrologic cycle has for centuries left people to consider water a free natural resource. Expansion of irrigated areas, followed by increasing demands from industrial and domestic water use sectors, stemming both from population growth and improved economic development, has changed society's views. These changes in water availability are felt hardest in the water scarce areas of West Asia and North Africa. Whereas no less than two decades ago the water issue (management and scarcity) was still considered the lone playing ground for hydrologists and engineers, it is now at the top of the priority list for planners and policy makers around the world.

Historically, the concept of system-wide (basin) efficiency of water use began to surface in the early 1990s. There were, however, some early gleams about it here and there (such as Bagely 1965 and Jensen 1967) but they did not cause notable impact. This is because water was not so scarce at that time and water problems were usually solved by increasing supply through the construction of more water infrastructures. Nowadays, the approach to water problems has been shifted from construction to management.

Water scarcity crises jumped to the forefront of global/international organizations' serious concerns in the early 1990s (Rio Declaration, Agenda 21, 1992). Among the main issues covered in the declaration are cooperation and participation, water economics, sustainable development, and food production. Before the Rio Declaration, the emphasis on water related issues was on availability of safe drinking water and sanitation. Emphasis on sustainable development and efficient use of water resources was explicitly and critically addressed in the First World Water Forum, 1997. Since then, water issues have become everybody's concern and worry. There has been a growing awareness that water will be one of the most critical natural resources in the twenty-first century.

There is considerable potential for efficiency improvement in water use in agriculture. But, this improvement in water use and management should be sustainable and environment friendly. Sustainability, in simple terms, implies continuity, long lasting, and preserving of the natural base of the development, whereas efficiency implies output per unit input. A shortcut to sustainable development is through integration and participation. Since sustainability is time dependent, there is no foolproof way to guarantee it. It may take decades to achieve (or may not achieve) it. However, sound indicators for sustainability play a paramount role in this regard.

From Efficiency to Productivity

Generally, the term "efficiency" reflects the ratio of output to input. However, constituents of the input and output components differ from one discipline to another and vary depending on the level and/or scale at which the word "efficiency" is being used. In irrigated agriculture, there are many efficiency terms; each has a specific use. Among these terms are Water Conveyance Efficiency, which reflects losses of water in the conveyance system; Water Application Efficiency, which reflects losses of water by runoff from the field and by deep percolation below crop root zone;

Water Distribution Efficiency, which reflects how uniform water is applied to the field; Water Storage Efficiency, which reflects the sufficiency of water stored in the crop root zone; and Irrigation Efficiency, which reflects the overall losses in irrigation. These are classical engineering efficiency indices commonly used by water engineers in the planning, design, evaluation, and management of water use/flow in irrigation projects. They are only applicable for use at irrigation project level or less. The inputs to them are merely water amounts (depth or volume)—there is no consideration for other types of returns such as crop yield or economic values.

With increasing water scarcity and the emerging necessity to optimize water use under limited availability of water, Water Use Efficiency (WUE), which reflects how good water is used in producing crops, takes the lead. The term Water Use Efficiency (WUE) has been defined in various ways by hydrologists, physiologists, and agronomists depending upon the emphasis that one wishes to place on certain aspects of the problem. This term is an outgrowth of a series of older terms which are more or less related to the water use by crops, such as: duty of water, consumptive use, transpiration ratio, and transpiration coefficient. Viets (1962) used the term WUE to characterize the ratio of crop yield (biological or economical) to crop water use (transpiration or evapotranspiration). WUE, in general, refers to the amount of plant material produced per unit of water used and, therefore, can be expressed as follows:

$$WUE = \frac{Y}{W} \qquad (22.1)$$

The numerator Y may represent total plant material produced, aboveground dry matter, or economic yield (grain or straw or both). The denominator W may represent: water transpired by the plant, evaporation from soil and plant plus transpiration (evapotranspiration), or water applied to the field (rainfall plus irrigation if any). The component Y is usually expressed as a mass (g, kg, and ton), while the component W is expressed as a unit volume of water, m^3 or ha-mm. WUE is not a physical engineering efficiency, which is usually dimensionless and has a maximum value of 100 percent. The *efficiency* term for this biological ratio has caused some concern and confusion with irrigation and water resources efficiency terms (Oweis et al. 1999). Here, WUE generally refers to the entire growing season.

Although *efficiency* is a measure of output per unit input, all the outputs or all the inputs involved in crop production are rarely considered. Plant growth and yield are more strongly related to transpiration than to evapotranspiration. WUE is sometimes evaluated per unit of water transpired (WUET), or per unit of irrigation water applied (WUEI). Often, WUE is based on evapotranspiration (WUEET). The difference is important since suppression of soil evaporation and prevention of weed transpiration can improve the WUEET, however, it need not improve the WUET, which is a measure of crop performance. Furthermore, it is important that *water quality* (and not only quantity) be taken into consideration when dealing with WP and water use efficiency issues.

Irrigation Water Use Efficiency (IWUE$_{ET}$) is a measure of the increase in the crop production (biomass or marketable component) relative to the increase in water consumed when irrigated, over the consumption under nonirrigated conditions as follows (Burman et al. 1981):

$$IWUE_{ET} = (Y_i - Y_n)/(ET_i - ET_n) \tag{22.2}$$

where:
  Y_i = mass of marketable crop produced with irrigation
  Y_n = mass of marketable crop produced without irrigation
  Et_i = mass of water used in ET by the irrigated crop
  Et_n = mass of water used in ET by the nonirrigated crop

The concept of WUE has proved useful in both experimental and field studies of crop water use in dryland and irrigated agriculture, however the term WUE is inappropriate to the intended concept behind it because a maximum established (by theory or observation) does not exist for reference and that is why WP is a more appropriate term than WUE, but it is still not as widely used/accepted as WUE.

It should be well understood that the issue of water productivity is multidisciplinary and scale or level-dependent. In their detailed analysis "Accounting for Use and Productivity of Water," Molden et al. (2003) put forward an interesting question (which crop and which drop?) when discussing the banner "More Crop per Drop"—a slogan used recently by the Secretary General of the United Nations as a response to the threatening crises of water scarcity. WP is addressed at different scales (plant, farm, project, and basin levels) and a conceptual framework for better understanding of WP and water accounting across scales is introduced. It has been pointed out that the highest WP at one scale does not necessarily result in the highest WP at another scale. Economic productivity and opportunity cost of water make the undertaking far more complex.

THE WATER ACCOUNTING FRAMEWORK

It starts by defining the boundaries of each scale or domain of use and the inflow and outflow of water across these boundaries. Components of water inflows and outflows are classified into various water accounting categories. A clear distinction is made between water depletion and water consumption. Water depletion is the use or removal of water (from a domain, particularly a basin) that renders it unavailable for further use. Water may be depleted by evaporation, flows to sinks (such as sea or saline groundwater), or incorporation into products (such as bottling water).

Not all water diverted to a service or use is depleted. Water depletion is classified into beneficial and nonbeneficial. Beneficial water depletion is subclassified into process depletion and nonprocess depletion. Process depletion is the amount of water diverted and depleted to produce an intended good. For agriculture, it is water transpired by crops. In industry, for example, it includes the amount of water evaporated by cooling systems, or converted into a product. Nonprocess depletion includes diverted water that is depleted, but not by the process it was intended for, such as evaporation from the water surface in canals, deep percolation losses to saline groundwater, and outflow of irrigation water to sea.

WATER PRODUCTIVITY PERFORMANCE INDICATORS

A set of performance indicators, based on the above-mentioned water components, have been developed that can be used to analyze and evaluate the water productivity at different scales/levels. These performance indicators have overcome two shortages or limitations of the classical efficiency indices: (i) nonagricultural water uses are taken into consideration, and (ii) make interaction with other water users evident and explicit. However, the lack of reliable data, particularly at basin level, confronts the wide application of this framework and its related indicators.

The beasting obstacles are those related to subsurface water flows, recovery cost for outflows, uncertainty about the fate of outflows, and the change in their quality. Outflows (surface runoff and deep percolation) from fields or farms are charged as inflows to the project level, and those from projects to the basin level. If the total production of a farm is maintained the same but with less amount of water abstracted from the main water supply canal of the project, this means that water productivity of the farm has increased. If the saved water that remained in the canal was not used by another user within the system, but rather flowed to sink (depleted), the WP of the project has not improved. Therefore, micro water saving may not result in better WP at a macro level. By the same token, higher WP does not necessarily result in greater economic efficiency.

The following four performance indicators are proposed by Molden et al. (2003):

PW of irrigation water: PWirrigation = Production/irrigation water amount (kg m^{-3})

PW of inflow water: PWinflow = Production/net inflow water amount (kg m^{-3})

PW of depleted water: PWdepleted = Production/depleted water amount (kg m^{-3})

PW of processed water: PWprocessed = Production/processed amount of water (kg m^{-3})

It should be pointed out that PW*irrigation* is not applicable for basin level, because production for the whole basin includes nonirrigated areas and such as forests and natural grazing.

ECONOMIC WATER PRODUCTIVITY

Water Resource Use Efficiency (RUE) considers all the components of production inputs: land, water, climate, nitrogen, cropping system, and others. When one resource is limiting yield, other nonlimiting resources are usually used less efficiently. If the limitation to yield is alleviated, all resources will be used more efficiently until another resource becomes limiting. Since water is often the most limiting resource, any improvement in water RUE will necessarily lead to enhanced use of other resources. From the perspective of a national economy, a key goal for water use productivity is to improve net economic returns per dollar invested in water use, favoring investment in the urban and industrial sectors. However, such a view fails to adequately recognize the social and environmental benefits from using water in agriculture.

The use and definition of water use efficiency and/or water productivity terms is still controversial. FAO relates water use efficiency to the level of performance of irrigation systems from the source to the crop and defines it as *the ratio between estimated plant requirements and the actual water withdrawal* (FAO 2003b). This definition is more or less closer to the classical irrigation efficiency. On average, this water use efficiency in the developing countries is estimated, by FAO, to be around 38%. FAO considers that improving irrigation efficiency is a slow and difficult and site specific process that largely depends on the local water scarcity situation. It may be expensive and requires willingness, know-how, and actions at various levels. Technology permits accurate water application to crops in the optimum quantity and timing. The application of advanced technology, however, depends on the investment and capacity as well as an economic incentive to make it worthwhile.

It should not be surprising that water use efficiency and water productivity issues make only slow progress and draw little attention whenever water is cheap (or highly subsidized), plentiful (such as in Latin America), or it has no other user to compete on. Among the serious causes for current inefficient use of water in most countries are that water interest, use, control, monitoring, and management are divided among many governmental and nongovernmental agencies, divisions, and bodies, simply because it has multiple uses that concern a wide range of disciplines and interests. There is lack of coordination and harmony among these divisions. There are numerous forces involved ranging from individuals to administration, political, and economics. This of course creates great difficulties in obtaining data and in improving the use and management of water resources. The increasing pressure and competition on water should bring a new and comprehensive and integrated approach to improve the current management of water resources.

RESPONSES TO IMPROVE WATER PRODUCTIVITY

Deficit Irrigation

When water is limiting irrigation, the rules of scheduling should be modified for improved water productivity. In intensive irrigation development, all efforts including research and advancement in technology development are geared towards achieving maximized yield per unit of land. However, in water-scarce areas, water, not land, is the most limiting factor to improved agricultural production. Accordingly, maximized yield per unit water (water productivity or water use efficiency), and not yield per unit of land, is a more viable objective for on-farm water management under such conditions. Fortunately, the two tracks, water and land efficiencies, are parallel for some distance, but not all the way. Irrigating for less than maximum yield per unit land (deficit irrigation) could save substantial amounts of water for irrigating new lands and hence producing more food from the available water. Deficit irrigation is not the only practice that has shown good potential, but other ways are available to modify water management principles to achieve more water-efficient practices. Research is needed to develop the required soil–water–plant–atmosphere relations under such conditions. New guidelines for crop water requirements and irrigation scheduling to maximize water productivity are yet to be developed for the important crops in the dry areas.

CHANGING CURRENT LAND USE

Due to increased water scarcity, globalization, and climate change, current land use and cropping patterns should be modified if more food is to be produced from less water (see chapter 4). Water is likely to be the major constraint and new land use systems that respond to external as well as internal factors must be developed based on available water. This should include adopting water efficient crops, varieties, and sound combinations of crops in the farming system.

PRECISION IRRIGATION

Improved technologies that already exist may at least double the amount of food produced from present levels of water use, if applied in the field. Implementing precision irrigation, such as micro- and sprinkler irrigation systems, laser leveling, and other techniques contributes to substantial improvement in water application and distribution efficiency. It is true that water lost during conveyance and on-farm application is not an absolute loss from a basin perspective, but its quality may deteriorate and its recovery comes at a cost. To account for these losses, the size of the irrigation system will significantly increase and this again comes at a very high cost. Policies to implement and transfer these technologies are vital. There is a need to provide farmers with economic and more efficient alternatives to on-farm water management practices with incentives that can bring about the needed change. There is a great scope for improving the productivity and use efficiency of water in the region. Research findings have shown that substantial and sustainable improvements in water productivity are attainable, but can only be achieved through integrated natural resources management approaches. On-farm water use efficient techniques if coupled with improved irrigation management options, better crop selection and appropriate cultural practices, improved genetic makeup, and timely socioeconomic interventions will help to achieve this objective. Conventional water management guidelines designed to maximize yield per unit area need to be revised for achieving maximum water productivity instead. Appropriate policies on farmers' participation and water cost recovery are necessary for adopting improved management options.

a) Increasing the productivity per unit of water consumed and or depleted;
through:

> *Changing crop varieties* to new crop varieties that can provide increased
> yields for each unit of water consumed, or the same yields with fewer
> units of water consumed.
>
> *Crop substitution* by switching from high- to less-water-consuming
> crops, or switching to crops with higher economic or physical produc-
> tivity per unit of water consumed.
>
> *Deficit, supplemental, or precision irrigation.* With sufficient water
> control, higher productivity can be achieved using irrigation strategies
> that increase the returns per unit of water consumed.
>
> *Improved water management* to provide better timing of supplies to
> reduce stress at critical crop growth stages leading to increased yields
> or by increasing water supply reliability so that farmers invest more in
> other agricultural inputs leading to higher output per unit of water.
>
> *Optimizing non-water inputs.* In association with irrigation strategies
> that increase the yield per unit of water consumed, agronomic practices
> such as land preparation and fertilization can increase the return per
> unit of water.
>
> *Policy reform and public awareness.* Policies related to water use and
> valuation should be geared towards controlling water use, reducing
> water demand, safe use and disposal of water, and encouraging the col-
> lective approach in using and managing water by users. These policies
> must be balanced, workable, and feasible, otherwise they will be dif-
> ficult to implement and/or enforced.

b) Reducing nonbeneficial depletion; by:

> Reducing evaporation from water applied to irrigated fields through
> specific irrigation technologies such as drip irrigation, or agronomic
> practices such as mulching, or changing crop planting dates to match
> periods of less-evaporative demand.
>
> Reducing evaporation from fallow land, decreasing the area of free
> water surfaces, decreasing non- or less-beneficial vegetation, and con-
> trolling weeds.
>
> Reducing water flows to sinks—by interventions that reduce irrecover-
> able deep percolation and surface runoff.
>
> Minimizing salinization of return flows—by minimizing flows through
> saline soils or through saline groundwater to reduce pollution caused by
> the movement of salts into recoverable irrigation return flows.
>
> Shunting polluted water to sinks—to avoid the need to dilute with fresh-
> water, saline, or otherwise polluted water should be shunted directly to
> sinks.
>
> Reusing return flow.

c) Reallocating water among uses:

> *Reallocating water from lower- to higher-value uses.* Reallocation will
> generally not result in any direct water savings, but it can dramatically
> increase the economic productivity of water. Because downstream

commitments may change, reallocation of water can have serious legal, equity, and other social considerations that must be addressed.

Tapping uncommitted outflows.

Improving management of existing facilities to obtain more beneficial use from existing water supplies.

A number of policy, design, management, and institutional interventions may allow for an expansion of irrigated area, increased cropping intensity, or increased yields within the service areas.

Possible interventions are reducing delivery requirements by improved application efficiency, water pricing, and improved allocation and distribution practices.

Reusing return flows through gravity and pump diversions to increase irrigated area.

Adding storage facilities. Infrastructures to store and regulate the use of uncommitted outflows, which is usually the case during wet years, could be considered so that more water is available for release during drier periods. Storage may take many forms including reservoir impoundments, groundwater aquifers, small tanks, and ponds on farmers' fields.

CASE STUDIES FROM THE MIDDLE EAST

SUPPLEMENTAL IRRIGATION IMPROVES RAINFED AGRICULTURE

Rainfed agriculture emerges as a potential key to sustainable development of water and food. Rainfed agriculture produces about 60 percent of total world cereals. Future forecasts under the Business As Usual (BAU) scenario show that rainfed agriculture will continue to play a major role in cereal production, contributing half the total increase in cereal production over the period between 1995 and 2025 (Rosegrant et al. 2002). Improving water management and crop productivity in rainfed area would significantly contribute to food security and relieve considerable pressure on water resources.

Rainfed agriculture worldwide is practiced on approximately 80 percent of the agricultural lands, while the remaining 20 percent is under irrigation. This rainfed area percentage varies widely among regions: from approximately 95 percent in the SSA tropics to 65 percent in Asia. Yield of rainfed agriculture is often low, around 1 t/ha, in semiarid tropics (Rockstrom 2001). For cereal, it is only 0.85 t/ha in the SSA and around 1.40 t/ha in the West Asia and North Africa (WANA) region. These numbers are way below the potential productivity. The rainfed cereal productivity in the developed countries is about 3.17 t/ha. In China, it is 3.59 t/ha, which is higher than the irrigated cereal productivity both in SSA (2.16 t/ha) and that in WANA (3.58 t/ha).

There is ample evidence that the low productivity of rainfed agriculture in the dry areas of the developing countries is not only because of the vagaries of the climate but due more to poor performance related to management aspects. Rainfed agriculture in the water-scarce tropic and a large part of the dry areas is often related to the intensity of rainfall (with large spatial and temporal variability) rather than to low cumulative volume of rainfall. The key challenges lie in mitigating the risk of the intra-seasonal dry spells in order to improve water productivity The technologies that have been proven to be successful in this regard are water harvesting, supplemental irrigation, and a combination of both. Equally important is addressing farmers' perceptions of risk and adapting these technologies, management strategies, and policies that address their concerns.

Simulation models for forecasting world food future and water use anticipate that basin water productivity in the developing countries will increase for both irrigated and rainfed agriculture in order to cope with the sharp increase in food demand. Using 1995 data as a base, it is anticipated that WP in the developing countries will increase from 0.45 kg/m^3 to 0.55 kg/m^3 by 2025 for rainfed cereals (excluding rice) and from 0.56 kg/m^3 to 0.93 kg/m^3 for irrigated cereals over the same period. This represents an increase in WP of 22 percent and 66 percent for the rainfed and irrigated cereals, respectively.

Precipitation in the dry rainfed areas, especially in the Mediterranean-type climate, is characterized by low annual amount, unfavorable distribution over the growing season, and great fluctuation among years. Except in limited areas and exceptional years, rainfall amount in the dry areas is much lower than crop water requirements for economic production. Furthermore, the distribution of the rain in these areas is irregular, unpredictable, and does not usually match crop needs (Figure 22.4). The suboptimal distribution of rainfall coupled with great variations from year to year make predictions very difficult.

As a result of unfavorable rainfall characteristics, soil moisture does not satisfy crop needs over the whole season. I will take wheat as an example here. In the wet months (Dec. to Feb.) stored rain is ample, crops sown at the beginning of the season are in early growth stages, and water extraction rate from root zone is very low. Usually little or no moisture stress occurs during this period. However, in early spring, crops grow faster, demanding a high rate of evapotranspiration and soil moisture depletion. Usually, at that time chances of rain become little while soil moisture drops below critical levels. Thus, a stage of increasing moisture stress starts and continues until the end of the season. This moisture stress occurs in all Mediterranean-type rainfed areas with no exception but varies in its starting time and severity.

As a result, rainfed yields are very low in all countries of the region. Potential yields are much higher and are attainable if soil moisture stress during dry spills is alleviated. Supplemental irrigation is an effective response to this problem. This is the addition of essentially rainfed crops of small amounts of water during times when rainfall fails to provide sufficient moisture for normal plant growth, in order to improve and stabilize yields.

Research results from ICARDA and other institutions in the dry areas as well as harvest from farmers showed substantial increases in rainfed crop yields in response to the application of relatively small amounts of water. This increase covers cases with low as well as high rainfall. Average increases in wheat grain yield under low, medium, and high annual rainfall in Aleppo reached about 400 percent, 150 percent, and 30 percent using amounts of SI of about 180, 125, and 75 mm, respectively (Figure 22.5). When rainfall is low, more water is needed but the response is greater,

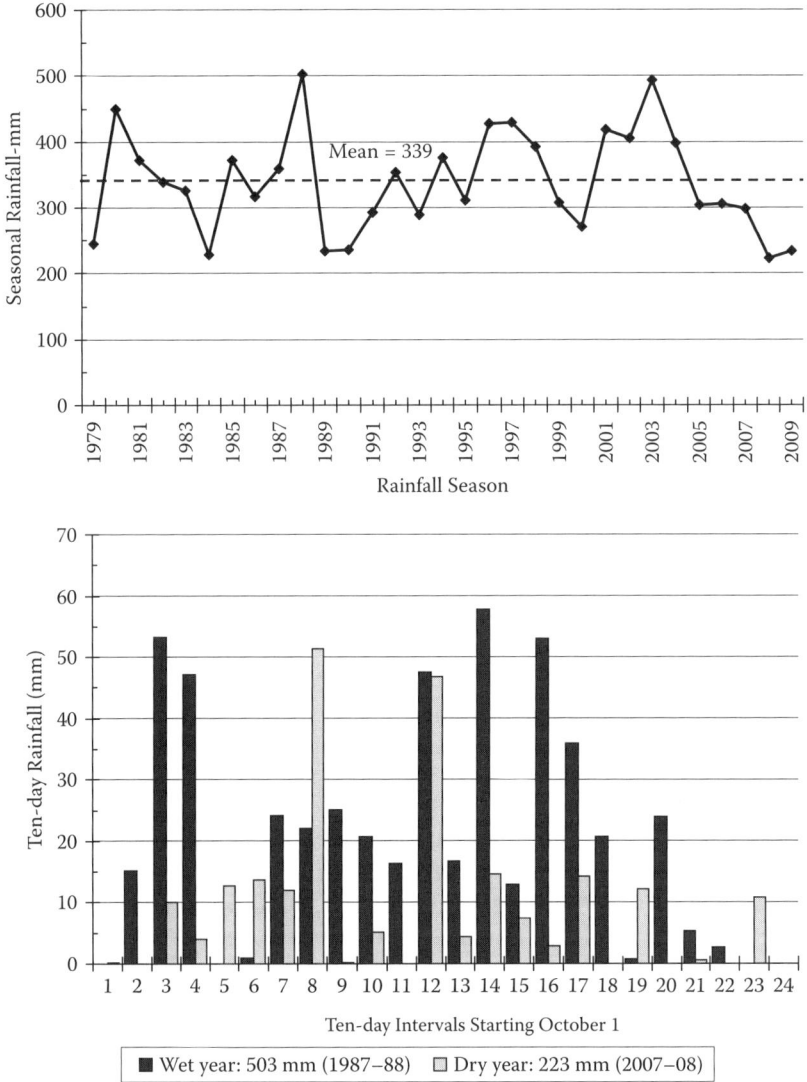

FIGURE 22.4   Rainfall characteristics at ICARDA research station, Tel Hadya, Aleppo, Syria (1979–2009).

and increases in yield are remarkable even when rainfall is as high as 500 mm. The response was found to be higher when rain distribution over the season is poor.

In Syria, average wheat yields under rainfed conditions are only 1.25 t/ha and this is one of the highest in the region. With SI the average grain yield was up to 3 t/ha. In 1996, over 40 percent of rainfed areas were under SI and over half of the 4 million tons national production was attributed to this practice. Supplemental irrigation does not only increase yield but also stabilizes farmers' production. The coefficient of

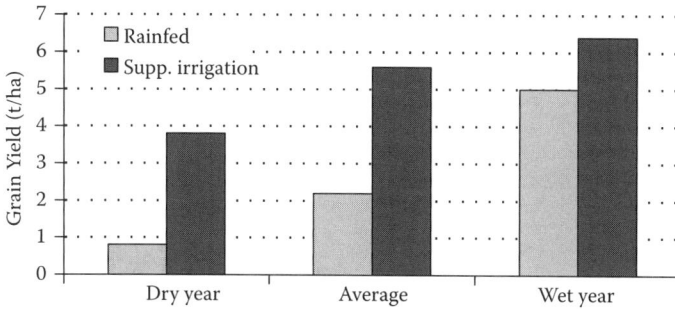

FIGURE 22.5 Impact of supplemental irrigation on wheat grain yield for a dry year (234 mm rainfall with SI of 183 mm); average year (316 mm rainfall with SI of 120 mm); and wet year (504 mm rainfall with SI of 75 mm). (Adapted from Oweis T. 1997. Supplemental Irrigation: A Highly Efficient Water-Use Practice. ICARDA, Aleppo, Syria. 16 pp.)

variation in rainfed production in Syria was reduced from 100 percent to 10 percent when SI was practiced.

Average WP of rain in producing wheat in the dry areas of WANA ranges from about 0.35 to 1.00 kg grain/m^3. However, water used in supplemental irrigation can be much more efficient. ICARDA found that a cubic meter of water applied at the right time (when crops suffer from moisture stress) and good management could produce more than 2.5 kg of grain over the rainfed production. This extremely high WUE is mainly attributed to the effectiveness of a small amount of water in alleviating severe moisture stress during the most sensitive stage of crop growth. The stress usually causes a collapse in the crop development and seed filling and reduces the yields substantially. When SI irrigation water is applied before such conditions occur, the plant may reach its high potential.

In comparison to the productivity of water in fully irrigated areas (rainfall effect is negligible) we find greater advantage with SI. In fully irrigated areas with good management, wheat grain yield is about 6 t/ha using a total amount of 800 mm of water. This makes WP about 0.75 kg/m^3, one-third of that under SI with similar management (Figure 22.6). This suggests that water resources may be better allocated to SI when other physical and economic conditions are favorable.

Deficit irrigation is an optimizing strategy under which crops are deliberately allowed to sustain some degree of water deficit and yield reduction. One important merit of deficit supplemental irrigation is the greater potential for benefiting from unexpected rainfall during the growing season due to the higher availability of storage space in the crop root zone. Results on wheat, obtained from field trials conducted in a Mediterranean climate in northern Syria, reported significant improvement in SI water productivity at lower application rates than at full irrigation. Highest water productivity of applied water was obtained at rates between one-third and two-thirds of full SI requirements, in addition to rainfall.

Research in the WANA region has shown that applying only 50 percent of full supplemental irrigation requirements causes a yield reduction of only 10 to 15 percent. Farmer-managed field plots were established to demonstrate this finding. It was

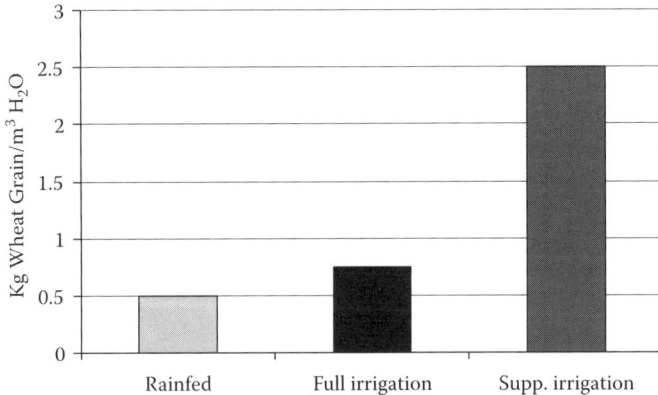

FIGURE 22.6 Productivity of one cubic meter of green water (rainfed), of blue water (full irrigation), and of blue water as supplemental irrigation for a basically rainfed wheat. (Adapted from Oweis T. 1997. *Supplemental Irrigation: A Highly Efficient Water-Use Practice.* ICARDA, Aleppo, Syria. 16 pp.)

observed that farmers tend to over-irrigate their wheat fields. When there is not enough water to provide full irrigation to the whole farm, the farmer has two options: to irrigate part of the farm with full irrigation leaving the other part rainfed, or to apply deficit irrigation to the whole farm. Assuming that under limited water resources only 50 percent of the full irrigation required by the farm would be available, the option of deficit irrigation was compared with other options. The results show that a farmer having a 4-hectare farm would on average produce 33 percent more grain from his farm if he adopted deficit irrigation for the whole area, than if full irrigation was applied to half of the area (Figure 22.7). The advantage of applying deficit irrigation

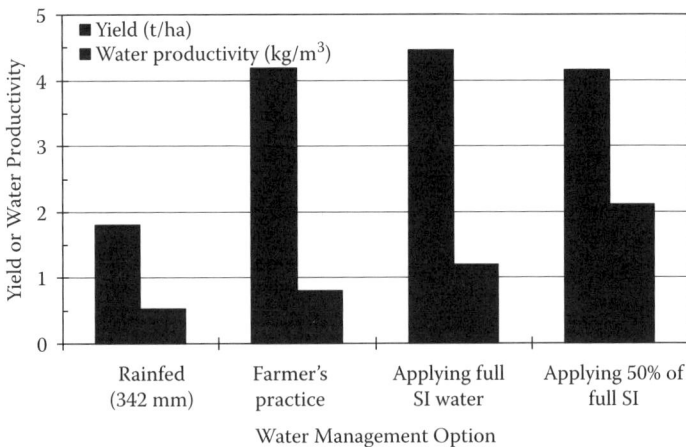

FIGURE 22.7 Yield and water productivity for wheat under different water management options for a four-hectare farm in northern Syria.

increased the benefit by over 50 percent compared with that of the farmer's usual practice of over-irrigation.

In the highlands of the WANA region, frost conditions occur between December and March and put field crops into dormancy. Usually, the first rainfall, sufficient to germinate seeds, comes late resulting in small crop stand when the frost occurs in December. Rainfed yields as a result are much lower than when the crop stand pre-frost is good. Ensuring a good crop stand in December can be achieved by early sowing and applying 50 mm of supplemental irrigation in October. SI given at early sowing dramatically increases wheat yield and water productivity. Applying 50 mm of SI to wheat sown early has increased grain yield by more than 60 percent, adding more than 2 t/ha to the average rainfed yield of 3.2 t/ha (Ilbeyi et al. 2006). Water productivity reached 4.4 kg grain/m^3 of consumed water (Figure 22.8). These are extraordinary values compared to water productivity values of wheat of 1–2 kg/m^3.

SI alone, although it alleviates moisture stress, cannot ensure highest performance of the rainfed agricultural system. It has to be combined with other good farm management practices. Of most importance is fertility, particularly in the Mediterranean region where nitrogen is usually the main deficiency. Absence of nutrient deficiency greatly improves yield and water use efficiency. Other areas may have different deficiency levels of N or deficiencies in other elements (Figure 22.9). It is always important to eliminate these deficiencies to get potential yield and WP.

The use of supplemental irrigation is an example of a concurrent change in both management practice and water-responsive cultivars to increase water productivity. The proper varieties need first to manifest a strong response to limited water applications, which means that they should have a relatively high yield potential. At

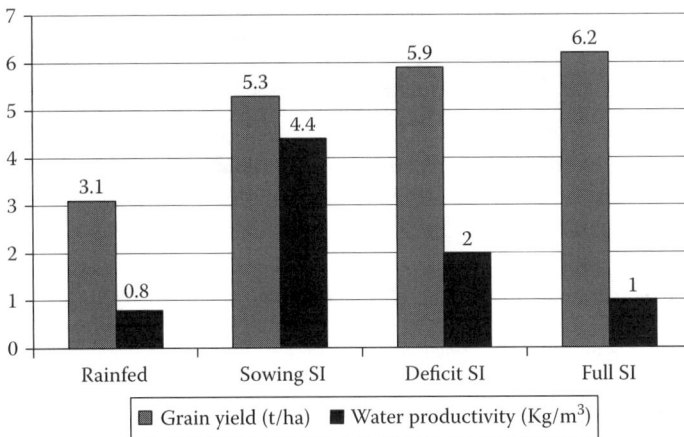

FIGURE 22.8   Wheat grain yield and water productivity at the central Anatolia plateau of Turkey's highlands under three types of SI management (Sowing SI: 50 mm of SI at sowing only; Deficit SI: sowing SI plus deficit SI during spring; Full SI: sowing SI plus full SI during spring) compared to rainfed case. (Adapted from Ilbeyi, A., Ustun, H., Oweis, T., Pala, M., Benli, B. 2006. Wheat water productivity and yield in a cool highland environment: Effect of early sowing with supplemental irrigation. *Agricultural Water Management* 82:399–410.)

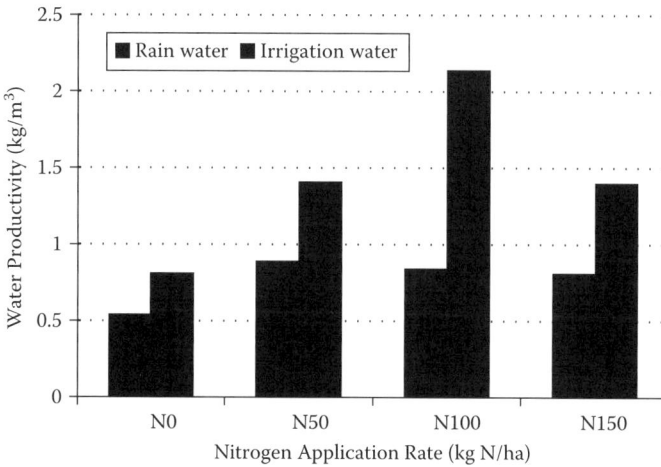

FIGURE 22.9   Different levels of nitrogen in northern Syria. Gains in water productivity for wheat grain.

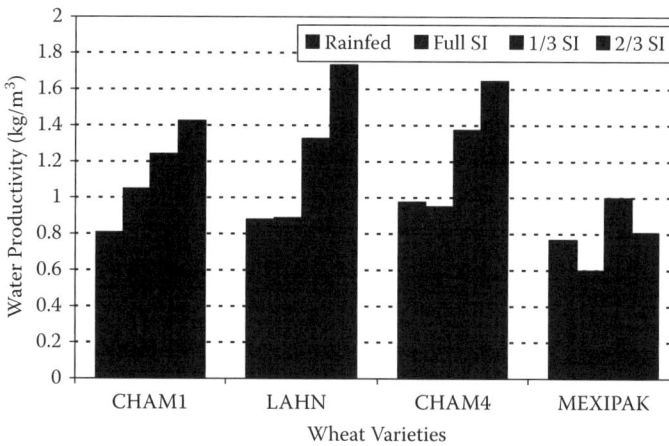

FIGURE 22.10   Average 4-year rainwater and supplemental irrigation (SI) water productivity for durum and bread wheat varieties grown in northern Syria. (Cham1 and Lahn are durum; Mexipak and Cham4 are bread wheat varieties). (Oweis, 2001, unpublished data.)

the same time, they should maintain some degree of drought resistance, and hence express a good plasticity. Figure (22.10) shows the variations in the response of two durum and two bread wheat varieties to various water management options

Using both Mendelian breeding techniques and modern genetic engineering, new crop varieties can be developed that can increase the water use efficiency while maintaining or even increasing the yield levels. For example, through breeding, winter chickpea and drought-resistant barley varieties that use substantially less water have been developed. The chickpea crop is traditionally sown in the spring. As a consequence, terminal drought stress occurs causing low yields. This was avoided by early planting with cultivars developed by ICARDA that are cold tolerant. On-station as well as on-farm trials have demonstrated that increases in yield and water productivity of 30 to 70 percent are possible by adopting early sowing. Currently, winter chickpea is spreading fast among the farmers in the WANA region.

In the winter rainfall environment of the WANA region, delaying the sowing date will prevent crop germination and seedlings establishment because of the rapid drop in air temperature starting generally in November. In the lowlands of the Mediterranean region, where continuous cropping prevails as pure cereal or cereal–legume rotations, mid-November was found to be an optimum sowing time for cereals. Every week delay after this time results in a 200–250 kg ha^{-1} yield decrease. If the onset seasonal rain is delayed, early sowing can be realized with the help of supplemental irrigation systems.

Methodologies to help farmers determine the appropriate supplemental irrigation management have been developed. Since rainfall amount cannot be controlled, the objective is to determine the optimal amount of SI that results in maximum net benefit to the farmers. Knowing the cost of irrigation water and the expected price for a unit of the product, maximum profit occurs when marginal product for water equals the price ratio of water to the product (Figure 22.11). The figure shows the optimal amount of SI to be applied under different rainfall zones and various price ratios.

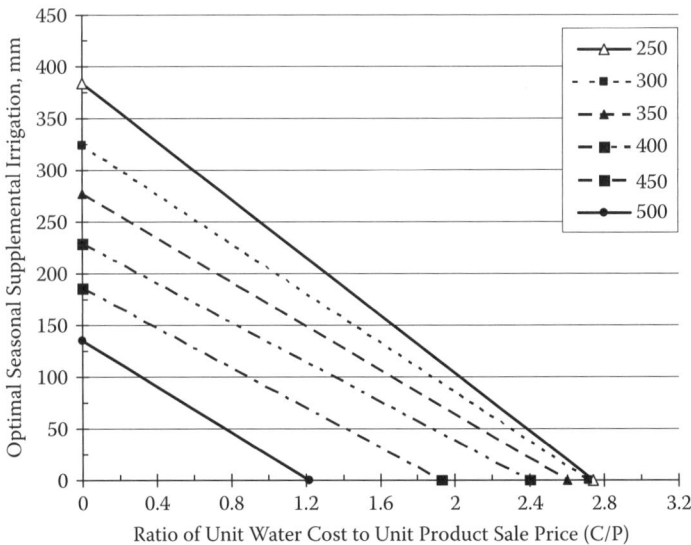

FIGURE 22.11 Optimal seasonal depth of supplemental irrigation of durum wheat at Aleppo, Syria versus the ratio of unit water cost to unit wheat grain sale price for different probable seasonal rainfall (mm).

Improved Water Productivity in Irrigated Agriculture

The WANA region contains vast deserts—about two-thirds of the region receives less than 100 mm annual rainfall and about three-quarters of WANA receives less than 200 mm annual rainfall (Table 22.1). If water (surface or subsurface) is available, irrigation is practiced. In these dry areas, irrigation agriculture produces most of the food. However, water productivity is generally low and soil and water quality are continuously declining. Soil salinity and water logging are other problems in WANA irrigated areas. The root causes for these chronic problems in the irrigated land are poor on-farm water management and lack of sufficient drainage and salt leaching. Despite its scarcity of irrigation water in the region, it is still abused and wasted. It has been reported that farmers, on average, apply 60 percent more water than the crops need (Oweis et al. 2000).

In most WANA countries, water for irrigation is generally managed by public sectors where local institutions are very weak. Policies of water allocation and use do not provide incentive for water savings. Expanding irrigated areas without additional water resources poses another source of pressure for more efficient use of water. The return for irrigation water in the region is low, and varies from one crop to another. However, substantial improvements in water productivity can be achieved by irrigation management and adjusting cropping patterns.

Higher water productivity (yield per unit of water) is usually associated with higher land productivity (yields per unit of land). This parallel increase in yields and WP, however, does not continue positively all the way. At some high level of yield incremental yield increase requires higher amounts of water to achieve. This means that water productivity starts to decline as yield per unit land increases above certain levels. This means that the amount of water required for producing the same amount of wheat at high yield levels is higher than the water requirement at lower levels. This is mainly due to increased losses of evaporation, etc. In this case and when water is more limiting than land it would be more efficient to produce lower yield while the saved water may better be used to irrigate new land than to produce maximum yield with excessive amounts of water at low water productivity. For summer crop under full irrigation, deficit irrigation also improves WP. This, of course, applies only when water, not land, is limiting resource and without sufficient water to irrigate all the available land.

The association of high water productivity values with high yields has important implications for the crop management for achieving efficient use of water resources in water scarce areas. Policies for maximizing yield should be considered carefully before they are applied under water-scarce conditions. Guidelines for recommending irrigation schedules under normal water availability may need to be revised when applied in water-scarce areas.

The identification of appropriate crops and cultivars with optimum physiology, morphology, and phenology to suit local environmental conditions is one of the important areas of research within cropping system management for improved water productivity. Under water scarcity many crops may be phased out of economical production. In many countries of WANA, inefficient crops are still being produced with very high water costs.

Irrigation is the artificial means to secure water necessary for supplying the soil with moisture needed for plant production. The source of irrigation water may be surface water, groundwater, recycled agricultural drainage water, and treated wastewater. The major land degradation problems in irrigated agriculture in the region include: soil salinization, water logging, and declining water productivity. The root causes of these problems are poor on-farm water management and lack of sufficient drainage and salt.

On-farm methods and techniques for applying irrigation water to the land include surface, sprinkler, and micro-irrigation (mainly drip). Surface irrigation, the traditional method of applying water to soil, is still the dominating method of irrigation in the region. On the average, about one-fourth of the cultivated area (fruit trees and permanent field crops included) is irrigated. More than 85 percent of the irrigated areas are under surface irrigation.

Inefficiency is an inherent trait in surface irrigation, due to the limitations in the basic mechanism of this method of water application. All local technical reports from different research institutions and governmental agencies have agreed that 15 percent of the water diverted from the water supply source to the project is lost through the open conveyance system by seepage and evaporation. Furthermore, 53 percent of the water reaching the farms is lost through the surface irrigation water application process by uncontrolled surface runoff at the fields' tails, deep percolation along the surface irrigation run, evaporation, and operational waste in the farm water conveyance system. Therefore, open conveyance systems coupled with traditional surface irrigation result in losing, on average, about 60 percent of the water resources. Water resources management on basin level disagrees with this analysis, claiming that these water losses are still inside the basin and be can recovered and reused within the basin, therefore they are not real losses. The real water losses from the basin are only evaporation and any water that enters salt sinks (i.e., severe quality deterioration).

The use of more efficient on-farm water application techniques, namely sprinkler and drip, must be promoted. Governments should encourage and support farmers, through loans and subsidies, to convert from surface irrigation to sprinkler and drip irrigation. There is a need to provide farmers with economic alternatives to the current practices that lead to wastage of water, and with incentives that can bring about the needed change.

The problem of using groundwater for irrigation in the dry areas is the overexploitation of this natural resource. Pumping groundwater in excess of natural replenishment of water to the aquifer endangers sustainability of the development, which depends on this water. Thousands of wells in the study region are drying out each year. Groundwater mining in the region is a serious problem that must be carefully considered, taking into account quantity and quality as well as legal and institutional aspects. Some countries have already exhausted their groundwater reserves and the quality, especially at the coastal areas, is ruined by saltwater intrusion.

Farmers in the region under consideration seldom have the means to monitor soil water depletion. Schedules based on soil moisture tension in the active root zone are useful but the farmers are unable to use them for want of availability of equipment and/or understanding. The challenge to agricultural water research is to develop simple and inexpensive tools and methods that are easy to implement and are easily understood by farmers and project managers. Methods based on soil water measurements and on plant-stress indicators present some difficulties, particularly for farmers. Tensiometers and plant indicators provide information on the irrigation date only. The farmer still needs information on how much water to apply. Farmers are always criticized for being wasteful and applying excessive water. Actually, the root cause behind this inefficient practice is that the farmer does not know how to measure water flow or quantity, therefore he is applying more than what is *needed* in order to be on the safe side.

In long-established irrigated areas, lands are threatened by rising water tables and salinization. Deterioration of water quality (mainly due to intensive use of chemicals and fertilizers) and quantity add to the problem. Despite today's understanding and the availability of preventive and remedial technologies, the problems of water logging and salinity continue to spread in the region. As a result, the productivity of land and water resources is steadily declining and sustainability is at risk, if not already collapsed, as indicated by land abandonment and migration to nearby cities.

Irrigation water in dry areas usually carries large amounts of dissolved salts. Therefore, hundreds of millions of tons of salt are added to soil by irrigation water each year in the study area. When water, but not salts, is removed from soil by plant roots absorption and soil surface evaporation, the salt is left over in the soil profile. This process is repeated each irrigation, thus concentration of harmful salts in the plant root zone accumulates with time and will build up to a fatal degree. This of course is one of the root causes resulting in salinization and land degradation, which threatens the sustainability of irrigated agriculture. It is well known that the protective measure to this danger is providing proper drainage and enough salt leaching.

Drainage is the cornerstone to sustainable irrigated agriculture. Provision of efficient drainage (especially, subsurface) with adequate leaching in any irrigated agriculture in the arid region is of prime importance and the key for sustainable development. In an attempt to cut down on the initial cost of developing an irrigated agriculture project, the drainage system and related facilities are often deleted from implementation and thus not constructed with the excuse that it is not necessary right now or in the near future. Quite often, the drainage system is already there in the project, but badly neglected with no proper maintenance, thus it is practically nonexistent. There are numerous examples for each of these cases in the WANA region. Overlooking or purposely avoiding, for cost reasons, the inclusion of drainage facilities in irrigated agricultural projects has resulted in agricultural disasters in the region due to water logging and soil salinization.

WATER HARVESTING FOR THE DRIER ENVIRONMENTS—STEPPE

The drier environments, *the steppe* or rangeland, or, as called in West Asia, *Al Badia*, occupy the vast majority of the dry areas of the world. The disadvantaged people generally live there. The natural resources of these areas are subject to degradation and the income of the people who depend mainly on grazing is continuously declining. Due to harsh conditions, people are increasingly migrating from these areas to the cities, with associated high social and environmental costs. Precipitation is low,

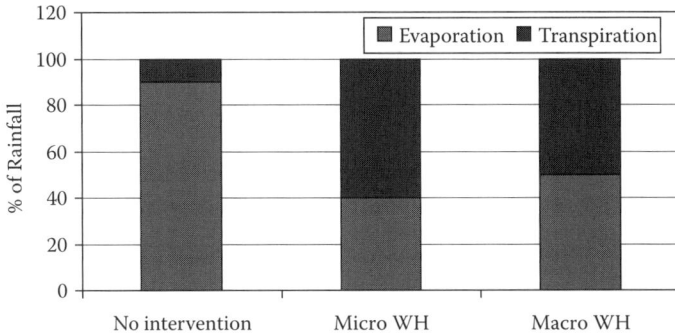

FIGURE 22.12 Percent beneficial use (transpiration) of rainwater under different interventions.

usually less than 200 mm, compared to crop basic needs. It is unfavorably distributed over the crop-growing season and often comes with high intensity. As a result, direct rainfall in this environment cannot support economical dry farming.

In the Mediterranean areas, rain usually comes in sporadic, unpredictable storms and is mostly lost in evaporation and runoff, leaving frequent dry periods during the crop-growing season. Part of the rain returns to the atmosphere, directly from the soil surface by evaporation after it falls, and part flows as surface runoff, usually joins streams, and flows to swamps or to "salt sinks," where it loses quality and evaporates; a small portion joins groundwater. Lack of vegetative cover, degraded soils, and the absence of proper management result in losing about 90 percent of rainwater by evaporation back to the atmosphere or by flow to salt sinks (Figure 22.12). Moreover, the lost water causes serious erosion and accelerate desertification. Water harvesting is the most effective technique to efficiently utilize this water. This simple intervention allowed runoff water to concentrate in smaller areas and be stored in the plants' root zone, supplying needed water to crops and changing the whole scene.

This is the practice of concentrating rainwater, by depriving part of the land of its share of rainwater (which is usually small and nonproductive) and adding it to the share of another part (usually smaller in area) through runoff for beneficial use. The concept of water harvesting differs from that of soil water conservation in which no land is intentionally deprived of its share of rainwater.

Water harvesting systems are broadly classified into two major types:

- The microcatchment systems where surface runoff is collected from a small catchment area and where sheet flow prevails over short distance. Runoff water is usually applied to an adjacent agricultural area to be stored in the root zone and used directly by plants or stored in a small reservoir for later use. The target area may be planted with trees, bushes, or with annual crops. Farmer has the control within his farm over both the catchment and the target areas.

- Macrocatchment systems are characterized by having runoff water collected from a relatively large catchment. Often, the catchment is a natural range, steppe land, or mountainous area. Catchments for these systems are mostly located outside the farm boundaries where farmers have no control over them. Generally, runoff productivity per area is much lower than for microcatchments. Water is often stored in surface or subsurface reservoirs but may also be stored in the soil profile for direct use by crops.

There are several water harvesting (WH) systems, used in WANA. Small farm reservoirs are especially popular in providing water for irrigation and for livestock. This is a macrocatchment WH system that is low in cost and can be very efficient. Harvested water in rainfed areas may be used for supplemental irrigation. Harvesting rainwater for supplemental irrigation is popular in Turkey, Syria, and Tunisia. The research work showed great potential in improving water-use efficiency in the production of winter and summer crops such as wheat and vegetables. However, problems include slitting, evaporation, seepage, and management. Social issues include water rights, farmer capacity, and land ownership. Upstream–downstream balance and efficiency of water collection and use are among the most important issues. A cistern is another WH system that has served human consumption and agriculture for centuries and is still widely used in many areas like northern Egypt.

Jessour is a traditional system in Tunisia and still supporting figs and olives. Preserving the systems and improving their agricultural return are subjects for several research projects currently implemented. Another macrocatchment WH system is the water spreading bunds that are constructed to help spread the water from a stream that passes through the farm to various parts of the farm. They are usually used for field crops. Among the widely used microcatchment WH systems are the contour ridges, usually used for shrubs and field crops. They are formed along the contour by maintaining good catchment surface conditions for runoff and downstream fertile cultivated areas. They are suitable for mechanized agricultural practices but require high precision in layout and implementation. A *Negarim* WH system is small rectangle or diamond-shaped basins surrounded by small earth bunds. Part of the area is used as a catchment where the lower part receives runoff water and contains the crop. Negarim is used for growing trees or bushes.

Runoff strips WH is a technique that is applied on gentle slopes and used to support field crops, where production is usually risky or has low yield level. The farm is divided into strips along the contour. One strip is used as a catchment and the one downstream is a cropped one. The same cropped strips are cultivated every year.

Semicircular bunds WH is constructed by forming a bund along the perimeter of semicircles, thus creating a basin behind the bund. The radius of the basin ranges from 2.0 to 5.0 meters used for bushes or trees and sometimes for field crops. It can be constructed manually or (now) by machines. Mechanizing the construction made it possible to implement it on very large scale in steppe areas. They are suitable for rangeland development and fodder production.

In the Syrian *badia*, a special implement mounted on a tractor (vallerani) was used to construct semicircular bunds mechanically. The implement was able to provide over 40 ha of bunds per day, creating over 5,000 bunds of varying size and spacing.

It is ideal when labor is not available or when it is costly. Operating cost is around US$100/ha, including planting. Shrubs planted in the semicircular bunds showed a survival rate of over 90 percent, compared to 10 percent without water harvesting. They survived three consecutive years of drought and supported local community with grazing in most difficult times. Currently only less than 10 percent of the rainwater is utilized in transpiration. By microcatchment beneficial rainwater can be increased to 60 percent and with macrocatchment to 40 percent.

Contour-bench terraces microcatchment WH are constructed on very steep slopes to combine soil and water conservation with water-harvesting techniques. Cropping terraces are usually built level and supported by stone walls to slow down the flow of water and control erosion. They are frequently used to grow trees and bushes, but rarely used for field crops. The historic bench terraces in Yemen are a good example of this system. Since they are constructed on steep mountainsides, most of the work is done by manual labor.

Rooftop water harvesting collects and stores rainwater from the roofs of houses, large buildings, greenhouses, courtyards, and similar impermeable surfaces, including roads. Most of the rain can be collected and stored. How the harvested water will be used depends on the type of surface used and how clean it is, as well as users' needs. Such systems provide a low-cost water supply for humans and animals in remote areas. Although mainly used for domestic purposes, this technique also has agricultural uses.

There are, however, several limitations to widespread use of water harvesting in the region. Among the most important are the appropriate selection of site, method, and crop; the design, data, and engineering work; the precision in implementation; the integration with other components of the system; and the annual maintenance. One of the greatest limitations to adopting water harvesting works is the lack of clear land ownership, which can be public, tribal, or other forms. Changes are required to provide incentive for people to invest in this promising water management intervention.

Direct benefits of water harvesting are not high. Most of the benefits are environmental and social. Farmers get only small portion of the total benefits and this may not be attractive for investment. The public shares the benefits in improved environment, less migration, and social stability, and this requires that they also share the cost. It is vital that users are involved from the planning stage of development. Institutions should be strengthened and integrated watershed management approaches should be used to resolve upstream–downstream conflicts.

WP IMPROVEMENT AND THE ENVIRONMENT

Now, it has been globally understood and accepted that *environment* is a water using sector, which is strongly linked to the sustainability of water resources development and management. This is a complex issue for both rich and poor countries. It is technically challenging and often entails difficult trade-off among social, economic, and political considerations. One of the important achievements in the water resources issue in recent years is the clear and firm inclusion of environmental water requirements in the picture of the water resources assessment and budget at various levels: global, regional, and basin-wide.

Strategies to reduce poverty should not lead to further degradation of water resources and ecology. Since we accept that environment is one of the water-using sectors, a valid question that should be answered is *what is the fair share for it*? Estimates of irrigation withdrawal in 2025, taking into consideration environment, anticipate that irrigation withdrawals need to be reduced by 7 percent from the 1995 level in order to sustain ecosystems. It looks as though that trade-off between environmental water needs and that for food production will be unavoidable. There are many options that have been proposed for solving this complex and paradox problem. The two most promising options are increasing water productivity and upgrading rainfed systems through the implementation of highly efficient tested techniques and improved water management practices—all at lowest environmental cost.

POLICIES AND INSTITUTIONS

The critical issue in improved water management in the area is lack of appropriate policies and poor implementation. Changing from supply to demand management require incentives and change in the attitude of the people. Valuating water is essential if efficiency is to improve. Sociopolitical constraints do not allow water pricing, but alternatives to pricing can be developed. Water trading through goods is an old practice. It can be used in countries with extreme water scarcity to reduce inefficient water use but agricultural practices of rural communities should be protected.

Water management institutions such as user associations and community cooperatives are weak in the region and need strengthening. They should be allowed to participate in the decision making regarding water issues. Their capacity is also poor and training is essential to improve skills and participation. Linkages between various organizations and disciplines would help integration and exchange of experiences.

CONCLUSIONS

Water is a finite resource and certainly at the heart of human and ecosystem development. It is unequally distributed among regions, countries, and basins. The annual share of available water per capita across the world varies from a few hundred cubic meters to more than 15,000 cubic meters. This dramatic difference in water availability greatly impacts the way people are valuing and using water. In the WANA region, where poverty generally prevails, water is becoming more scarce due to the high rate of population rise and reallocation of water from agricultural uses to other water-using sectors, primarily domestic and industry.

Increasing water productivity is the more viable option to cope with scarcity. Substantial increase in WP requires a shift in thinking to increase water productivity in agriculture substantially through integrated and participatory water resource development and management. It is essential that substantial changes be made in the way water is managed to help alleviate poverty, promote economic growth and overcome potential conflicts. Changes needed include:

- The emphasis from land to water with new guidelines for water management.
- Water allocation to more water-efficient techniques.
- Current land use, cropping patterns, and germ plasm to be more water-efficient.
- The way water is valuated to truly reflect the scarcity conditions.
- Trade policies to import goods with high water demand for production.
- The attitude toward regional cooperation.
- Policies to address water scarcity issues.

REFERENCES

Bagely, J. M. 1965. Effects of competition on efficiency of water use. *Journal of Irrigation and Drainage Engineering Division of the American Society of Civil Engineers* 91 (IR 1): 69–77.

Burman, R., P. Nixon, J. Wright, and W. Pruitt. 1981. Crop Water Requirements. In: *Design and Operation of Farm Irrigation Systems*, M.E. Jensen (ed). American Society of Agricultural Engineers Monograph No. 3. St Joseph, MI. ASAE.

Cai, X., and M. Rosegrant. 2003. World water productivity: current situation and future options. In: Kijne, W.J., Barker, R., Molden, D. (eds.), Water Productivity in Agriculture: Limits and Opportunities for Improvement. CABI Publishing, Wallingford, U.K. pp. 163–178.

El-Beltagy, A., Hamdi, Y., El-Gindy, M., Hussein, A., Abou-Hadid, A., 1997. Dryland farming research in Egypt: strategies for developing a more sustainable agriculture. American Journal of Alternative Agriculture, 12 (3).

FAO. 2003a. Review of world water resources by country. Water Reports 23. FAO, Rome, Italy.

————, 2003b. Agriculture, food and water. FAO, Rome.

Ilbeyi, A., Ustun, H., Oweis, T., Pala, M., Benli, B., 2006. Wheat water productivity and yield in a cool highland environment: Effect of early sowing with supplemental irrigation. Agricultural Water Management 82: 399–410.

Jensen, M., Swarner, L., Phelan, J., 1967. Improving Irrigation Efficiencies. In: Irrigation of Agricultural Lands. American Society of Agronomy Monograph No. 11. Madison, Wisconsin, USA. PP 1180.

Kally, E., 1994. Cost of inter-regional conveyance of water and costs of seawater desalination. In Water and Peace in the Middle East. Issac and Shuval (eds). Elsevier Science B.V., Amsterdam, The Netherlands.

Margat, J., Vallae, D., 1999. Water Resources and Uses in the Mediterranean Countries: Figures and Facts. Blue Plan, UNEP- Regional Activity Center.

Molden, D., 1997. Accounting for water use and productivity. SWIM paper 1. Colombo, Sri Lanka, IWMI.

Molden, D., Murray-Rust, H., Sakthivadivel, R., Makin, I., 2003. A water-productivity framework for understanding and action. Pages 179–197. In: Kijne, W.J., Barker, R., Molden, D. (Eds.), Water Productivity in Agriculture: Limits and Opportunities for Improvement. CABI Publishing, Wallingford, U.K.

Oweis T., 1997. Supplemental Irrigation: A Highly Efficient Water-Use Practice. ICARDA, Aleppo, Syria 16 pp.

Oweis, T., Hachum, A., Kijne, J., 1999. Water Harvesting and Supplemental Irrigation for Improved Water Use Efficiency in the Dry Areas. SWIM Paper 7. Colombo, Sri Lanka: International Water Management Institute.

Oweis, T., Shdeed, K., Gabr, M., 2000. Economic assessment of on-farm water use efficiency in agriculture: methodology and two case studies. 76 p. (En). United Nations, New York, USA.

Qadir, M., Oster, J., 2004. Crop and irrigation management strategies for saline-sodic soils and waters aimed at environmentally sustainable agriculture. Science of the Total Environment 323, 1–19.

Rockstrom, J., 2001. Green water security for the food makers of tomorrow: Windows of opportunity in drought-prone savannahs. Water Science and Technology 43 (4): 71–78.

Rosegrant, M. W., Cai, Ximing, Cline, S. A., 2002. World Water and Food to 2025. IFPRI, Washington, D.C.

Scott, C., Faruqui, N., Raschid-Sally, L., 2004. Wastewater use in irrigated agriculture: Management challenges in developing countries. In: Scott, C.A., Faruqui, N.I., Raschid-Sally, L. (Eds.) Wastewater Use in Irrigated Agriculture: Confronting the Livelihoods and Environmental Realities. CABI Publishing, Wallingford, UK, pp 1–10.

Seckler, D., Molden, D., Barker, R., 1999. Water Scarcity in the Twenty One Century. Water Brief 1, IWMI, Colombo, Sri Lanka.

UN/WWAP (United Nations/World Water Assessment Programme), 2003. UN World Water Development Report: Water for People, Water for Life. UNESCO, Paris.

Viets, F. G., 1962. Fertilizers and the Efficient Use of Water. Advances in Agronomy. 14: 223–264.

World Resources Institute. 2003. Earth trends, the environmental information portal. http://www.earthtrends.wri.org/

Epilogue

Cambodia's Institutional Frameworks for Ecocultural Restorative Redevelopment
Introduction to Optimistic Management Models Useful for Iraq

CONTENTS

Introduction 485
Integrated Water Management 487
Community Fisheries Management 489
Sustainable Livelihoods Management 491
Summary 492
Conclusion 493
References 493

> The optimist proclaims that we live in the best of all possible worlds; and the pessimist fears this is true.
>
> —**James Branch Cabell**

INTRODUCTION

Many ecocultural parallels were drawn between Iraq and Cambodia and Peru in chapter 14. These countries do share another final, dark similarity that offers, however, an optimistic contrast to coverage in the daily newspapers or television reports about the stagnant situation in Iraq. As Iraq sometimes seems to teeter on the edge

485

of slipping into civil war, it is hard to think of a time in the future when the situation will stabilize enough to allow for the further restoration of the marshes within such a framework as that proposed by the practitioners and theorists brought together in France (2011). And as to the restorative redevelopment of the marshlands and their inhabitants along the lines of the sort of examples gathered from around the world and laid out in the pages of the present book ... well, that simply seems too far away to be realistic. But look around that world and it is possible to see situations where such commensurate feelings of despair have been replaced by those of buoyant optimism. It was only a decade and a half ago, for example, when unrest and violence had effectively ended international tourism in both Peru and Cambodia. Today, the turmoil in both counties has abated and their governments are actively rebuilding their formerly devastated rural communities at the same time as welcoming millions of tourists (and their money) to visit their spectacular historical and natural wonders. In the long view, Iraq is a place where citizens and environments have rebounded time and again from the ravages inflicted by capricious nature or despotic rulers (France 2007). And things will certainly improve there yet again.

Based on hydrological and ecological similarities, the huge wetlands in Vietnam's Mekong River Delta have been offered as a model for the management of the Iraqi marshlands. But Vietnam offers other useful, in this case sociological, comparisons to Iraq as well. Today, three decades after more than two decades of war, the level of international and national attention brought to studying and managing the Mekong River Delta in Vietnam is a real success story. But, to interpret this in the spirit of Cabell's pessimistic quip above, three decades seems such a long, long time.

Chapter 14 focused on introducing ecotourism development concerns in Cambodia. Here I wish to outline a handful of restorative redevelopment plans and projects that are currently underway in Cambodia *less than a decade* after its international and internecine strife finally ended. In this respect, there may be no more apt or more optimistic model for what the future could hold for Iraq. But first, a reminder about the sorrowful history of modern Cambodia.

Cambodia carries the unfortunate moniker of being the "sick man of Asia." Given its turbulent history, it is a marvel that the country even exists at all, for there has been no place on the planet in recent times that has been devastated by human carnage to the same extent as Cambodia. And if Cambodia and its people can rebound enough to work with foreign donors on restorative redevelopment, the situation will be relatively easy in Iraq (with the major proviso, however, that a nonfundamentalist government of some form or another is finally accepted in Baghdad).

For much of the twentieth century, Cambodia was occupied by foreign powers. It wasn't until 1953 that it gained full independence from France. Cambodia's struggle to remain neutral during the Vietnam War of the 1950s and 1960s ended with the U.S. bombing campaign that dropped over a million tons of ordnance in three thousand raids, driving the North Vietnamese deeper into the country and leading to the formation of the Khmer Rouge. What followed was the genocidal reign of Pol Pot (the famous "killing fields"), a Vietnamese invasion and occupation, a civil war, and a protracted guerrilla war, all facilitated by the self-centered meddling of Thailand, China, and the United States. In the end, after nearly three decades of turmoil, more than 3 million Cambodians were dead (a quarter of the population), and the landscape which had been ravaged by clearcutting in an attempt by the Khmer Rouge to increase agricultural production was largely empty of both inhabitants and infrastructure. Today, the level of poverty in Cambodia remains high and the standards of living low by comparison with the rest of Southeast Asia. The environment is, however, on its way to recovery, natural "rewilding" having been found to be superior in this regard to anthropogenic replanting.

The following three examples of aid programs, all internationally supported (the Asia Development Bank [ADB] partnered with donor countries, the United Nations' Food and Agriculture Organization partnered with the Asia Forest Network, and the Mekong River Commission partnered with Helsinki University of Technology and the World Bank), are only a small sample of the myriad development projects presently underway in Cambodia. These ongoing projects were selected for introduction here as being representative of the sort of useful approaches that might one day be applied in Iraq's own restorative redevelopment.

INTEGRATED WATER MANAGEMENT

Varis, Kummu, and Keskinen (2006) advance a useful umbrella model of Integrated Water Resources Management (IWRM) as being the most viable way for Cambodia to develop its water resources of the Tonle Sap Great Lake. In 1995 Cambodia joined the newly formed Mekong River Commission, whose mandate is to work toward "a balance between the economic, social, and environmental decisions and development" within "an economically prosperous, socially just and environmentally sound Mekong River Basin." And Varis et al. believe the way to bring this about is through basin-wide IWRM which aims at "developing democratic governance and promotes balanced development in poverty reduction, social equity, economic growth and environmental sustainability."

In particular, IWRM is a multidisciplinary approach that addresses the following issues (Varis, Kummu, and Keskinen 2006):

- *Environment*, through hydrology, chemistry, biology, ecology, and erosion
- *Economy*, through traditional livelihoods, industry, modern agriculture, fisheries and forestry, services and tourism, and the informal sector
- *Social concerns*, such as equity, empowerment, polarization, marginalization, and poverty
- *Participation*, including education and capacity building such as universities, administration, and public awareness; village surveys of local actors; stakeholder links; and communication and workshops
- *Governance*, through linking central government to the local level, links between sectors, international actors, nongovernmental organizations (NGOs), and legislation and conventions

Along these lines, the ADB, recognizing that "threats to the Tonle Sap lake requires an integrated, cross-sectoral, cross-boundary, and multi-disciplinary institutional framework that operates at the basin level," set about to establish a Tonle Sap Basin Management Organization whose justification is explained as follows:

Basin management is not about command-and-control. It is about harmonizing the activities of a multiplicity of natural resource managers, and users so that their actions consider impacts on resource flows and the needs of others. Achieving this requires the creation of institutional processes through which these flows, needs, and impacts can be identified and shared by the full range of stakeholders. Critical to this is the establishment of a structure that provides for planning and management at the basin level to be undertaken holistically in the context of effective dialogues between planners, managers, and users of natural resources, between different levels of government, and across administrative boundaries. River parliaments in India and polder water boards in the Netherlands are two examples of how societies integrate management of land, water, and biotic resources. Where and how such organizations can be established depends on the socio-political context, legal and institutional arrangements, the economic situation and, not least, the resource characteristics of particular basins. Therefore, the form and span of authority of such organizations differ. But effective basin management organizations are characterized by their ability to coordinate planning, contribute to its implementation, and provide a neutral body to monitor achievement and report to stakeholders. (ADB 2005)

The ADB (2005) developed a technical assistance framework in order to plan the development of the institutional framework for integrated basin planning and management for the Tonle Sap. The overarching goal was the sound management of natural resources and the environment of the Great Lake with the following performance indicators and targets:

- Adopt policies based on integrated basin planning and management.
- Increase the participation of stakeholders in decision making.
- Improve integrated management and delivery of water.
- Protect water quality and preserve aquatic ecosystems.
- Invest in the water sector.

Wisely, using a similar strategy of intent to that described in the present book, the ADB (2005) did not set out to reinvent the wheel for Cambodia but took a lateral approach and investigated the particulars of other models of basin management organizations, including the Comisión Nacional del Agua (CONAGUA) in Mexico, various river authorities in Germany and the United Kingdom, the Tennessee Valley Authority in the United States, as well as a suite of commissions in China, Vietnam, France, Australia, and the United States, in addition to those mentioned previously in the Netherlands and India. All successful basin management organizations were found to have the following attributes (ADB 2005):

- Its area of responsibility is clearly defined.
- The technical and financial inputs that are necessary to its operations are provided for.
- The stakeholders and their forms of participation are defined.
- It is a public institution.

And in order to act on its responsibilities of integrated basin planning and management, the organization must include the following:

- A widely representative body in which public administrations, user associations, and NGOs are represented
- An executive, with the relevant financial resources, in charge of the direct implementation of actions or making them possible by funding
- A multidisciplinary technical team having the necessary technical means for achieving efficient management and monitoring of land, water, and biotic resources
- An organization in charge of natural resource use policing and monitoring

COMMUNITY FISHERIES MANAGEMENT

Evans, Marschke, and Paudyal (2004) targeted a single floating village community as a demonstration project about bottom-up comprehensive resource management in such a way as to enable communities to have a clear understanding and strong consensus about its contents, thereby allowing the plan to be fully implemented by the community. Kompong Phluk is made up of almost three thousand people inhabiting three small villages, with fishing being the main occupation for most. Following analysis of the history of traditional resource management, the authors worked with locals to develop several community organizations for the more effective and sustainable management of their forests and fisheries. The Community Fisheries Management Plan (CFMP) included the following objectives:

- To protect and manage flooded forests for a regular supply of daily needed forest products and provide habitat to fish for spawning and nourishing
- To conserve flooded forests to provide shelter for aquatic life (conserve aquatic biodiversity)
- To conserve forests to protect villagers from storms
- To develop sustainable fishing practices for livelihood improvement of fishers

And the following strategies were developed for implementing the CFMP:

- Strategy of controlling illegal activities (enforcement and patrolling)
- Strategy for resource protection (protection areas and fish sanctuary)
- Strategy for sustainable resource harvests (fisheries and forestry)
- Strategy for dry-season agricultural practices (replacing rice with mung bean cultivation)
- Strategy for ecotourism development (establishing a visitor center, building resting huts in the flooded forest, and building a floating fish culture cage)
- Strategy for institutional development (identifying options for income generation, setting up an office and communication system, and using transparent accounting)

Because such a system of community management planning was new in Cambodia, a key principle became "learning by doing" (Evans, Marschke, and Paudyal 2004). The management-planning steps and activities included the following:

1. Preparation of fieldwork
   - Gather and review existing information.
   - Identify any traditional management systems.
   - Collect background visual material.
2. Meeting with community
   - Meet with all committee members, village council chiefs, and other local stakeholders.
   - Review the management-planning process with the preparation team and all stakeholder representatives, including community elders.
3. Training the team
   - Discuss the key concepts of CFMP and its importance for development.
   - Cover roles and responsibilities of community related to management planning.
4. Socioeconomic conditions
   - Discuss the socioeconomic profile of the village.
   - Discuss issues, problems, and illegal activities.
   - Discuss traditional uses and practices of resource management.
5. Participatory resource mapping
   - Map out important locations and their uses.
   - Divide areas into blocks and define block objectives.
   - Verify and finalize blocks with Global Positioning System (GPS) mapping.
6. Natural resource inventory
   - Prepare simple inventory methods to define resource status within blocks.
   - Collect and analyze data.
7. Discussions of field results
   - Review results of inventory.
   - Discuss management options according to blocks and activities.
   - Discuss operating budget for implementation of CFMP.
   - Discuss benefit sharing, fish sanctuaries, and controlling illegal activities.
   - Define roles and responsibilities of stakeholders for implementation.
8. Drafting a management plan
   - Compile each component of CFMP into an agreed format and prepare an annual action plan.
9. Finalizing a management plan
   - Review the draft plan and discuss with different interest groups.
   - Organize a local stakeholder workshop.
   - Prepare the final plan, incorporating comments.
10. Approving community fisheries management
    - Gain approval of plan from village council.
    - Seek approval from regional government levels.

SUSTAINABLE LIVELIHOODS MANAGEMENT

Iraqi Minister of Water Resources, Abdul-Latif Jamal Rasheed, titled his introduction to the 2004 conference "The Future of the Iraqi Marshlands." As an interesting bookend, the incredibly comprehensive management plan of the ADB called the Tonle Sap Initiative uses as its motto the phrase "Future solutions now" (ADB 2006).

With over a third of its population living below the poverty line, Cambodia ranks 130th out of 173 countries on the Human Development Index (ADB 2006). And despite Cambodia being rich in natural resources, it is difficult for local villagers to make a living there. Because the health, integrity, and restoration potential of the environment are closely tied with the well-being of its human inhabitants (Millennium Ecosystem Assessment 2005), it is necessary to address the issues of poverty in restorative redevelopment projects. The ADB (2003, 2006) bases its Tonle Sap Initiative on a program of addressing livelihood assets, which are divided into five core types of capital:

1. *Human capital* is the total of skills, knowledge, experience, labor ability, and health that enable people to pursue different livelihood strategies.
2. *Social capital* consists of the active connections among people, including trust, mutual understanding, and shared values and behaviors.
3. *Natural capital* refers to both renewable and nonrenewable resources, as well as environmental services, from which livelihoods are derived.
4. *Physical capital* provides the basic infrastructure (transport, shelter, water supply, sanitation, energy, etc.) needed to support livelihoods.
5. *Financial capital* comprises the monetary resources available to enable people to actualize their livelihood plans.

As described by the ADB (2003), "The livelihoods approach is a way of thinking about the objectives, scope, and priorities for development. It seeks to develop an understanding of the factors that lie behind peoples' choice of livelihood strategy and then to reinforce the positive aspects and mitigate against the constraints or negative influences. Its core principles are that poverty-focused development activity should be people-centered, responsive and participatory, multi-level, conducted in partnership, sustainable, and dynamic." The project design comprises a feasibility study, preliminary engineering, cost estimates, technical and socioeconomic analyses, an environmental analysis, a social impact assessment, and a study of initial benchmark indicators.

With a goal of asset accumulation, the ADB (2003) set some objectives of what the areas of direct and selected indirect support might be to accomplish this task:

- Human capital
  - Support to health, education, and training infrastructure
  - Support to health, education, and training personnel
  - Support for the development of relevant knowledge and skills
- Social capital

- • Support to improve the internal functioning of groups in terms of leadership and management
- • Support to extend the external links of local groups
- • Natural capital
  - • Support to conserve natural resources and biodiversity
  - • Support for the provision of services and inputs for agriculture, forestry, and fisheries
  - • Reform of organizations that supply services to those involved in resource extraction
  - • Changes in institutions that manage and govern access to natural resources
  - • Improvements in environmental legislation and enforcement mechanisms
  - • Support to market development to increase the value of agriculture, forestry, and fisheries products
- • Physical capital
  - • Support to service provision
  - • Support to infrastructure provision
- • Financial capital
  - • Support to the development of financial services organizations

The ADB identified a set of entry points for sustainable livelihoods that represent an early attempt to achieve synergies between priorities identified by community representatives in a survey and the types of activity that would build on strengths and opportunities (ADB 2006):

- • Establishing village development funds
- • Strengthening community-based management of natural resources
- • Improving agronomic practices and small-scale irrigation
- • Developing postschool literacy for women
- • Supporting self-help groups

And focusing on the entry points most germane to the subject of this book, the purpose of strengthening community-based natural resource management was to generate transparency and equity into issues of access to land and other natural resources that create conflict and dispute in villages, and to improve the sound management of those resources. And the purpose of improving agronomic and irrigation practices was to achieve optimum production from local crop varieties, especially rice, and to make agricultural output more reliable with the rehabilitation of existing or the construction of new small-scale water management projects (ADB 2006).

SUMMARY

The point of this epilogue is to provide an overview of various initiatives that are currently underway in the restorative redevelopment of Cambodia. Those familiar with the methods and models that have been utilized in international development may well be unsurprised by many of the approaches summarized above. My intent here is not to present these approaches for their novelty but rather to provide a frame of reference for the future restorative redevelopment of Iraq in terms of highlighting these particular models for the optimism that they convey in terms of being applied for Cambodia, a country whose recent history of conflagration far exceeds that in Iraq. In other words, though the situation on the ground in Iraq may sometimes look bleak, if such interventions in restorative redevelopment can be underway in Cambodia, they, or some version like them, should be able to be adapted to and adopted for Iraq or any other country devastated by conflict or natural disaster.

CONCLUSION

As the present book in general, and this epilogue in particular, demonstrates, ecological restoration, if it is to play a role in international development in terms of helping to transform and renew treasured places instead of just physical spaces, and human lifestyles instead of merely animal lives, needs to grow into the multifaceted discipline that I refer to as "restorative redevelopment." By paying close attention to the issue of natural assets (as opposed to just natural capital), restorative redevelopment deals with issues of environmental justice and thus navigates a path from despair to hope in what Boyce, Narain, and Stanton (2007) recently referred to as "reclaiming nature," the new vision of positive environmentalism for the twenty-first century—in those years long after any particular worries about "2012" would have been forgotten.

REFERENCES

Asia Development Bank. 2003. *Technical assistance to the Kingdom of Cambodia for preparing the Tonle Sap sustainable livelihoods project*. Manila: Asia Development Bank.
———. 2005. *Establishment of the Tonle Sap Basin Management Organization*. Manila: Asia Development Bank.
———. 2006. *Future solutions now: The Tonle Sap Initiative*. Manila: Asia Development Bank.
Boyce, J. K., S. Narain, and E. A. Stanton. 2007. *Reclaiming nature: Environmental justice and ecological restoration*. London: Anthem Press.
Evans, P. T., M. Marschke, and K. Paudyal. 2004. *Flood forests, fish and fishing villages: Tonle Sap, Cambodia*. Rome: Food and Agricultural Organization.
France, R. L. 2011. *Restoring the Iraqi marshlands: Potentials, perspectives, practices*. Sussex, UK: Sussex Academic Press.
France, R. L., ed. 2007. *Wetlands of mass destruction: Ancient presage for contemporary ecocide in southern Iraq*. Winnipeg, MB: Green Frigate Books.
Millennium Ecosystem Assessment. 2005. *Ecosystems and human well-being: Synthesis*. Washington, DC: Island Press.
Varis, O., M. Kummu, and M. Keskinen. 2006. Special issue: Integrated water resources management on the Tonle Sap Lake, Cambodia. *International Journal of Water Resource Management* 22.

Index

A

Aboriginal Land Rights Act (1976), 51
ADB. *See* Asia Development Bank (ADB)
Agricultural development, 127
  environmentally sensitive, 74
  guidelines, 94
  in Pantanos de Centla, 175
  potholes and, 160
Agricultural pollution, 169
Agricultural water, 444, 455–456
  decline, 456
  drainage, 68, 458
  increasing productivity, 460–465
    accounting framework for, 463–464
    economic approach to, 464–465
    evolution of concept of, 461
    from efficiency to, 461–463
    performance indicators for, 464
  research, 477
Al Wathba Nature Reserve, 445, 446
Alexander River Restoration Project, 26, 34
American Society for Ecological Restoration, 2
Angkor, 234–235, 245
  archaeological site, 262
  as hydraulic city, 230, 234
  fishing industry, 247
Angkor Archaeological Park, 230
Angkor Centre for Conservation of Biodiversity,
  263, 264–265
Angkor Wat, 228
Animal(s), 299
  as kinfolk, 351
  biological diversity, 135
  effect of agricultural drainage on, 68
  protein intake, 247
  rehabilitation, 263
  sanctuaries, 277
APSARA. *See* Asian Pacific Self-Development
  and Residential (APSARA)
Aquifer mining, 89, 90
Archaeological tourism, 332
Asian Development Bank (ADB), 268, 487
  goal of asset accumulation, 491
  management for Tonle Sap, 488
    livelihood approach to development, 491
Asian Pacific Self-Development and Residential
  (APSARA), 247
Audubon Society, 223
Azraq Management Plan, operational objectives,
  9
Azraq Oasis

aquifer mining, 89, 90
building foundations from Arab colonization
  period, 88
groundwater withdrawal, 91
history, 87
interpretive signage, 91–94
paleolithic artifacts in, 86
population history, 87
proximity to Qusar Amra, 110
rebirth of, 90
rehabilitated wetlands in, 96, 97
success of, achievement indicators for, 103,
  108, 115
tourist lodge, 109
visitor center, 98
  amenities, 104–108
  pedagogic displays at, 99–101
Azraq Wetland Reserve, 103
Aztec civilization, 34–35

B

Babylon, restoration of, 35
Bed and breakfast inns, 312, 316–317
Biological nutrient removal, 415
Biological oxygen demand (BOD), 379, 380
Biotoxic concerns, 134
BOD. *See* Biological oxygen demand (BOD)
Bolivia, 272
Bounce relationships, 143

C

California-Florida model, for wastewater reuse,
  393
Cambodia, 245. *See also* Tonle Sap Great Lake
  Chong Khneas in, 248–249, 259
    accommodating variable water levels
     for passenger port and fish market,
     273–275
    as shipping port, 268
    building typologies in, 271
    description of, 265
    ecological park near, 276–277
    living conditions in, 250
    location, 265
  foreign occupation of, 486
  history of modern, 486
  Khmer Rouge of, 486
  offshore oil reserves, 257
  poverty, 486, 491
  water resource management in, 487–489

issues addressed by, 487
  economic, 487
  educational, 487
  environmental, 487
  governance, 487
  social, 487
Canadian International Development Agency, 387
CCPRD. *See* Clark County Parks and Recreation Department (CCPRD)
Cement mine, 161
CFMP. *See* Community Fisheries Management Plan (CFMP)
Chesapeake Bay, 168–169
Chicago Wilderness Project, 23
Chong Khneas, 248–249, 259
  accommodating variable water levels for passenger port and fish market, 273–275
  as shipping port, 268
  building typologies in, 271
  description of, 265
  ecological park near, 276–277
  living conditions in, 250
  location, 265
Chong Khneas Environmental Improvement Project (CKEIP), 268
CKEIP. *See* Chong Khneas Environmental Improvement Project (CKEIP)
Clark County Parks and Recreation Department (CCPRD), 120, 124
Clark County Wetlands Park, 117, 120
  goals, 225
  key to success, 222, 225
  master plan, 121, 222
    implementation of, 225–226
  planning process, 223
  programming and land use, 224
CLIS. *See* Council of Leading Islamic Scholars (CLIS)
Cluster treatment systems, 422–424
Co-management model, 50–52
Coiba, Panama, 298–303
Colorado River Basin, 437
Community Fisheries Management Plan (CFMP), 489–490
  objectives, 489
  planning steps and activities, 490
    community meeting, 490
    consideration of socioeconomic conditions, 490
    discussion of field results, 490
    drafting management plan, 490
    fieldwork preparation, 490
    finalizing management plan, 490
    natural resource inventory, 490
    participatory resource mapping, 490
  seeking approval government approval, 490
  team training, 490
  strategies for implementation, 489–490
Competitive discriminator, as tool in conservation marketplace, 160
Confederated Salish and Kootenai Tribes (CSKT), 348, 359
  partnership in design, 360
  tubal preservation staff, 369
Conservation International, 216
Conservation projects, 132
Conservation *vs.* restoration, 15
Contaminants, 134
Control power, effect on indigenous culture, 56–57
Council of Leading Islamic Scholars (CLIS), 396
Craft production, 309–311, 313
Crop irrigation, 393
Cultural landscapes, prairies, 369
Cultural tourism, 53
Cultural values
  as foundation for design decisions, 348
  impact of landscape changes on, 352
  supporting, 365
Culverts, 119, 151, 363–364, 365

D

Dana Guest House, 313
Dana Nature Reserve, 305
  as model for Iraq, 308
  as paradise for tourism, 307
  craft production, local, 309–311, 313, 314
  growth of, 312
  participation of local people in project, 309
  RSCN management, 307
  size, 308
  socioeconomic model for, 309
  surrounding villages, 308–309, 310, 311
    bed and breakfast inns in, 316–317
    irrigation channels for restored fruit orchards, 315, 316
    RSCN guest house in, 321–322
  tourism, 312, 335
Dana Village, 308–309, 310, 311
  as location for shooting films, 333
  as victim of its success, 332
  bed and breakfast inns in, 316–317
  irrigation channels for restored fruit orchards, 315, 316
  RSCN guest house in, 321–322
Deficit irrigation, 465–466
Degradation, 5
Delta Marsh, 165, 166, 167
Denitrification, 380, 407
Desalinization, 456

Development, effect on indigenous culture,
　56–57
Ducks Unlimited, 158
　case studies, 163–169
　financial resources, 159
　management approach, 160–163
　priority areas, 160
　regional management prioritization, 159
Dupont Nylon Treatment Wetlands, 376

E

Earth Watch Institute, 211
　Conservation International and, 216
　engaging people as key concern, 212–213
　funding, 214
　mission, 211
　model, 213–215
　Pantanal wetlands as case study, 215–217
　projects, 215
　　corporate participation in, 217
　　research, 214
　tourist economies, 214
Ecocultural landscapes, 127
Ecological engineering, 391
Ecological restoration, 196
　creation of ecological citizens and, 24
　defined, 22
　defining issues, 22
　historical fidelity and, 2
　holistic approach to, 3
　importance of, 22–23
　integrating cultural values with, 348
　model, 153
　practitioners, 2
　question, 19
　short comings, 2
　therapeutic gardening and, 25
Ecotechnology, 391, 402–408
　development of, 419
　goals of, 402
　land use, 402
　purely passive systems, 405, 407
　semipassive systems, 407
Ecotourism, 41, 261
　archaeological tourism and, 332
　as conservation strategy, 227, 263, 324–325
　as hope and threat, 262
　at Prek Tol Core Area, 260, 261
　benefits and challenges, 262
　center, 77
　challenges in development of, 128, 177, 262
　companies, 291
　goal, 74
　in Dana Village, 310, 311
　in Jordan, 310
　negative aspects, 260

planning alternatives for, 302
　program, 206
　　in Iraq, 305
　responsible, 260
　role of farmers in, 81
　strategies for development, 227, 263, 489
　tour operators, 332
Ecotoxicological concerns, 134
Ecumenical restoration, 1
Energy reuse, 407, 408
Environmental engineering, 391
Environmental planning, public policy, 303
Environmental problems, 29
Environmental tourism, 53, 54
Ethnic tourism, 53
Eutrophication, 292
Evapotranspiration, 185
Everglades Nutrient Removal Project, 385

F

Fair Oaks Farm, 146, 148, 149
Fallowing land, in salinity management, 436
Federal Emergency Management Agency, 138
Federal Highway Administration, 359
Fertile Crescent, 441–443
Feynan Eco Lodge, 326–328
First World Water Forum, 461
Fish production, 430
Flathead Indian Reservation, 348
　boundaries, 350
　introducing visitors to, 368
　land ownership and land use patterns on, 355
　location, 349
　mammal species, 349
　prairies as cultural landscapes on, 369
Florida Everglades, 385, 387–388
Forced Bed Aeration™, 406
Fort Whyte Nature Centre, 161, 162

G

GEF. *See* Global Environmental Facilities (GEF)
Global Environmental Facilities (GEF), 90
Global Positioning System (GPS), 168
GPS. *See* Global Positioning System (GPS)
Green infrastructure, 385
Greenhouses, 148
Growth-peak-decline model, 54

H

Hazor, ruins of Bronze Age, 66
Historical tourism, 53, 54
Homo sapiens as *Homo reparans,* 15
Hula Nature Reserve, 68–71
　trail system, 71

wetland remnants in, 70
Hula Swamp, 65
Hula Valley
    Arab population, 67, 68
    history, 66
    irrigation history, 67
Hula Valley Restoration Project
    as example for Iraqi marshlands, 82–83
    conversion to agricultural fields, 68
    goals, 71, 74
    green parking lot, 72
    major goals of, 71
    observation area, 73
    observation tower, 76
    results of, 76–80
    scientific approach to, 79
    visitor center, 72
    walkways, 74–75
    water buffalo at, 73
Hydraulic(s), 381–382
    loading rates, 381
    modeling, 122
Hydrology, 151, 381–382
    modeling, 122

I

India, 387
Indigenous culture, 50–52
    central control conservation model and,
        48–50
    co-management model and, 52–53
    governance and, 53
    tourism and, 53
    transformation, 48
Indigenous land
    development stages on, 54–56
        exploration, 55, 56
        involvement, 55, 56
        prime development, 55, 56
Indigenous people, 50
Indigenous sovereignty model, 52–53
Integrated Revitalization Guide, 33
Integrated Water Resources Management
        (IWRM), 487–489
    issues addressed by, 487
        economic, 487
        educational, 487
        environmental, 487
        governance, 487
        social, 487
International Conservation Plan, 159
Iraq
    agricultural history, 443–444
    as conservation strategy, 227, 263
    ecotourism in, 305
    restoration in

funding of, 35
    international support for, 35
    technology transfer approaches to, 387
    Warsaw effect and, 30
Iraqi marshlands, 17, 21, 163
    as tourist attraction, 47, 53–54
    ecological significance, 157
    Hula Valley Restoration Project as example
        for, 82–83
    restoration, 22, 132
        master plan for, 41
        questions for consideration in, 41–45
        recommendations for, 445–446
        salt considerations in, 444
        strategies for, 40
        study design, considerations for, 44–45
    variability of, 40
Irrigation, 393
    agricultural, 75, 397
    center pivot systems, 145, 147
    channels in Dana Village, 315, 316
    defined, 477
    flumes, 134
    groundwater, problem of using, 477
    in Hula Valley, 393
    landscape, 397
    methods, 477
    overflow, 386
    precision, 466–468
    projects associated with U.S. 93 corridor, 356
    renovated channels, 312, 315, 316
    reuse, 401 (*See also* Wastewater reuse)
    spray, 395
    supplemental, 468–475
    wastewater for (*See* Wastewater reuse)
Isla de los Uros, 283, 284, 285
Isla Taquile, 279
    demonstration of indigenous agricultural
        methods and craft making on,
        280–281
    terrace farming, 279–280
    tourism on, 282–283
Islam, 392–393, 395–397
Island of Coiba, 299

J

Jessour, 480
Jewish National Fund, 34
Jordan. *See also* Dana Nature Reserve
    protected areas in, 324
    sprawl of luxury hotels, 334
    water resources, 87–90
Jordan Rift Valley, 332
Jordan River, 69

K

Kankakee Sands, 145–153
  restorative process, 145–151
    drainage infrastructure, 147
    greenhouses in, 147
    mapped ditch systems, 146
    mapped soil types, 148
    phasing program for implementation,
      149, 150
    seeding and planting in, 149
    soil preparation specifications in, 149
    tile networks, 148
    vegetation management in, 149
Khmer Empire, 228, 233–234
  hydroengineering projects, 229
  wildlife carved on temples, 256–257

L

Lagoons, 2, 179, 227, 405
  aerobic, 404
  anaerobic, 404
Lake Agassiz, 139
Lake Agmon, 77, 78
Lake Hula, 65
  as area for migratory birds, 67
  draining of, 67, 68
    problems associated with, 70
Lake Texcoco, 184
Lake Titicaca, 272–293
  archaeological remains, 272, 278
  floating islands, 291
  history, 272
  Isla Taquile in, 279
  location, 272
  reed marshes in, 288
Land use, 224, 402
  changes, increased water productivity and,
    466
  planning, 298
Las Vegas Valley, 119
Las Vegas Valley Water District, 120
Las Vegas Wash, 222, 225
  adaptive management, 123
  committee, 123
  environmental problems, 121–122
  geography, 118
  grade, 122
  park, 118–119
  protective efforts, 120–121
  restoration efforts, 122–123
  uniqueness, 119
Las Vegas Wash Development Committee, 120
Livelihood(s)
  assets, 491
    capital, 491

financial capital, 492
human capital, 491
natural capital, 491–492
physical capital, 491–492
social capital, 491–492
sustainable management of, 491–492
Living landscape projects, 132
  water use considerations, 133

M

Mekong River, 228
  catfish, 228
  hydroelectrical dams on, 257
Mekong River Basin, 487
Mekong River Commission, 487
Mekong River Delta, 486
Mesopotamia, 34
Mexico. See Pantanos de Centla; Xochimilco
Millennium Ecosystem Assessment 2005, 2
Montana's highway 93, 348–370
  curvilinear design, 361, 362
  flexible lane configurations, 361
  landscape of, 349–350
  redesign
    approach to, 358
    client as partner in, 358
    context-sensitive approach to, 360
    farming and irrigation projects associated
      with, 356
    honoring "spirit of place" in, 360, 367
    memorandum of agreement for, 359
    preservation of rural character as
      consideration in, 367
    recommendations for, 363–364
  revegetation guidelines, 365–367
  wildlife connectivity across, 351

N

National park
  conservation model, 48
  control power relationships in development,
    49
  indigenous people and, 50
  traditional model for, 48, 54
Native Americans
  Confederated Salish and Kootenai Tribes,
    348
    history of, 350
    relationship between people and place,
      350–351
    seasonal calendar of, 33
    tribal lands, 348
Neal Smith National Wildlife Refuge, 220
Nevada Water Authority, 123

New Mexico Environment Department Reuse
    Workgroup, 395
Ninepipe wetland complex, 357
Nitrification, 407
Nitrogen, 379–380

O

Olentangy River Wetland Reserve, 387
Oremet titanium mine, 438
Organochlorines, 134
Ottoman village, ruins of, 335
Outfall pipes, 205
Owens Lake project, 439–440

P

Pantanal wetlands, 164–165, 215–217
Pantanos de Centla, 174–184
    biodiversity, 175
    environmental threats, 175
    forested islands in lagoon, 179
    grazing livestock, 182
    history and description, 174–175
    lessons for Iraqi marshlands, 206
    location, 174
    nature center, 180–181, 184
    observation tower, 184
    park management, 175–177, 179
    size, 174
    transport into wetlands, 178
    visitor center, 183
    wetland living in, 180
Parks Watch International, 176
Peru, 283
Petra, 334, 335
Phnom Kraom, 265, 270
Phnom Penh, Cambodia, 268
Plankton ecology, 403
Planting design, 382–383
Pollution, agricultural, 169
Pothole ecosystems, 160
    agricultural development and, 160
Prairies, as cultural landscapes, 369
Precision irrigation, 466–468
Prek Toal Bird Sanctuary, 259, 260
Prek Toal Core Area, 259
    ecotourism visitor center and park
        management office, 261
    handling of security issues in, 263
    survival, 260
Pressure sewers, 398
Puno, Peru, 272, 289, 290, 291

Q

Qusar Amra, 110–114

R

Rain forests, 160
Rainfed agriculture, supplemental irrigation and,
    468–475
Rainwater harvesting, 459
Reblindness, 31–33
Recreational tourism, 53, 54
Red River
    elevations, 138
    ice jam problems, 139
    profiles, 138
    spring flood potential, 139
Red River basin, 136–145
    beach ridge areas, 138
    glacial lake plain, 138
    glacial moraine, 138
    lake-washed till plain, 138
    landforms, 138
    major landforms, 137
    profiles, 138
    run-off travel time, 139, 140
    timing zones, 141
    understanding watershed context for
        restoration of, 136–139
Regional Flood Control District, 120
Rehabilitation, 5
Restoration
    as big business, 22
    conservation *vs.,* 15
    design, 4
        ecology *vs.,* 2, 21
    economics of, 22, 30
    growing importance of, 22
    improvement *vs.,* 17
    integrative value, 33–36
    living landscape *vs.* conservation projects,
        132
    models
        co-management, 50–52
        indigenous sovereignty model, 52–53
    moral issues intrinsic to, 23–24
    of hydrology, 151
    of Iraq marshlands, 17
    principles for scaling up, 135–136
    projects, 15
    proposals for, questions regarding, 18
    relationships and, 24–25
    study designs, 40
        stakeholders and, 40
        technical research team and, 40
    success of, 17
    sustaining, 80–82
    training, 31
    writ large, 18
Restorative development, 3
    defined, 5

framework, 4
  underlying concepts, 6
Revegetation guidelines, 365
Revitalization, integrated, 33
Ring Water Basin, 165, 168
Roads
  as cultural constructs, 348
  as ecological entities, 348
Rock inflow dissipators, 78
Rocky Mountain plateau people, historic range
    of, 352
Royal Society for the Conservation of Nature
    (RSCN), 81, 87, 283
  Azraq Oasis Handicraft Center, 102–103
  campsite, 318–320
  crafts and products linked to conservation,
    309–312
  Dana Nature Reserve management, 307
  esteem and regard for, 312
  guest house in Dana village, 321–322
  in Jordan, 214
  national plan for protected areas, 307
  plans, 332
  work with local schools, 95, 98
RSCN. See Royal Society for the Conservation of
    Nature (RSCN)
Runoff travel time, 139, 140

S

Sacramento Valley Basin-wide Management
    Plan, 440
Safe Drinking Water Act, 121
Salinity
  fish production and, 430
  management, 436–440
    changing crops and, 436
    examples of, 437–440
      implications for Iraq, 441–448
    fallowing land and, 436
    in California, 437–438
    leaching and flushing in, 436
    reducing consumptive use in, 436
    seasonal, 438
  plant productivity vs., 435
  sulfate as component of, 433
Sam Veasna Center for Wildlife Conservation,
    263
Schistosomiasis, 401
Sea of Galilee, 65, 77
Selenium removal, 386
Septic tank effluent gravity sewers (STEGs), 398
Settled-sewage collection system, 420, 422–423
Shallow surface ditches, 145, 147
Sheet flow, maintenance of, 382
Siem Reap, Cambodia, 228, 230, 237, 248
  development, 257, 263

tourist industry, 269
Single-home options, 417
Site rehabilitation, 5
Society for Ecological Restoration, 2, 23
Soil salinization, 432
South America
  rainforest areas, 160
Southwest Wetlands Consortium, 121
STEGs. See Septic tank effluent gravity sewers
    (STEGs)
Steppe, water harvesting in, 478
Sulfide toxicity, 433
Sweetwater Wetlands, 376, 377, 386

T

Taybet Zaman village, 334
  transformation of ruins into hotels, 336–338
Technical and Science Advisory Committee
    (TSAC), 141
Tenochtitlan, 34, 35
Terrace farming, 279
Tile networks, 148, 149
Tonle Sap Basin Management Organization,
    487–488
Tonle Sap Biosphere Reserve (TSBR), 259
Tonle Sap Great Lake, 228–271, 232–233
  as UNESCO Biosphere Reserve, 259
  bamboo and net fencing to contain fish in
    flooded forest, 232–233
  core areas for conserving biodiversity,
    259–271
  floating villages, 230, 251
    community activities in, 240–245
    fishing as major source of income, 231
    health conditions in, 245
    homes, 236–237
    merchants and craftsmen of, 231
    pens for livestock, crocodiles and fish,
      238–239
    poverty of, 245
    replica of, 258
    scenes of life in, 252–255
    television and communication towers in,
      238–239
    water and land access, 271
  increase in surface area, 229
  Khmer ruins surrounding, 233
  Mekong River and, 228
  rice fields in flood plains surrounding, 231
  social ecology, 266–267
  submerged forest in, 230
  tourism, 237
Tourism. See also Ecotourism
  Dana Nature Reserve, 312
  guidebooks, 307
  jobs related to, 324

lessons learned from, 330
reed boat for, 286
regulating and monitoring, 333
Tonle Sap Great Lake, 237
type
    cultural, 53
    environmental, 53, 54
    ethnic, 53
    historical, 53, 54
    indigenous culture and, 53, 56–57
    recreational, 53, 54
TSAC. *See* Technical and Science Advisory
    Committee (TSAC)
TSBR. *See* Tonle Sap Biosphere Reserve (TSBR)

U

U.S. Agency for International Development
    (USAID), 18, 39, 41
U.S. Army Corps of Engineers, 34
U.S. Department of Interior, 220
U.S. Environmental Protection Agency, 392
U.S. Fish and Wildlife Service, 119, 220
U.S. Highway 93. *See* Montana's highway 93
UNESCO, 35, 127, 192
UNESCO Biosphere Reserve, 259, 268
United Nations Educational, Scientific and
    Cultural Organization. *See* UNESCO
United Nations Environment Programme, 445
USAID. *See* U.S. Agency for International
    Development (USAID)

V

Valley Sanitation District Wetland, 381

W

Wadi Dana, 314
    archaeological ruins in, 322–324
Wakodahatchee wetlands, 385
Walnut Creek National Wildlife Refuge. *See* Neal
    Smith National Wildlife Refuge
Warsaw effect, 30
Wash Coordination Committee, 222
Wastewater
    dischargers, 118
    flow, 121
    impure, 396
    industrial, 376
    lagoons, 405
    outfall pipes for treated, 205
    pathogen removal, 401
    quantity, 121
    salinity treatment, 375
    treatment plant, 192, 386
Wastewater reuse, 391–409

California-Florida model, 393
centralized, 392
conventional technologies, 418
culture and, 391–397
decentralized, 181, 387, 397–399
    cost effectiveness of, 397
    defined, 397
    location of system for, 399
    odor control for, 399
    pressure sewers in, 398
    scale of, 398
    STEGs in, 398
    underground, 400
    vacuum sewers in, 398–399
economics, 394, 396
ecotechnology and, 402–408
energy consumption, 407, 408
fatwa on, 396
for crop irrigation, 393
for vertical-flow wetland treatment, 400
infrastructure, 414–424
    models for, 414–415
    single-home, 417–420, 421
    village-scale, 420, 422–424
Islam and, 392–393, 395–397
no risk philosophy, 394
obligate internal pathogens and, 401
opposition to, 394
package treatment system, 401
promoting, 397
public health and environmental concerns,
    399–402
rationale for, 397
salinity of, 401–402
semipassive systems, 422–423
single-home options, 417
standards for, 393
subsurface-flow wetlands for, 405–407, 419
    pathogen removal, 405
surface-flow wetland for, 404–405
    pathogen removal from, 405
technology, 387
theoretical framework, 415, 416
to drinking standard, 394
WHO standards for, 393
Wastewater, zooplankton predation and, 403
Water
    accounting framework, 463–464
    agricultural, 455–456
        drainage, 458
        increasing productivity of, 460–465
            accounting framework for, 463–464
            economic approach to, 464–465
            evolution of concept of, 461
            from efficiency to, 461–463
            performance indicators for, 464
    annual renewable share, 454

applied, 434
as carrier medium, 419–420
balance, 381
consumptive use, 434
developing new supplies of, 456–457
    cost of, 456–457
distribution to, 382
drainage, 434
in dry areas
    scarcity, 452–454
    water resources in, 453
marginal-quality, 457–458
overdrainage, 435
pricing, 460
productivity
    environmental considerations of, 481–482
    increasing, 460–468
        accounting framework for, 463–464
        case studies from Middle East,
            468–483
        deficit irrigation and, 465–466
        economic approach to, 464–465
        evolution of concept of, 461
        from efficiency to, 461–463
        in drier environments, 478–481
        in irrigated agriculture, 476–478
        land use changes and, 466
        performance indicators for, 464
        precision irrigation and, 466–468
    land productivity and, 476
    policies and institutions for, 482
regional transfer of, 458–459
runoff travel time, 139
scarcity
    for agriculture, 455–456
    in dry areas, 452–454
    options for coping with, 456–460
tower, 201
treatment and restoration, 375–388
    case studies in benefits of, 385–387
virtual, 459–460
Water Application Efficiency, 461
Water Conveyance Efficiency, 461
Water Distribution Efficiency, 462
Water harvesting
    direct benefits of, 481
    limitations to use of, 481
    rooftop, 481
    runoff strips, 480
    semicircular bund, 480
    systems, 479–481
        contour-bench terraces microcatchment,
            481
        macrocatchment, 480
        microcatchment, 479, 481
Water management, integrated, 487–489
Water Resource Use Efficiency, 464–465

Water Use Efficiency, 462
Watershed
    diversity, 143
    hydraulic performance, 143
    outlet design characteristics, 143, 144
    ratio, 143
    restoration, 33
Western Transportation Institute, 365
Wetland(s), 160
    aerated, 406
    as green infrastructure, 385
    as multipurpose park, 377
    as outdoor laboratories, 376
    cement mine reclaimed for, 161
    constructed, 404
    contaminant removal processed, 381
    decentralized wastewater use and, 404
    denitrification, 380, 407
    design and performance data, 378
    ecological risk to, 378
    hydraulic loading rates, 381
    industrial effluents in, 378
    natural treatment processes, 378–380
        BOD and, 379
        design considerations in, 379
        pH and, 379
    nitrification, 407
    nitrogen assimilation, 379–380
    number of constructed, 378
    pH, 379
    public use, 383–385
    questions about treatment, 378
    restoration, 133
        as area for migratory birds, 67, 152
        ecotoxicological and biotoxic concerns,
            134
        hydrological framework for considering,
            140–143
        plant seeding in, 152
        Red River basin, 136–139
            design and siting criteria for, 142
            understanding watershed context for,
                136–139
        rewetting dewatered and, 134
        risks to wildlife during, 134
        water rights and diversion for, 133
        water treatment and, 375–388
    sulfide toxicity, 433
    surface-flow, 376, 404–405
    technologies, 376
    tidal flow, 407
    vector control of, 378
    water movement, 382
WHO. *See* World Health Organization (WHO)
Wild Jordan Café, 325, 329–332
Wildlife
    crossing, 364–365

guidelines, 363
   structures for, 364
cultural significance of, 351
roads as impediments to, 348
threats to, 306
vehicular deaths, 354–355
World Commission on Protected Areas, 325
World Health Organization (WHO)
   standards for wastewater reuse, 393
World population, 413

X

Xochimilco, 184
   amenities, 202–204
   boardwalks, 187–188
   boats for hire, 193
   cultural significance, 191–192
destruction, 191
ecological and recreational parks, 197–198, 204
environmental restoration and regeneration of natural capital, 192–193, 195–196
flowers, 197, 198, 199
history and description, 184–185 190
lessons for Iraqi marshlands, 206, 209
villas, 195
visitor center, 189
   biodiversity dioramas, 190
   waterways at, 186
water tower, 201

Z

Zooplankton predation, 403